The Technology of Reservoir Characterization

Jia Ailin Cheng Lihua

Petroleum Industry Press

内 容 提 要

本书介绍了精细油藏描述内容、技术发展状况，论述了精细油藏描述的步骤方法，并针对每一步骤的技术方法进行阐述，包括地层结构描述、沉积微相描述和储层定量参数描述等内容。既有方法理论的介绍，也有实践经验总结。

本书可供从事油田开发后期精细油藏描述及提高采收率研究的科研人员、工程技术人员和高等院校相关专业师生参考使用。

图书在版编目（CIP）数据

精细油气藏描述程序方法 = The Technology of Reservoir Characterization：英文 / 贾爱林，程立华著. — 北京：石油工业出版社，2022.3
ISBN 978-7-5183-5327-9

Ⅰ.①精… Ⅱ.①贾… ②程… Ⅲ.①油气藏-油气勘探-英文 Ⅳ.①P618.130.8

中国版本图书馆 CIP 数据核字（2022）第 062260 号

出版发行：石油工业出版社
　　　　（北京安定门外安华里 2 区 1 号　100011）
　　网　址：www.petropub.com
　　编辑部：(010) 64523708
　　图书营销中心：(010) 64523633
经　销：全国新华书店
印　刷：北京中石油彩色印刷有限责任公司

2022 年 3 月第 1 版　2022 年 3 月第 1 次印刷
787×1092 毫米　开本：1/16　印张：19.25
字数：590 千字

定价：160.00 元
（如出现印装质量问题，我社图书营销中心负责调换）
版权所有，翻印必究

Preface

After nearly 30 years of development, the reservoir description technology has developed from a sub-discipline of oilfield development geology into a backbone discipline and has further evolved into a reservoir description technology adapted to the early stage and entire process of oilfield development, and a fine reservoir description technology specially intended to improve reservoir description accuracy and enhance the recovery ratio of old oilfields in their late development stage. The so-called fine reservoir description refers to the finer research on the relationships of reservoir geology with oil and water using the dynamic and static data of all development wells and seismic data after the completion of the initial development well pattern of an oilfield than the evaluation in its early development stage. Fine reservoir description includes two important prerequisites: one is the refinement of research data, and the other is the refinement of research depth.

Nothing has a completely programmed model, and the same is true for fine reservoir description, but there is a regular understanding of the development of everything and the solution of problems. It is for this reason that, based on many years of research on reservoir description, we have completed monographs such as "Techniques and Methods for Establishing Geological Models in the Evaluation Stage of Reservoirs", "Fine Research Methods for Reservoirs", "Fine Reservoir Description and Geological Modeling", etc., and a large number of related papers; moreover, we attempt to compile a booklet about procedures and methods of fine reservoir description, introducing the key steps of fine reservoir description and expounding the working and technical points that must be completed in each step, as well as the data and technical methods to meet the requirements of the work.

With the continuously increasing oilfield development degree and the further enrichment of research data, the fine reservoir description at present has undergone a very large change compared with that ten years ago, mainly as follows: in terms of data, due to the continuous promotion of 3D seismic and even 4D seismic in old oilfields, the prediction and description of interwell reservoirs are more accurate; compared with the previous general injection and general production, separate layer injection and production profiles are refined to every single layer, which provides more reliable dynamic data for development geology, especially for the research on reservoir rhythm characteristics and reservoir continuity and connectivity. In addition, the development geological depiction of genetic units of well profiles and the inside of genetic units has further improved the understanding and research accuracy of reservoirs.

This book, which introduces the procedures and methods of fine reservoir description. Consists of ten chapters, involving the main technical means and development status of fine reservoir description, the establishment of reservoir geological models, and the fine description of reservoir distribution characteristics and remaining oil distribution, etc. This book was compiled referring to a lot of research results in combination with our years of experience in reservoir description, and focusing on how to solve problems in the specific reservoir description work in oilfields, also quoting a large number of research examples, especially the specific research objects involved in the scientific research undertaken by the authors in the past. Here, the authors would like to sincerely thank the people who undertook the research content involved in the book.

During the process of the project research and the writing of this book, we received consistent support and guidance from Academician Yuan Shiyi, Deputy Chief Engineer Zhong Taixian, and Director Luo Kai. Thank then for that. In addition, we would also like to thank the senior engineers such as Liu Wenling, Yang Huidong, He Dongbo, Guo Jianlin, *et al.* who participated in the project research! We would like to express our sincere gratitude to Academician Han Dakuang, Professor Qiu Yinan and Professor Pan Xingguo for their kind guidance on many technical issues. Thanks again to the colleagues who participated in this research work with us for many years, and the people who worked hard for the publication of this book!

Contents

Chapter 1	**Overview of Fine Reservoir Description**	(1)
Section 1	Development Status Quo of Main Technologies	(1)
Section 2	Development Direction of Fine Reservoir Description	(15)
Chapter 2	**Contents and Steps of Fine Reservoir Description**	(19)
Chapter 3	**Regional Background Description**	(29)
Section 1	Content and Methods of Regional Background Description	(29)
Section 2	Alluvial Fan Glutenite Bodies	(33)
Section 3	River Sand Bodies	(38)
Section 4	Lacustrine Sand Bodies	(47)
Chapter 4	**Fine Stratigraphic Structure Description**	(63)
Section 1	Division and Correlation of Reservoirs by the "Cycle Correlation and Hierarchical Control" Method	(63)
Section 2	High-Resolution Sequence Stratigraphy Analysis Technology	(70)
Section 3	Fine Stratigraphic Structure Establishment Example	(86)
Chapter 5	**Description of Sedimentary Microfacies of Reservoirs**	(97)
Section 1	Basic Methods of Sedimentary Microfacies Description	(97)
Section 2	Sedimentary Facies Research Examples in Dense Well Pattern Areas	(102)
Chapter 6	**Establishment of Reservoir Geological Knowledge Database**	(115)
Section 1	Main Content of Reservoir Geological Knowledge Database	(115)
Section 2	Establishment Methods of Reservoir Geological Knowledge Database	(116)
Section 3	Establishment of Reservoir Geological Knowledge Database Using Geological Outcrop Data	(119)
Section 4	Establishment of Reservoir Geological Knowledge Database Using the Method of Dense Well Pattern Area Anatomy	(133)
Chapter 7	**Prediction of Interwell Reservoirs with Seismic Data**	(143)
Section 1	Reservoir Prediction with Seismic Attributes	(143)
Section 2	Logging-Seismic Joint Inversion	(154)
Section 3	Seismic Frequency Division Technology	(159)
Section 4	Development Seismic Technology	(163)
Chapter 8	**Description of the Internal Architecture of Genetic Units**	(166)
Section 1	Theoretical Methods of Reservoir Architecture Analysis	(167)

Section 2　Reservoir Architecture Research Steps and Examples ·················· (175)

Chapter 9　Description of Reservoir Physical Parameters and Fluids ················ (197)

　　Section 1　Reservoir Heterogeneity Description Technology ····················· (197)

　　Section 2　Description of Remaining Oil Distribution ································ (220)

　　Section 3　Natural Gas Distribution Description ···································· (236)

Chapter 10　Geological Models of Reservoirs ·· (252)

　　Section 1　Classification of Geological Models of Reservoirs ···················· (252)

　　Section 2　Geological Modeling Methods ··· (255)

　　Section 3　Fine Geological Modeling Steps ·· (269)

　　Section 4　Deterministic 3D Reservoir Modeling ···································· (272)

　　Section 5　Well-Seismic Joint Stochastic Simulation and 3D Reservoir Modeling ······ (284)

References ·· (295)

Chapter 1 Overview of Fine Reservoir Description

Reservoir description originated early and developed rapidly in China in the 1990s. The geological evaluation contens and technical means of oil reservoirs and gas reservoirs can be in common use; therefore, it was called oil reservoir description in the past, which actually covers the content of gas reservoir description, but is not sufficiently targeted. In recent years, with the continuous expansion of the natural gas development scale, more and more attention has been paid to gas reservoir description, and especially the development of unconventional gas reservoirs has further increased the depth of gas reservoir description research. Therefore, for the middle and late stages of oil and gas field development, multi‐disciplinary and multi‐professional forces in geology, geophysics, reservoir engineering, etc. have been concentrated to conduct comprehensive researches on the fine reservoir description, and the outstanding progresses have been made.

Section 1 Development Status Quo of Main Technologies

There are always two constant themes in the process of oil and gas field development, namely, understanding reservoirs and stimulating reservoirs. The main task of reservoir description is how to understand the actuality of reservoirs. The development of this technology is mainly driven by two aspects. Firstly, the requirements for an understanding of geological bodies faced are getting higher and higher. In terms of exploration, high-quality integral traps and reservoirs have almost been explored globally, and the exploration difficulties in blocks to be explored and discovered have increased sharply. In terms of development, a large number of oil and gas fields have entered the late stage of development, and the extensive development methods are far from meeting the needs of production. Therefore, it is necessary to understand and study reservoirs on a finer scale than before to find out the objects for further tapping the potential and realizing the stable production of old oilfields. Secondly, new reservoir description technologies and methods make it possible to achieve these goals. Logging and seismic technologies have been rapidly developed in recent years, and the content that can be studied and the research accuracy have been greatly expanded and improved, directly promoting the accuracy of reservoir description research. The improvement of the computer application level provides new ideas for the comprehensive study of multi‐disciplinary information on the same data platform. Among the technologies, quantitative geological research technology, prototype geological model‐building technology, and high-resolution sequence stratigraphy research technology are developing fastest. Under the dual promotion of these technologies and methods, the reservoir description technology has transformed to comprehensive developments from qualitative to quantitative, from macro to micro, from a single discipline to multi-discipline and multi-specialty.

Ⅰ. Fining and quantification of geological research

The core issue of geological research in reservoir description is to describe and predict the heterogeneity of reservoirs at all levels as accurately as possible. (1) The study at layer series scale shall describe the sandstone density, stratification coefficient, interlayer permeability heterogeneity, the spatial configuration of different levels of reservoirs, barriers distribution, etc.; (2) The study at single-layer scale shall describe the in tra layer sedimentary rhythm, in tra layer heterogeneity, interlayers distribution, etc.; (3) The study at sand body scale shall focus on describing the geometry, continuity, connectivity, and the directionality of macroscopic permeability of sand bodies; (4) The study at pore scale shall focus on the description of pore types, pore-throat configuration, pore structure, etc.

At present, the fine and quantitative geological research is mainly manifested in three aspects. (1) The scale of research is getting smaller and smaller; (2) The basic unit is getting smaller and smaller; (3) The trend of quantification and prediction is becoming more and more obvious. This development trend is also to meet the requirements of production. Especially the in-depth tapping of the potential of old oilfields will be very difficult without fine geological research.

(Ⅰ) **Fine geological research**

The fine geological research is manifested in the scale of stratigraphic units, the detail level of structural interpretation, the detail level of diagenetic evolution, and the development process of pore structures. In terms of the division of stratigraphic units, the stratigraphic division taking layer groups as units cannot meet actual needs, and even the division at a single sand body scale cannot meet the requirements of production. At present, it is necessary to conduct in-depth research on specific issues such as the structure, rhythm characteristics, and interlayer distribution within a single sand body. Figure 1-1 is the schematic diagram of detailed reservoir research level by level.

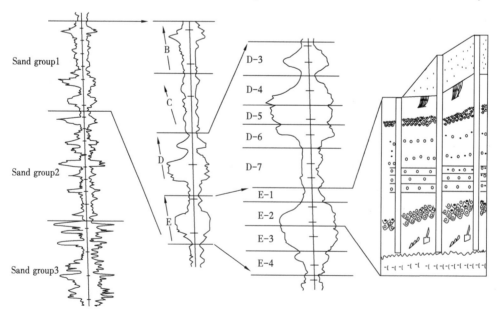

Figure 1-1   Schematic diagram of fine reservoir research

Finally 7 units with different structures and grain sizes have been divided in detail within a channel sand body in the fan delta's nearshore channel. The lithology from bottom to top: massive fine conglomerates (containing carbonized tree trunks)—brecciform sandstones with large trough cross-beddings—coarse sandstones with parallel beddings—massive medium sandstones. After such a detailed study, the distribution of underground fluids under certain development conditions can be known; in addition, since the internal change characteristics of the sand body are known in advance, the possible distribution characteristics of the remaining oil and gas after development to a certain extent can be predicted accurately before development.

Daqing Oilfield has conducted a more in-depth study of the fine description of sand body structures through the subdivision of sedimentary units and the research of sedimentary microfacies.

1. Internal architecture of thick oil layers

There are seepage barrier layers inside the sand bodies of some thick oil layers and between the rhythm sections of a single sand body. These seepage barrier layers are related to the sedimentary environment and diagenesis, and can be roughly divided into four types.

Type Ⅰ interface: the mudstone or calcareous layer between outer surface layers, the interface thickness ≥0.2m.

Type Ⅱ interface: the calcareous layer or physical interlayer between effective layers, the interface thickness ≥0.4m.

Type Ⅲ interface: the calcareous layer or physical interlayer within the effective thickness, 0.1m ≤ interface thickness ≤ 0.4m.

Type Ⅳ interface: the permeability grading interface or water-flooded interface superimposed on two phases of sandstones, the interface thickness <0.1m.

On the basis of the subdivision of sedimentary units, Daqing No.6 Production Factory has realized the hierarchical cycle subdivision of thick oil layers based on the principle of analytic hierarchy process and has accurately described the internal architecture of sand bodies and its influence on seepage, thereby laying a solid foundation for fine description of the distribution of remaining oil and gas in thick layers.

2. Research on identification methods for single channels in composite sand bodies

There are many sand bodies deposited in the delta distributary plain facies in the Xingnan Development Area of Daqing Oilfield. These sand bodies are mostly superimposed by single channel sand bodies. Daqing No. 5 Production Factory distinguishes between single channels of different phases vertically and traces single sand bodies horizontally. Based on the modern sedimentary theory and guided by sedimentary models, the paleogeographic environment is restored, the facies belt position, the scale of a single channel, and the distribution direction of the facies belts in the research target layers are ascertained, and the distribution map of single-channel facies belts is accurately drawn.

The identification of a single channel in a composite sand body accurately depicts the heterogeneity of reservoirs and the connection relationship between sand bodies, ascertains the main factors affecting the polymer flooding development effect and the characteristics of sand body production, which is of much significance to the selection of stimulation wells and layers and the guidance to the adjustment of water injection schemes.

3. Microstructure research

Structural interpretation is no longer an issue of structural shape and fault combination etc. in the tapping of the potential of old oil and gas fields. In old oilfields, the general well pattern density reaches 200-300m, and the density of individual well patterns used in experimental areas can reach 50m or less. Moreover, with the continuous improvement of seismic interpretation accuracy, it is necessary to further improve the interpretation accuracy of subtle structural changes and faults with very small fault throw.

4. The development of diagenesis research towards 3D direction

The study of diagenetic evolution has gone far beyond the simple definition of diagenetic stages, but shall be conducted from the process of diagenesis. ① Within the scope of a basin, it is necessary to comprehensively study the thermal force, burial, mineral composition, and rock types of the basin under the control of different structural units and different fluid fields. ② Within the scope of a reservoir, it is necessary to study the diagenetic order, mineral transformation, fluid types in the reservoir, and the deposition mechanism of genetic units under the conditions of unified structures and fluid fields.

The research on diagenesis has developed from one-dimensional diagenesis research to two-dimensional diagenesis research and is developing towards the three-dimensional diagenesis simulation. In particular, it needs to be explained that the development of new analytical technologies and microscopy technologies in recent years has made it possible to refine the study of reservoirs and diagenesis.

(Ⅱ) **Quantification of geological research**

Reservoir geology has developed from general description to semi-quantitative description for more than 10 years and has achieved considerable success in many aspects. With the rapid development of computer technology and the requirements of other related disciplines for geology, the current reservoir geology has been far from meeting the needs, and the development of quantitative geology is bound to become inevitable.

Quantitative geology is to quantitatively characterize the shape and characteristics of geological bodies, various proportional relationships between sand bodies, and various parameters inside sand bodies. The research methods of quantitative geology mainly include the anatomy of field outcrops, the study of dense well pattern areas in oil and gas fields, the physical simulation and numerical simulation of sedimentary systems, etc.

During the 9th Five-Year Plan period, the fine and quantitative geology research on the reservoirs in the two outcrops—Luanping fan delta and Datong braided river was completed. According to the characteristics of the sedimentary systems in the two outcrops, the research on physical simulation and numerical simulation of the same sedimentary system has been carried out. The actual underground conditions of oil and gas fields with the same sedimentary system have been anatomized, and the research results of the three aspects have been combined. The research on fan deltas, braided rivers, and quantitative geology has been greatly developed. A series of quantitative research results have been summarized, including the morphological characteristics of sand bodies, the size ratio and configuration relationship of sand bodies, the quantitative data of sand bodies, the physical property distribution law within sand bodies, sand body prediction methods, etc. In addition, geological knowledge databases have also been established, including lithology and lithofacies library, environmental type library, scale library, interlayer type library, sand body statistics library, prototype model library, etc. Moreover, preliminary research methods for quantitative geology of continental clastic reservoirs have been established.

In the process of quantitative geology research, except for the geological parameters (e.g., sedimentary system, sedimentary environment, etc.) that cannot be expressed numerically, all other parameters shall be expressed as quantitatively as possible. For example, the previous understanding of fan deltas basically remained on the types of microfacies contained in a sedimentary body with a three-layer structure, and the proportion and size of various microfacies could not be pointed out. Moreover, within the distribution range of a width-to-thickness ratio prediction volume, only the width-to-thickness ratio parameter of various rivers was used, while the microfacies units inside a fan delta couldn't be accurately predicted. Through research, a large number of quantitative geological laws have been summarized in this aspect, thus guiding the prediction of similar underground reservoirs.

The development of a series of quantitative statistics and prediction technologies will lay a certain foundation for the refined development of reservoir research.

II. The role of logging technology in reservoir description has become increasingly remarkable

The role of logging technology in reservoir description has become increasingly remarkable. Logging technology has always been regarded as the eyes of the underground conditions of oil and gas fields and is one of the most important means to understand underground conditions in the process of oil and gas field development. Especially at present, China's old oilfields have entered the stage of high water cut, water-flooded layers and low-resistivity oil layers are the main potential tapping targets, and fractured low-permeability reservoirs are also a difficult problem in logging interpretation and evaluation. It is particularly important to develop new logging technologies and make full use of traditional logging means for further research, mainly involving the following aspects.

(I) **Traditional logging means are still the main tools for reservoir description**

For general sandstone and mudstone formations, the main problems to be solved by logging include division of permeable layers, effective layers and barriers; interpretation of the types and properties of fluids in reservoirs, namely oil, gas, and water layers; interpretation of reservoir parameters, including porosity, permeability, and oil saturation. It shall be said that if a correct interpretation chart that conforms to the characteristics of the reservoir is established based on the four-property relationship, logging means can completely meet the needs of the interpretation accuracy of the other eight parameters except for permeability.

The current conventional logging methods cannot be directly used to obtain permeability, but permeability can be obtained indirectly through other parameters. Generally, permeability is calculated as per porosity and shale content and is also directly calculated as per porosity. At present, the common practice of various oil companies in the world is to firstly establish a relationship between porosity from core analysis and permeability data, and then use this relationship to obtain permeability as per the porosity from logging interpretation. Although this method has been widely used and proved to be practical through practice, it is very difficult to control its error to be lower than 30%. We all know that the factors affecting permeability are very complex. Even in the same reservoir, the difference in the permeability of reservoirs with the same porosity is of the order of magnitudes. In addition that the interpretation of permeability needs to be further improved, there are still some problems that cannot be solved very well through conventional logging, mainly including the three aspects. (1) Due to the problem of logging interpretation resolution, it is still difficult to interpret thin layers, and the needs of old oil and gas fields for tapping the potential of thin layers and poor layers cannot be met. (2) There are still many problems in the quantitative interpretation of water-flooded layers in open hole wells and cased wells, and the changes of formations and fluids during the development process cannot be interpreted very well. (3) The logging interpretation technology for fractured reservoirs and low-permeability reservoirs needs to be improved urgently to meet the actual needs of production. To solve these problems, many new logging technologies and methods have emerged in recent years.

(Ⅱ) **New logging technologies have been developed and applied to a certain extent**

In response to the above-mentioned difficult problems, new logging methods and technologies have emerged. These technologies can solve some problems to a certain extent. Among them, the most representative ones include the following.

1. Imaging logging technology

The imaging logging tool is the fifth-generation logging tool and a major technological breakthrough in logging. It can provide high-resolution real wellbore conditions in reservoir description. It can be used to determine stratigraphic dip, detect fractures, pores and holes, quantitatively describe thin layers, and determine the location of faults and other geological features. To a certain extent, it has the function of replacing coring, and can also make up for the deficiencies of conventional coring in determining bedding structure and fracture orientation. The highest resolution of imaging logging can reach 0.5cm, and the deposition of large gravels can be observed through imaging logging. Main imaging logging series include array induction imaging logging, azimuthal lateral imaging logging, formation micro resistivity scanning imaging logging, dipole S-wave imaging logging, ultrasonic imaging logging, etc.

2. NMR logging

NMR logging is the fastest-growing logging technology in recent years. NMR logging is a new method that can measure the free porosity of reservoirs, and is not affected by lithology, shale content and fluid salinity. In reservoir characterization, NMR logging can quantitatively and reliably obtain the data of porosity, pore size distribution, permeability, and bound or movable liquid saturation; moreover, NMR logging can detect the recovery ratio and effective thickness of reservoirs, and is used to calculate the recoverable reserves of oil and gas fields. Currently, the

NMR logging technology has begun to enter the practical stage. The rapid development of this technology not only makes it possible to quickly measure reservoir characteristics on site but also makes it possible to provide cheap and high-quality measurement results.

3. Logging while drilling (LWD)

LWD is another breakthrough in logging technology, and the content of LWD is constantly expanding, including formation resistivity, GR, neutron, density, acoustic, etc. LWD was used by Occidental Petroleum in the early 1990s in the drilling of information wells and horizontal wells. The biggest advantage of LWD is that it can measure the true conditions of original formations and fluids, and does not take up drilling time; in addition, during the drilling of horizontal wells, the drilling trajectory can also be adjusted in time according to the changes of formations.

4. Modular dynamic formation test (MDT)

MDT is designed based on the seepage principle. The tool can acquire multiple fluid samples and flow velocity data each time it is run in a hole, and there is an optical indicator to monitor the properties of the reservoir fluid entering the sampling chamber. The main functions of MDT in reservoir description include direct measurement of formation pressure and permeability, direct finding of oil and gas, evaluation of production capacity of reservoirs, etc.

5. Through-casing logging

Through-casing logging includes pulsed neutron capture logging, C/O ratio logging, gravity logging, and through casing resistivity logging. Especially for old oil and gas fields, through-casing resistivity logging provides a powerful means for dynamic reservoir description. It overcomes the shortcoming that conventional resistivity logging can only be performed after a new well is drilled, but cannot obtain effective resistivity in cased wells. By conducting through-casing resistivity logging in old wells and comparing the through-casing resistivity logging data with the initial logging data, the water-flooded degree, remaining oil, water flooding efficiency, etc. can be understood; in addition, the connectivity of sand bodies and the heterogeneity of reservoirs can be further understood.

Of course, these new logging technologies cannot be widely used as conventional methods at present. In addition, most of the old wells in old oil and gas fields use old logging series, and thus it is still a realistic task to further improve the interpretation accuracy of conventional logging series. Furthermore, these new logging technologies will also play an important role in calibrating and improving conventional methods.

(Ⅲ) **The potential of conventional logging data has been deeply tapped**

The emergence of new logging technologies has improved the ability to understand the true underground conditions. But at present, their cost is too high and it is very difficult to apply them on a large scale in production. It is very important to make full use of conventional logging data to solve special reservoir problems. At present, the successful application is to use conventional logging information to identify and study fractures. The main basis for using conventional logging data to identify fractures is: the conductivity of fluid-containing fractures, the heterogeneity and anisotropy of fractured formations, and the relationship between fracture development and lithology, etc. Pore structure index method, porosity logging combination method, and fracture logging indicator method are some of the better applications. Using these methods, the porosity and permeability of fractured reservoirs can be calculated. These methods have good practical value.

III. The fining of development seismic research makes fine prediction of reservoirs possible

The development seismic technology is a discipline of reservoir geophysics that emerged in response to the needs of production practice. It is a seismic observation, processing, and interpretation technology for specific small-scale reservoirs that require fine structure description based on seismic exploration. It can be understood as the seismic technology used in the development stage and is also called reservoir seismology or hydrocarbon reservoir seismology. It is a technology developed on the basis of exploration seismology, which makes full use of observation methods and attribute processing technologies for oil and gas reservoirs. By closely combining multi-disciplinary data involving drilling, logging, petrophysics, oil and gas field geology, reservoir engineering, etc., the development of seismic technology is a discipline for transverse prediction of reservoir characteristics and complete description and dynamic monitoring of them during the development and production of oil and gas fields. The main technologies include seismic target processing, 3D continuous processing, prestack depth migration, high-resolution seismic exploration, seismic attribute analysis and hydrocarbon detection, coherence cube analysis, quantitative seismic facies analysis, comprehensive seismic interpretation and visualization, seismic inversion, reservoir feature reconstruction and feature inversion, AVO analysis and inversion, 3D AVO, cross-well seismic, 4D seismic, multi-wave and multi-component, etc. The connotation of development seismic includes two parts, namely, static reservoir description and dynamic management of reservoirs.

For seismic reservoir prediction, in addition to the need to further improve the interpretation accuracy of general sandstone and mudstone formations, there are also challenges in the prediction of special types of reservoirs. The main types include thin limestone, conglomerate, mudstone, eruptive volcanic rock, intrusive rock, weathering crust and intrusive rock alteration zone reservoirs, etc. The common feature of these reservoirs is that they are characterized by fractured reservoir spaces and dissolution spaces. Conventional seismic inversion can only obtain the information of acoustic velocity (or acoustic impedance). When reservoirs have obvious acoustic characteristics or wave impedance characteristics, very good effects can be obtained from conventional seismic inversion. However, many complex reservoirs and especial fractured reservoirs often have no obvious recognizable acoustic (acoustic impedance) characteristics, and conventional seismic inversion is helpless for this. For this kind of complex reservoirs, the reservoir feature reconstruction technology has been developed. The principle of this technology is: aiming at specific geological problems, based on the relationship of rock physics, a reservoir characteristic curve is reconstructed from numerous logging curves so that the reservoir has obvious characteristics on this curve. Then, the reservoir characteristics are extrapolated beyond the well point through reservoir characteristic inversion to obtain characteristic reconstruction data volumes for reservoir description.

Currently, cutting-edge seismic technologies mainly include 4D seismic, cross-well seismic, P-wave azimuth AVO and multi-wave azimuth AVO, multi-wave and multi-component, etc. The 4D seismic technology is mainly aimed at changes in reservoirs and fluids during development. The cross-well seismic technology is for fine reservoir research and reservoir management during the development of oil and gas fields. The P-wave azimuth AVO and multi-wave azimuth AVO technology is a new development of AVO technology, and is mainly used for direct identification of reservoirs. The application of the multi-wave and multi-component technology mainly involves two aspects such as (1) imaging and (2) lithology and fluid identification, and is mainly for the study of complex geomorphology and accumulation process.

With the vigorous promotion of 3D seismic, both seismic resolution and seismic application scope have been unprecedentedly improved. Especially the combination of seismic technology with logging technology and sedimentology research technology has enabled the discipline to achieve considerable development. On the basis of 3D seismic data, technologies such as logging-constrained inversion, 3D visualization, etc. have made great progress in further improving the resolution and application scope of seismic data. For example, the 3D visualization technology is no longer a simple display, but an interpretation technology, which can be used to evaluate the type and distribution of reservoirs in the early stage of exploration. The logging-constrained seismic inversion technology is to calibrate seismic attributes on well points. Inferred from this, the recognition accuracy of thin layers can be greatly improved within a certain distance range.

IV. The high-resolution sequence stratigraphy analysis method has been applied

As a new branch of geology, sequence stratigraphy has been greatly developed in recent years. Especially the relatively large-scale sequence stratigraphy research methods are basically mature. However, high-resolution sequence stratigraphy for development is currently in the development stage.

The high-resolution sequence stratigraphy analysis is to carry out stratigraphic division and correlation based on the theory of formation process-response sedimentary dynamics, outcrop data, core data, logging data and high-resolution seismic reflection profile data, as well as the A/S (accommodation space/sediment supply) ratio and base level cycles. This theory breaks through the limitations of traditional sequence stratigraphy derived from passive continental margin marine strata, and is more applicable to high-precision division and correlation of continental strata. Therefore, since its introduction to China in the mid-1990s, this theory has greatly promoted the development of domestic high-resolution sequence stratigraphy, forming a relatively complete theory of continental sequence stratigraphy research, and achieving many successful applications.

The basic principles and methods of high-resolution sequence stratigraphy include base-level principle, volume division principle, facies differentiation principle and cycle isochronous correlation rule. The basic theory is: due to the changes in the ratio of accommodation space of sediments to sediment supply (A/S) in a base-level cycle process, volume distribution of sediments in the same depositional system tract or facies tract occurs so as to result in changes in the preservation degree of sediments, the accumulation pattern of strata, facies sequence, facies type, and rock structure. These changes are a function of the location in the base level cycle process and accommodation space. Therefore, the stratigraphic distribution form of isochronous stratigraphic units controlled by the base level cycle is regular and predictable.

The core of high-resolution sequence stratigraphy is how to identify multi-graded base-level cycles and multi-graded stratigraphic cycles in stratigraphic records. There must be traces that can reflect the time experienced in response to the grade of the base-level cycle in each grade of the stratigraphic cycle. Its sedimentary characteristics and stratigraphic characteristics are mainly vertical changes in the physical properties of a single phase, phase sequence changes, phase combination changes, cycle superimposition pattern changes, and the geometry and contact relationship of strata.

In short, high-resolution sequence stratigraphy is currently still in the stage of introduction and development, but according to the analysis of current development trends, it may be an important development direction of sedimentology and stratigraphy.

V. Establishment of reservoir prototype models and geological knowledge databases for different types of sedimentary systems

The core of reservoir description is to establish a reservoir model. To establish an accurate reservoir model, a set of practical mathematical prediction methods and software are needed, and the distribution law of various parameters of reservoirs also needs to be grasped, that is, a reliable reservoir prototype model and a geological knowledge database need to be established. Therefore, the reservoir prototype model and geological knowledge database are important foundations for fine reservoir prediction and modeling, and an important manifestation of the quantification and digitization of fine reservoir description.

The so-called prototype model is a reservoir geological model that is similar to the simulated target reservoir, has sufficiently dense control points, and is described in detail. Reservoir geological knowledge database refers to the series parameters that are highly summarized after a large number of studies, can qualitatively or quantitatively characterize the geological characteristics of different types of reservoirs, and have universal significance. The choice of prototype model is based on two basic principles. (1) The sedimentary characteristics of the prototype model are similar to those of the simulated target area. (2) There are conditions for dense sampling, and the density of sampling points must be much greater than that of well points in the simulated target area. The establishment of the reservoir geological knowledge database is generally completed in the following three ways. (1) Fine measurement of field outcrops and modern sedimentary sand bodies. (2) Physical simulation and numerical simulation of reservoirs. (3) Fine anatomy of dense well pattern areas in oil and gas fields.

For outcrop areas and modern sedimentary areas, 3D sand body structure measurement can be carried out, and dense sampling and the measurement of rock physical properties (porosity, permeability) can be performed in 3D space. The density of the sampling grids can reach meter level or even centimeter level. Therefore, a very fine 3D reservoir geological model can be established and used to guide underground reservoir prediction. This is the best way to obtain the reservoir prototype model and geological knowledge database. There have been many successful examples at home and abroad, involving a variety of sedimentary types, including meandering rivers, braided rivers, fan deltas, deltas, etc., thereby providing new research ideas and methods for the fine description of reservoirs and the establishment of prediction models. The most typical foreign example is the Gypsy profile research and Yorkson profile research conducted by BP, which has greatly enriched the geological knowledge base of the fluvial facies prototype model. In China, the examples are represented by the establishment of the fan delta and braided river prototype models and geological knowledge databases. For the two outcrops in Datong and Luanping, detailed description, measurement, sampling, etc. were performed; in addition, drilling engineering and logging engineering etc. were also carried out, the internal structure characteristics and change laws of Luanping fan delta and Datong braided river were fully anatomized in detail, and the geological knowledge databases of the two types of outcrop reservoirs were established. Moreover, the prediction methods of the two types of sand bodies have been summarized using the stochastic modeling technology. In the case of 200 well spacing, the prediction accuracy can reach 78.3%, which is a major step forward in the refinement and prediction of outcrop reservoirs in China, and a model for the establishment of China's reservoir prototype models and geological knowledge databases.

In the dense well pattern area of mature oil and gas fields and especially the dense well pattern area with paired wells, prototype models can also be established, but their accuracy is lower than that of those in outcrop areas or modern sedimentary areas, but they can be used to guide the study of stochastic modeling is relatively sparse well pattern areas. Yin Taiju and Zhang Changmin (1997) established a downhole geological knowledge database by anatomizing the dense well pattern area of Shuanghe Oilfield for providing guidance to the establishment of a reservoir skeleton model. Chen Cheng (2006) also established a prototype skeleton model of the fan delta front by using the data of the dense well pattern area of Shuanghe Oilfield and obtained a quantitative geological knowledge database for the scale and shape of underwater distributary channel sand bodies. Currently, there are relatively few research achievements in this aspect. Compared with field outcrop and modern sedimentary research methods, the study of reservoir geological knowledge databases for dense well pattern areas is more difficult, and a lot of research work is still needed.

In general, the research on reservoir prototype geological knowledge databases has made some progress, and the degree of research on fan deltas, braided rivers, and meandering rivers is the highest. However, geological bodies are complex and changeable, and there are also plenty of sedimentary types. The current research achievements cannot satisfy the requirements of understanding complex and diverse geological bodies. On the one hand, it is necessary to carry out research on various types of sedimentary bodies and establish the corresponding reservoir geological knowledge database. On the other hand, it is necessary to further broaden the content of the geological knowledge database, enhance its operability, and give full play to the guiding role of the geological knowledge database in the characterization of underground reservoirs.

VI. Application and development of geostatistics and stochastic simulation technology

Geostatistics generally includes three basic components: correlation analysis of spatial functions, Kriging estimation and stochastic simulation. The correlation analysis of spatial functions refers to the analysis of various functions and covariance functions, and is the basis of Kriging estimation and stochastic simulation.

The Kriging method is a smooth interpolation method, which is actually a special weighted average method. Compared with the traditional mathematical interpolation method, the Kriging method considers the spatial correlation of reservoir parameters caused by geological laws all the more, improves the estimation accuracy of reservoir parameters, and is currently the most widely applied deterministic interpolation method. Of course, the Kriging interpolation method also has its limitations and is difficult to characterize the subtle changes and discreteness of inter-well parameters (such as complex changes of inter-well permeability). Moreover, the Kriging method is also a local estimate method, which does not take into account the overall structure of parameter distribution to a sufficient degree. When the reservoir continuity is poor, the well spacing is large, and the well spacing distribution is uneven, the estimate error is large. Therefore, the inter-well interpolation point given by the Kriging method is a certain value, but it is not the true value, but close to the true value. The magnitude of the error depends on the applicability of the Kriging method and objective geological conditions.

As one of the three major contents of geostatistics technology, stochastic simulation has attracted attention because of its great advantages in analyzing the heterogeneity and spatial uncertainty of geological phenomena. The Kriging interpolation method is actually a linear smoothing low-pass filter and an estimation of conditional mathematical expectation, and thus has a smoothing effect. The interpolation method conceals the degree of heterogeneity (i.e., discreteness), and especially the degree of heterogeneity of reservoir parameters (such as permeability) with obvious discreteness, so it is not suitable for the characterization of permeability heterogeneity. Stochastic models can reflect subtle changes in reservoir properties, which is particularly important for oil and gas field development and production.

Stochastic simulation technology is a combination of geology with mathematics and computer technology, and is a direction of development towards advanced computing technology. Stochastic simulation technology has developed rapidly in the past 15 years and is constantly improving.

An underground reservoir itself is definite, and has definite properties and characteristics at every location point, but randomness may appear when to recognize it. This is due to the following factors. (1) insufficient data and information. (2) The information itself is uncertain. (3) The geological laws of some reservoir properties have a certain degree of randomness. Stochastic modeling technology has been developed to reflect both the actuality of reservoirs as much as possible and their uncertainty that may exist.

The principle of stochastic modeling is: based on the information of known control points and the theory of random functions, the stochastic simulation method is used to give a variety of possible and equal probability implementations (prediction results) for the distribution and changes of the geological characteristic attribute parameters between well points, which are provided to development geologists for selection. That is, in stochastic modeling, the known structural statistical characteristics of a certain property of a geologic body are used to simulate the distribution of this property in an unknown area through some stochastic algorithms to make it identical with the known statistical characteristics, thus reaching the purpose of simulating the reservoir heterogeneity and then predicting the interwell parameter distribution.

To complete stochastic modeling, there are two key points. (1) How to master the inherent geostatistical characteristics of the modeling object (a certain parameter of a certain sedimentary type reservoir) correctly. (2) Choose a suitable simulation algorithm.

The geostatistical characteristics of various parameters of different types of reservoirs are mainly obtained through the establishment of prototype models. Carry out meticulous survey and measurement and intensive sampling of sedimentary outcrops of reservoirs, and then obtain a parameter distribution field with very high sample density (such as 0.5m×0.5m), which is the prototype model of such sedimentary reservoirs. The various geostatistical characteristics obtained represent the true features of such reservoirs and can be used to model similar sedimentary reservoirs downhole. Downhole data with dense well patterns in the late development stage can also be used as a prototype model for predicting similar reservoirs when the well patterns are sparse in the early development stage. The reservoir change information obtained from cluster wells drilled at the same well site and in different development layer series in some domestic old oil and gas fields is also very important geological knowledge. In the past ten years or so, domestic and foreign oil companies have proposed to return to the research on fine surveying of outcrops by investing huge sums of money, which is for this purpose. China has also done very good work for two outcrops in response to the fan deltas and channel sand bodies with strong heterogeneity in continental reservoirs, and has achieved some very meaningful work.

Represented by Stanford University of the United States and the Fontainebleau Geostatistical Center of the French School of Mines, some units developed many stochastic algorithms and formed some commercial software, such as STORM, RC^2, GOCAD, Herisim, Gridstat, etc. The domestic research team headed by Wang Jiahua developed a new stochastic modeling algorithm based on the actual production of oilfields, and proposed a random walk modeling method suitable for the widely distributed channel sand reservoirs in China, and compiled GASOR—the first stochastic modeling software system with China's independent copyright, which is being improved and upgraded continuously. RIPED has also formed its stochastic modeling software system. These simulation methods can be divided into a discrete method—object-based simulation method; a continuous method—pixel-based simulation method. They can also be divided into conditional simulations—methods that are faithful to the sampling point data; unconditional simulations—methods in which the known sampling points can also be changed.

More than 20 simulation algorithms have appeared abroad. The currently popular stochastic algorithms include Boolean simulation method, marked point process method, truncated Gaussian method, sequential Gaussian simulation, sequential indicator simulation, Markov-Bayesian simulation, fractal simulation, simulated annealing, etc. (Table 1-1).

Table 1-1 Classification of stochastic modeling methods (according to Wu Shenghe, 1997)

Simulation method	Algorithm and model / Stochastic model	Sequential simulation	Error simulation	Probability field simulation	Optimization algorithm (simulated annealing and iterative algorithm)	Model property
Target-based stochastic simulation	Marked point process (Boolean model)				Marked point process simulation (using annealing or iterative algorithm)	Discrete
Pixel-based stochastic simulation	Gaussian domain	Sequential Gaussian simulation, LU simulation	Turning band simulation	Probability field Gaussian simulation	(Optimization algorithm can be used in post-processing)	Continuous
	Truncated Gaussian domain	Truncated Gaussian simulation	Truncated Gaussian simulation	Truncated Gaussian simulation	(Optimization algorithm can be used in post-processing)	Continuous
	Indicator simulation	Sequential indicator simulation		Probability field indicator simulation	(Optimization algorithm can be used in post-processing)	Discrete/continuous
	Fractal stochastic domain		Fractal simulation		(Optimization algorithm can be used in post-processing)	Continuous
	Markov stochastic domain				Markov simulation (using iterative algorithm)	Discrete/continuous
	Two-point histogram				(Rarely used alone, mainly used for post-annealing treatment)	Discrete

The geological models obtained from stochastic simulation are lots of equiprobable geological models. How to apply them in actual work? They roughly have the following applications.

(1) Directly used as conceptual models to ensure that the basic features of main reservoir properties such as continuity, heterogeneity, etc. are correctly reflected. Such application is very successful in the early evaluation stage of oil and gas field development.

(2) Used to estimate the uncertainty of a geological model. The differences between multiple implementations themselves reflect the possible uncertainty.

(3) Selection under geological constraints. That is, development geologists select the most possible model based on geological knowledge and experience.

(4) Application of Monte Carlo style. Reservoir geological models are finally used for numerical simulation and prediction of production performance. The most optimistic, or the most pessimistic, or the most likely geological model is selected from multiple implementations, and the

production performance calculated from it also reflects the above three possibilities and is used as a basis for risk analysis and decision-making.

(5) Application of maximum probability. Choose the model with the maximum probability of occurrence among multiple implementations, and sometimes even apply the model averaged from multiple implementations into practice.

(6) Rapidly selected and simplified flow numerical simulation methods through development performance history fitting. At present, some oil companies have developed some simplified digital simulation software that can directly perform fast calculations without coarsening the geological model of millions of grids. Through rough screening, the optimal geological model is selected and then used in formal simulation development indexes. Stochastic modeling methods are currently booming.

The main development trends of multi-information integrated stochastic dynamic modeling technology are as follows. (1) vigorously carrying out outcrop surveying of various sedimentary reservoirs, and enriching prototype models and geological knowledge databases. (2) studying the adaptability of various simulation algorithms to different reservoirs and different geological targets. (3) developing new algorithms.

Section 2   Development Direction of Fine Reservoir Description

I. Comparison between domestic and foreign reservoir description technology levels

The historical process of the development of domestic and foreign reservoir description technologies is different, so they place emphasis on different specific problems to be solved, and the formed reservoir description technologies have their respective characteristics.

(1) The level of sedimentology is roughly the same at home and abroad. China's oil and gas fields are dominated by continental reservoirs, so China has formed its characteristic sedimentology theories and working methods in lake basin sedimentology and has formulated sedimentology research specifications for reservoir description in the petroleum industry, which have found very good applications in the development of oil and gas fields.

(2) In terms of quantitative geology, the domestic and foreign levels are close, and several quantitative geology and prototype model research bases have been established both at home and abroad. Foreign quantitative geology research results are represented by Gypsy profile in the United States, and domestic quantitative geology research results are represented by Luanping fan delta and Datong braided river outcrop. The establishment of quantitative geological knowledge databases provides a reference template for describing and predicting the spatial distribution of reservoirs more finely.

(3) Foreign companies are at the leading position in the development and application of new logging technologies. Domestic companies mainly import and utilize foreign logging technologies. In recent years, China has gradually formed its unique technologies for solving fracture fractures

through conventional logging and interpreting water-flooded layers and low-resistivity reservoirs.

(4) Foreign companies have a complete set of development seismic technology systems and take the lead in the development and application of new development seismic technologies, but there are still technical bottlenecks in prediction accuracy, and especially it is difficult to predict thin layers. Domestic companies have initially established their technical systems, but it is difficult to accurately predict thin reservoirs of less than 6m thick with these technical systems.

(5) Stochastic algorithm is one of the main development directions in geological modeling. Foreign companies have established a set of mature algorithm systems and formed relatively mature commercial software, while domestic companies mainly import and apply them.

(6) In terms of sequence stratigraphy, based on marine sequence stratigraphy, foreign companies have basically formed a relatively mature theoretical system. Some companies have also formulated the application specifications and manuals of sequence stratigraphy, and have achieved high levels in production and application. The initial research on sequence stratigraphy in China was based on the introduction of marine sequence stratigraphy theory. A series of research achievements have been made in the research and exploration of continental sequence stratigraphy based on the characteristic that continental deposits predominate in China, but there are still some theoretical problems of continental sequence stratigraphy that have not been solved. Especially in applications, different experts and scholars have formed different standards, which are difficult to unify, and bring certain difficulties to production and application.

II. Challenges in fine reservoir description

In order to meet the needs of tapping the potential of remaining oil and gas and enhancing oil recovery, finer geological models shall be established and the distribution of reservoirs shall be described on a smaller scale. Currently, a precision goal of transverse 100m-level × longitudinal centimeter-level has been proposed. To achieve this goal, far from solving problems with a single geological study, multidisciplinary comprehensive studies are needed. There are still many technical problems to be tackled in the series of fine reservoir description technologies.

(1) There are not enough types of outcrop prototype models. Some progress has been made in the establishment of reservoir prototype models, e.g., the establishment of prototype geological models of two sedimentary types such as fan delta and braided river, but for now, the types of outcrop prototype models established in China are not abundant enough.

(2) The application of high-resolution sequence stratigraphy in alluvial (fluvial) deposits is not mature yet. At present, high-resolution sequences are divided mainly based on logging curves, which still have certain difficulties and randomness.

(3) There is no effective method to evaluate the change of physical property parameters with the development process. The interpretation error of water flooded layers is large, and the coincidence rate is only below 70%.

(4) It is difficult to predict the thin layers in alternate layers of sandstones with mudstones.

(5) A true geological model of fractured reservoirs cannot be established yet for the description of fractured reservoirs. Only the faults at the seismic data interpretation level and the

minute fractures at the core slice data interpretation level can be effectively described. The middle-level fractures can only be predicted, and their parameters such as spacing, density, scale, etc. cannot be truly and comprehensively described.

III. Development direction of fine reservoir description

(I) **The research on reservoir sedimentology needs to be further deepened and developed**

In China's clastic reservoirs, onshore and underwater reserves account for half respectively. The research on high-resolution sequence stratigraphy of alluvial facies (mainly fluvial facies) is still one of the key points and difficult points. The study of reservoir heterogeneity needs to be strengthened. There is a need to strive to summarize the key heterogeneity characteristics and laws of different sedimentary types of reservoirs. A set of reservoir prediction methods for different sedimentary characteristics and different parameters must be developed to improve the prediction accuracy. It is necessary to continue the field outcrop study of different sedimentary types of reservoirs. By comprehensively using the research results of outcrops and dense well patterns, the geological laws of reservoir prototype models are summarized, and the geological characteristics and laws of the transformation from outcrops to underground are found. More realistic geological models shall be established to provide a scientific basis for the calculation of various parameters and decision-making. Moreover, it is necessary to carry out repeated experiments and tests on various understandings and methods to improve the practicability and reliability of technical methods.

(II) **Continued development of multidisciplinary synergistic research**

Since the 1990s, the technology of reservoir characterization has been moving towards the direction of multidisciplinary development. This is mainly due to the two factors below. On the one hand, as the depth of research continues to increase, any single discipline can no longer solve the new problems faced by reservoir research, and there is a must to take the road of multi-disciplinary comprehensive development. On the other hand, reservoir research is no longer a purely geological problem, but is to be applied as input data for reservoir engineering and numerical simulation. Therefore, these disciplines must be combined with each other in order to provide better and more accurate geological model input data, reservoir parameters, and numerical simulation parameters.

At present, due to the bottleneck problem of seismic resolution, it is very difficult to make a substantial breakthrough in the prediction of thin layers using seismic technology alone, and seismic means shall be combined with geologic means. Geological research shall provide geological models that can be predicted and applied in seismic research, so as to reduce the multiplicity and uncertainty of seismic prediction and improve prediction accuracy. Logging research shall mainly provide accurate parameters for special types of reservoirs (fractured reservoirs, low permeability reservoirs, low saturation reservoirs, etc.) based on general interpretation. The dynamic and static data of oil and gas fields shall be organically combined. Reservoir engineering and production performance data shall always be used to verify and modify reservoir description results and geological models and the geological models can be modified and corrected in real time and conform to actual underground conditions all the more.

(Ⅲ) Vigorous promotion and application of new technologies and new theories

The continuous development of the reservoir description technology is mainly driven by two aspects. On the one hand, the geological bodies we are facing are becoming more and more complex and are difficult to understand. A large number of oil and gas fields have entered the late stage of development, and the extensive development methods are far from meeting the needs of production. It is necessary to understand and study reservoirs on a finer scale than before in order to find out the objects for the further tapping of the potential and realize the stable production of old oil and gas fields. On the other hand, the continuous introduction of new technologies and methods makes it possible to achieve the above production requirements.

Logging technologies and seismic technologies have been rapidly developed in recent years, and the content that can be studied and the research accuracy have been greatly improved, directly promoting the improvement of the accuracy of reservoir description research. The development of computer application level has made it possible to study reservoirs on the same data platform by comprehensively applying multiple disciplines.

New technologies and methods that are constantly emerging provide new ideas for fine research. Among them, quantitative geological research technology, prototype geological model building technology and high-resolution sequence stratigraphy research technology are developing fastest. Under the dual promotion of these technologies and methods, the reservoir characterization technology has begun to develop from qualitative to quantitative, from macro to micro, from a single discipline to multi-discipline and multi-specialty.

The development and application of the following aspects shall be emphasized in the development of reservoir description technologies and methods in the future.

(1) Carry out theoretical and applied research on continental high-precision sequence stratigraphy, and look for stratigraphic boundaries that can be identified underground.

(2) Continue to promote and improve the application of prototype geological laws and geological knowledge databases in reservoir prediction.

(3) The development of new logging technologies and the development and optimization of a series of modeling algorithms.

(4) 4D seismic and multi-wave seismic are further developed on the basis of 3D seismic so as to solve the problems involving the prediction of remaining oil and gas distribution in old oil and gas fields and the prediction of complex reservoirs such as thin reservoirs, fractured reservoirs, etc.

Chapter 2   Contents and Steps of Fine Reservoir Description

Up to now, fine reservoir description has been comprehensively developed in terms of both individual technologies and comprehensive ideas and methods. This chapter systematically summarizes the working contents and steps of fine reservoir description from the aspects of its research content, research methods, research objectives and requirements, etc., and is divided into eight steps for the explanation. Each subsequent chapter details and concretizes the research content of each step.

The core of fine reservoir description is to achieve high-precision reservoir prediction and quantitative depiction of the internals of reservoirs. Therefore, the contents and steps of fine reservoir description are intended to anatomize and understand reservoirs hierarchically step by step, establish fine reservoir geological models, and predict remaining oil distribution centering on revealing the characteristics of reservoirs and comprehensively using multiple technical means based on the research idea "from macro to micro, and from overall to partial".

Ⅰ. Macroscopic regional background description

A geological process is an extremely complex event process, and the result of multiple factors such as climate, topography, water system, etc. A sedimentary process is complicated, but it is the result of natural actions, and there are some regular changes in sedimentary bodies. Therefore, through the study of the overall sedimentary background and sedimentary environment, this regularity can be grasped, and the development characteristics of sand bodies can be understood macroscopically so as to provide a basis for fine depiction of sand bodies.

Macroscopic regional background description focuses on the study of regional provenance characteristics and the types and characteristics of depositional systems, and the identification of sedimentary subfacies. The research methods adopted include provenance analysis, sedimentary facies analysis, and genetic mechanism analysis. Usually, in the fine reservoir description stage, the understanding of macroscopic regional background is relatively clear, there is no need to carry out systematic work, and the focus is to ascertain the control action of depositional systems based on the location of the fine description area. For example, the main reservoir of Daqing Oilfield is a large river-delta depositional system developed under a gentle slope background, the topographic slope of the basin bottom is about 1°, and the water bodies are shallow. This special sedimentary background determines that after entering the lake, the river can still maintain high energy, and the advancing distance to the lake basin water bodies is large, forming a constructive delta. In addition, the thickness of the sedimentary sand bodies is small, estuary dams are not developed, and there are obvious alternate layers of sandstones with mudstones. Specifically, the sand bodies in the anatomy area also have the characteristics of distributary channel sand bodies, and the distribution pattern of the barriers and interlayers also conforms to the sedimentary characteristics of distributary channels.

Similarly, typical low-permeation tight sand gas reservoirs are developed in He-8 Member of Shihezi Formation of Upper Paleozoic in Ordos Basin, which macro regional sediment is a set of braid river sedimentary system wide-developed under morass and swamp background. Pebbly sandstones, gritstones, medium sandstones, fine sandstones are developed, with a sand/gross ratio over than 50%, showing the characteristics of good continuity of sandstones interwells and continuous distribution in a large area. But, because of the large buried depth and strong compaction and sandstones tighten, only partial sandstones of coarser particles could become the effective reservoir with a proportion 25%-30% of the sandstone thickness. Therefore, under this regional background, it determined that, the effective scales of low-permeation tight sand gas reservoirs of Upper Paleozoic in Ordos Basin are small, low continuity and almost isolated distribution and low correlationableness interwells. The gas layer development structure feature of "sandstone-in-sandstone" should be fully taken into account during the fine description of different developing blocks.

Therefore, sedimentary backgrounds and sedimentary water systems determine the development scale and distribution law of reservoirs, the analysis of regional sedimentary backgrounds can provide a reference for further fine research on reservoirs, and the development characteristics of depositional systems must be understood.

II. Stratigraphic structure description

Fine reservoir description relies on the fine division and understanding of stratigraphic structure.

Stratigraphic division and correlation and fine structure interpretation are the basis for reservoir analysis. The distribution law of reservoirs can be truly reflected and the accuracy and reliability of reservoir prediction can be improved only by establishing a reasonable stratigraphic structure. The main methods of the fine stratigraphic division include cycle correlation and high-resolution sequence stratigraphy division, and the assisted technical means include core interface identification, logging curve comparison, fine seismic interpretation and development performance data analysis. Establishing an isochronous stratigraphic framework and ascertaining accurate structure distribution comprehensively using logging data, seismic data and development performance data are the main means to accurately establish stratigraphic configuration.

"Cycle correlation, hierarchical control" refers to the identification and correlation method of marker horizons and sedimentary cycles. The layer correlation method that began in the 1960s and 1970s in China adopted this cycle correlation method. In the 1980s, the fine correlation of oil layers for river-delta deposits gradually formed the fine reservoir correlation technology represented by that of the Daqing Oilfield. This technology has been widely used in tertiary oil recovery of thick reservoirs and increasing of well pattern density for thin poor reservoirs. This sedimentary cycle correlation method can be applied to both fluvial facies and lacustrine facies; although there is a certain diachronous phenomenon sometimes, it can be avoided to a certain extent by combining this method with various seismic means and logging means.

The application of high-resolution sequence stratigraphy is effective for the prediction and evaluation of reservoirs in the development stage, and is an important means of fine reservoir description. The principle of high-resolution sequence stratigraphy is: based on core data, logging data, and seismic data, the internal structural characteristics of strata of different levels are analyzed, and a high-resolution stratigraphic framework is established through the identification of base-level cycles and isochronous correlation; moreover, local fine stratigraphic correlation is performed according to the characteristics of low-level cycles, which can provide a basis for fine reservoir description. The identification of base-level cycles is applicable to lacustrine and marine strata but has certain limitations in alluvial facies. Therefore, the application of high-resolution sequence stratigraphy also requires specific analysis according to specific issues.

III. Description of sedimentary microfacies of reservoirs

The sand bodies of genetic units are identified, and the spatial distribution and superimposition patterns of the sand bodies are determined by subdividing sedimentary microfacies. The sedimentary microfacies analysis technology is the key technology in this step. Sedimentary facies research runs through the entire process of geological work and is also essential research content in the process of oilfield exploration and development. With the deepening of oilfield development, various accumulated data are gradually enriched, and the accuracy of sedimentary facies research is also higher and higher. In recent years, with the continuous advancement of outcrop analysis technology, dense well pattern anatomy technology, logging technology, and seismic technology, the geological theory of sedimentary facies research has continued to innovate, and the established reservoir deposition models have also become finer.

The subdivision results of sedimentary microfacies give sedimentary sand bodies the meaning of genesis. Therefore, the combined contact relationship of different microfacies from the perspective of sedimentology can be analyzed, the spatial distribution of sand bodies can be determined, and sand body superimposition patterns can be established with the aid of outcrop anatomy, modern sedimentary observation, etc., thereby building reliable reservoir deposition models which are used to provide guidance to reservoir prediction, finely depict the spatial distribution of sand bodies in different reservoirs, and reveal more subtle heterogeneous characteristics of reservoirs.

Taking Xingerzhong area of Daqing Oilfield as an example, the fine sedimentary microfacies research results show that the main reservoir is in the delta plain subfacies sedimentary environment, which can be further subdivided into five types of sedimentary microfacies such as distributary channel, branch channel, abandoned channel, overbank deposit (natural levee, splay), and distributary bay (Figure 2-1). As the main reservoirs, the distributary channel sand bodies are the focus of fine anatomy. This kind of distributary channel sand bodies can be developed either continuously or in isolation. In order to finely depict the distribution characteristics of such sand bodies, firstly abandoned channels are identified according to logging curve shape, sand body thickness change, sedimentary microfacies combination, etc., thus dividing the different channel sand bodies continuously developed into a single channel sand body as much as possible. on this basis, point sand dam sedimentary bodies are further identified, laying a foundation for the correlation of lateral accretion layers inside point sand dams and the fine description of the internal structure of sand bodies.

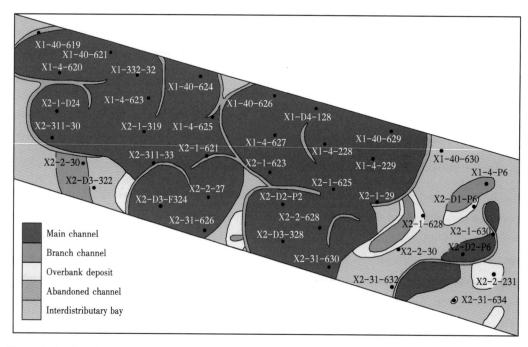

Figure 2-1   Distribution map of the sedimentary microfacies of Pu-I31 in Xingerzhong area of Daqing Oilfield

IV. Description of the geological knowledge base of reservoirs

The purpose of establishing the geological knowledge base of the prototype model is to determine the shape, scale (length, width, thickness, and proportional relationship), development frequency, and other parameters of the sand bodies of genetic units, and provide quantitative parameters for the prediction of interwell sand bodies. Seismic information is an important basis for interwell prediction of the reservoirs, but due to the limitation of the seismic resolution, only reservoirs with a large thickness (usually greater than 5m) can be predicted using seismic information. It is difficult to predict thin sand bodies and split thick sand bodies, while this is the focus of the remaining oil tapping work in the late stage of oilfield development. The key technology to solve this problem is the geological knowledge base technology of the reservoir prototype model.

The establishment of the geological knowledge base of the reservoir prototype model is an important aspect of fine reservoir description. Compared with the digitization of conventional reservoir description and display methods, the characterization of sand body characteristics is more digitized and quantitative, which can provide a quantitative parameter basis for interwell reservoir prediction and reservoir modeling. The geological knowledge base of the prototype model refers to the parameters that can quantitatively characterize the spatial characteristics, boundary conditions, and physical characteristics of various sand body genesis units, as well as various sedimentary models that can be qualitatively characterized, mainly including lithology and lithofacies library, sedimentary environment and sedimentary microfacies library, geometry library, physical parameter library, diagenetic library, etc. (Table 2-1). There are very typical examples at home and abroad for the study of the geological knowledge base of the prototype reservoir model. The first foreign example is the Gypsy profile invested and studied by BP, which has greatly enriched the geological knowledge base of the fluvial facies prototype model. For the two outcrops in Datong and Luanping in China, detailed description, measurement, sampling, etc. were performed; in addition, drilling and logging data were acquired and analyzed, the internal structure characteristics and change laws of Luanping fan delta deposits and Datong braided river deposits were fully anatomized in detail, and the geological knowledge bases of the two types of outcrop reservoir prototype models were established. Moreover, the prediction methods of the two types of sand bodies have been summarized using the stochastic modeling technology.

Table 2-1 One of a series of geological knowledge bases of Luanping fan delta

Genetic unit	Shape	Size		Barrier layer	Sedimentary structure	Grain size	Sorting property
		Width, m	Thickness, m				
Plain mudflow		1-3	10.0-30.0	Horizontal argillaceous siltstone	Massive bedding, large trough cross-bedding		
Braided channel		300-500	1.5-8.0	Horizontal mudstone, horizontal siltstone	Massive bedding, parallel bedding, tabular bedding		
Near-rock channel		200-1000	3.0-16.0	Horizontal mudstone, muddy argillaceous siltstone	Parallel bedding and cross-bedding (trough, tabular)		
Infralittoral channel		100-500	0.5-5.0	Horizontal mudstone	Parallel bedding, cross-bedding		
Natural levee		50-200	0.5-1.5	Wavy argillaceous siltstone	Parallel bedding, wavy bedding		
Mouth bar		30-100	2.0-4.0	Wavy argillaceous siltstone	Cross bedding		
Far sand dam		20-70	1.0-3.0	Wavy argillaceous siltstone	Small cross bedding		
Sheet sand		300-2000	0.5-2.0	Horizontal shale	Massive bedding, parallel bedding, wavy bedding		

Continued

Genetic unit	Shape	Size		Barrier layer	Sedimentary structure	Grain size	Sorting property
		Width, m	Thickness, m				
Slump turbidite		50-200	2.0-6.0	Horizontal shale, inclined mudstone	Massive bedding		
Overbank deposits		100-300	1.0-1.5	Horizontal mudstone	Wavy bedding		

The methods for establishing the geological knowledge base of the prototype model include outcrop fine anatomy, dense well pattern anatomy, modern deposition, and simulation test research. Combining these methods, learning from other's strong points to offset one's weakness, and mutual verification among them is the most ideal way.

However, underground geological bodies generally lack comparable outcrops and modern sedimentary bodies, and the most commonly available data are core data and dense well pattern data. Therefore, dense well pattern anatomy has become the most commonly used technology. The dense well pattern here can be a dense well pattern in the research area, or a comparable dense well pattern in mature oilfields. The geological knowledge base of the prototype model established by dense well pattern anatomy has lower accuracy than that established based on outcrop data or modern sedimentary data, but it can be used to guide reservoir prediction research in relatively sparse well pattern areas. At present, through the use of horizontal well data and production logging data, the accuracy and credibility of dense well pattern anatomy have been greatly improved.

V. Research on the prediction of seismic interwell reservoirs

Seismic technology has always been an important means of reservoir prediction and plays an important role in the exploration and development stage of oilfields. Digital fine reservoir description must use seismic data, which can provide both fine structure interpretation and a basis for interwell reservoir prediction. The core of the aforementioned four steps is the study of the sedimentary model of geological bodies. Under the guidance of the sedimentary model combined with the interwell reservoir distribution information provided by seismic data, interwell reservoir prediction can be performed more reliably. The key technology in this step is the development and application of fine seismic technology in the development stage.

Improving the interpretation accuracy of seismic data is a prerequisite for the effective application of seismic data in the development stage, and can provide a reliable basis for studying inter-well reservoir changes. In terms of thinking, the oil (gas) fields entering the middle and late stages of exploration and development are extremely rich in various basic data and the understanding of reservoir characteristics is also deep. Therefore, it is necessary to integrate as much existing data information and mature geological knowledge as possible in the process of seismic data processing and interpretation to form a fine seismic data processing and interpretation technology that uses rich well data as hard data and mature geological knowledge as soft ideas. In terms of technical means, the development and improvement of new seismic technologies provide effective means for finely depicting the geometric structure and physical parameter distribution of reservoirs during the development and production of reservoirs, and for dynamically monitoring fluid changes during the development process, e.g., vertical seismic profile (VSP), crosswell seismic, time-lapse seismic, etc. Although these technologies are not yet mature enough, they have shown great development potential in application practice.

The VSP technology is mainly used to interpret near-wellbore fine structures, describe lithology and obtain fracture parameters. This technology can cooperate with the interpretation of seismic data in the well area so as to provide parameters for the description and geological modeling of near-wellbore reservoirs. The time-lapse seismic technology is also called 4D seismic technology and time-delay seismic technology. The basic principle of this technology is to conduct dynamic monitoring of reservoirs by observing changes in seismic information of reservoirs over time. This technology plays a special role in finding remaining oil in old oilfields and dynamically monitoring reservoirs. The crosswell seismic technology can be used to predict the lateral changes of crosswell geological conditions through crosswell seismic imaging (including crosswell tomography and crosswell reflection wave imaging), and provide services for crosswell geological modeling and fine reservoir description. Field application results in several oilfields such as Daqing and Shengli etc. show that the crosswell seismic data can be used to distinguish strata with a thickness of 2-3m, and the amplitude resolution capability may reach 1m, especially for crosswell small faults (Figure 2-2).

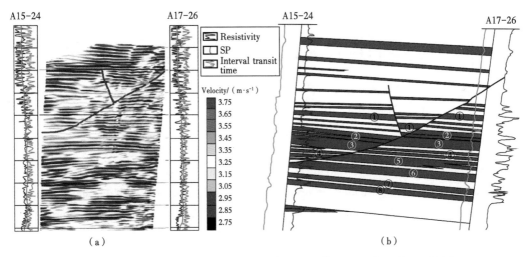

Figure 2-2 Application effect diagram of cross-well seismic data of an oilfield

In addition, the reprocessing and reinterpretation of the existing 3D seismic data of old oilfields is an important aspect of improving the resolution of seismic data, where the main method adopted is seismic frequency-division processing. The seismic reflections generated by geological bodies with different thicknesses correspond to different discrete frequency components in the frequency domain, thick layers correspond to low frequencies, and thin layers correspond to high frequencies. The frequency division technology mainly extracts geological bodies with different frequency components so as to obtain the complete geological characteristics of geological bodies through the combination of different frequencies. The seismic frequency division technology is suitable for the processing and interpretation of 3D seismic data. Its resolution is higher than the resolution capability which can be achieved by the conventional seismic main frequency technology, and it has some advantages in determining reservoir boundaries and predicting reservoirs.

VI. Description of single sand body architecture

As an oilfield enters the middle and late stages of development, the focus of digital fine reservoir description is not the continuity of reservoirs, but their separability. Various isolated reservoirs that are not swept by injected water are the main areas where the remaining oil is enriched. The establishment of a reservoir architecture model can further finely depict the internal structural characteristics of reservoirs, reflect the separability of reservoirs, and describe the lowest-level separator to the greatest extent. As shown in Figure 2-3, through the identification and correlation of lateral accretion layers, a single point sand dam is divided into multiple lateral accretion bodies, thereby revealing the internal separability of sand bodies more finely, and reflecting their internal heterogeneity.

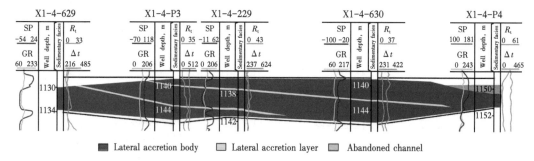

Figure 2-3　Anatomy of the internal architecture of single channel sand bodies in Block Xingerzhong of Daqing Oilfield

In this step, the reservoir architecture analysis technology is mainly used. Reservoir architecture analysis is also called reservoir architectural structure analysis and hierarchical structure analysis. In 1985, Miall et al. deeply studied the division of low-level interfaces in a depositional system and proposed a set of methods and theories for the study of hierarchical interface division and corresponding structural units. The hierarchical interface analysis method emphasizes the study of the hierarchy and structure of the system itself from the point of view of system theory, as well as the isochronism and discontinuity of deposition; therefore, the method is compatible with sequence stratigraphy upwards, and can be subdivided infinitely downwards while being always consistent with sedimentary genesis analysis. Based on this, scholars at home and abroad have extensively carried out research on reservoir structure units (reservoir architecture, reservoir architectural structure, and hierarchical structure), and conducted fine anatomy of reservoirs. At present, the degree of research on fluvial facies is the highest. In recent years, domestic scholars have carried out a lot of research. Jia Zhenyuan and Zhang Changmin introduced the research ideas and methods of reservoir architecture from different aspects. Although the terminology used is different, it belongs to the same conceptual system as the study of reservoir architecture, which reflects the hierarchy of the study of reservoir architecture. In addition, they pointed out the method of dividing and identifying architectural elements. Since then, Yin Taiju, He Wenxiang, Liao Guangming, *et al.* analyzed delta sedimentary bodies from the perspective of reservoir architecture analysis. Yin Taiju identified the architectural elements of fan delta sedimentary bodies. Guided by Miall's architecture analysis theory, Wu Shenghe, Yue Dali, Jiang Xiangyun, *et al.* conducted an architecture analysis of underground meandering river point dams on the basis of predecessors' analysis of modern deposits and outcrops of meandering river point dams. These research results show that the study of reservoir architecture from a static perspective provides a sedimentary basis and a new research approach for the further subdivision of strata and the study of reservoir heterogeneity, as well as new research ideas and methods for studying the internal characteristics of reservoirs in digital fine reservoir description.

Ⅶ. Description of physical property parameters of reservoirs and fluid distribution

The study of the previous six steps is intended mainly for the establishment of a reservoir skeleton model. The establishment of reservoir physical parameters and fluid distribution models is the difficult point of modeling work in the late stage of oilfield development. The key technology is the reservoir parameter evaluation technology.

The description of reservoir physical parameters includes quantitative interpretation of physical parameters, analysis of influencing factors, spatial distribution law analysis, and interwell prediction, as well as research on the evolution law of physical properties. The study of fluid distribution law refers to the qualitative and quantitative description of the type, nature, and saturation etc. of fluids contained in reservoirs. For oilfields entering the middle and late stages of development, the reservoir physical parameters and fluid distribution are more complicated. Especially for long-term waterflooding oilfields, due to the influence of factors such as reservoir permeability, rock cementation degree, production rate, crude oil viscosity, etc., the distribution of reservoir physical parameters changes greatly during water flooding. Generally, highly water-flooded zones gradually form relatively high permeability channels, i.e., large pore channels, which cause ineffective water circulation, seriously affect oil displacement efficiency, and lead to enrichment of remaining oil in the areas not swept by injected water; in addition, some low-permeability reservoirs will form blockages, reducing the oil phase permeability. Therefore, the formulation of remaining oil potential tapping measures must consider changes in reservoir physical parameters, especially the distribution and control factors of large pore channels. The study of the distribution of static reservoir physical parameters alone can no longer meet the requirements of the late stage of oilfield development. In recent years, oilfields have continuously strengthened research on the formation mechanism, identification methods, and distribution models of large pore channels in reservoirs, which is beneficial to EOR of oilfields by taking water plugging and profile control measures.

Digital fine reservoir description is mainly carried out for old oilfields in the middle and late stages of development. Therefore, changes in reservoir physical parameters and fluid distribution laws, especially the formation and distribution laws of large pore channels, are the research content worthy of attention.

VIII. Establishment of integrated reservoir geological models

The analysis of the previous steps provides a variety of parameters and constraints for reservoir geological modeling, thereby obtaining a clearer geological understanding. The final key step is to establish a reliable reservoir model through the optimization of reservoir modeling methods and the introduction of geological knowledge and to predict remaining oil distribution through numerical simulation. At present, a lot of reservoir description software integrates a variety of modeling interpolation methods, which can fuse various data and information. The software not only is mature in 3D display and human-computer interaction but also has made great progress in interwell reservoir prediction. In addition, the development of 3D visualization technology, virtual reality technology, and human-computer interaction integration technology has made reservoir modeling more digitized and automated and has also provided technical support for digital fine reservoir descriptions.

Chapter 3 Regional Background Description

Macroscopic regional background description does not refer to the description of geotectonics and climatic conditions etc., but the description of the development characteristics of depositional systems. What is more relevant to the fine reservoir description in the development stage is the control of sedimentary facies characteristics on genetic units. Therefore, the macroscopic regional background description here focuses on the description of the development characteristics of sedimentary facies, including their provenance properties, development characteristics, facies belt distribution, etc. This chapter focuses on the development characteristics of clastic sedimentary facies. There are mainly three common genetic types, namely alluvial fan sand bodies, river sand bodies, and lacustrine sand bodies. In different types of sand bodies, due to the differences in sedimentary subfacies and microfacies, there are certain differences in the distribution and storage performance of sand bodies.

Section 1 Content and Methods of Regional Background Description

I. Sedimentary background description

As an oilfield enters the development stage, the understanding of the macroscopic regional background is mature, so there is no need to carry out a lot of work, but it is necessary to master the provenance characteristics, topographical conditions, and hydrodynamic conditions of the research area. The research on this part of the content is the investigation and analysis of regional research results to a certain extent, without the need to carry out systematic research work. Here is a brief introduction to several research methods of provenance analysis.

Provenance research includes determining the material source direction of sediments, the location of the provenance area, provenance properties, and the transport path of sediments, which has reference significance for determining the distribution direction of depositional systems and the distribution law of genetic units. There are many provenance analysis methods. Conventional provenance analysis methods include heavy mineral method, clastic rock method and deposition method; in addition, fission track method, geochemistry method and isotope method are also commonly used research methods. The fundamental basis for these methods lies in some differences in the characteristics of rocks, mineral compositions and their combinations the development status of strata (including contact relations and sedimentary interfaces etc.), the lateral change and vertical superimposition of lithofacies, geochemical characteristics, etc. from different provenance areas.

The heavy mineral method includes single mineral analysis and combined mineral analysis. In addition to drawing planar distribution maps, mathematical methods such as cluster analysis

(R-type or Q-type), factor analysis, and trend surface analysis can also be used to study mineral combination characteristics, similarity and other indexes, so as to extract the information reflecting provenances.

The clastic rock method reflects the provenance area and transportation trend through the clastic composition and structural characteristics of clastic rocks. A statistical analysis can be made on the content of quartz, feldspar, and debris in sandstone samples from the selected horizon, then, triangle graphs are used for value projection, and the provenance type and transportation direction are determined according to the distribution of points. In addition, the provenance area can be identified according to the grain size change characteristics of clastic rocks, including the content of glutenites, the change law of median grain size, etc.

The principle of the deposition method is: based on the drilling data, logging data, seismic data and other data of a basin, the stratigraphic isopach map, sandstone thickness map, sand-strata ratio change chart and other related maps of a certain period are worked out through detailed stratigraphic correlation and division, and the relative position of the provenance area is inferred; in addition, the provenance area is made more reliable by means of lithology, composition, sedimentary body morphology, grain size, sedimentary structure (wavemark, cross-stratification, etc.), paleocurrent direction, plant micro-fossils, *et al*.

II. Description of depositional system and sedimentary facies

The method of describing depositional system and sedimentary facies types is mainly to determine the types of depositional system and sedimentary facies through the identification of facies markers, divide subfacies, and establish the corresponding sedimentary model suitable for the characteristics of the research area. Depositional system and sedimentary facies are two inseparable concepts.

Depositional system is an explanatory concept developed by American sedimentologists in the petroleum exploration practice of the Gulf of Mexico due to the needs of paleoenvironment reconstruction and basin analysis. Although this concept has been applied for many years in the petroleum geology field, especially in the petroleum geology field of Texas, the understanding of it is not consistent. The book "Depositional system: An Introduction to Sedimentology and Stratigraphy" published by Davis (1983, 1992) summarizes this important concept of sedimentology.

According to the definition by Scott and Fisher (1969), a depositional system is a combination of sedimentary environments held together by deposition. A deposition system may include multiple deposition environments, and each deposition environment has its characteristic sediments, biological combinations, and deposition. The basic constituent unit of a depositional system is a combination of facies representing a particular depositional environment and is called the framework component of the environment. Any depositional system is composed of some genetically related units. Miall (1984) pointed out that the concept of depositional system is an extension of Walther facies rule. According to Walther facies rule, the vertical relationship of sedimentary sequences is a reflection of their lateral relationship. For example, a delta depositional system consists of three parts such as prodelta, delta front, and above-water delta. Each part is a sedimentary facies that differs from the surrounding sediments in terms of lithology, sedimentary structure, and biological appearance. The three parts represent three different depositional environments. However, the lateral relationships and vertical sequences of the three parts are completely unified, and they together form a depositional system. Therefore, he defined a depositional system as a complete combination of a set of environments and their sedimentary products (i.e., sedimentary facies), which may be of unconformity up and down or defined by another depositional system that is completely unrelated in genesis. The sedimentary facies within a depositional system obey Walther facies rule, and their relationships are integrated. But there is no continuity of deposition or its products between deposition systems. Delta deposits can cover both littoral sediments and neritic sediments in the process of progradation, which is a very good example.

Posamentier *et al.* (1988) deemed that a depositional system is a 3D combination of lithofacies linked by deposition or sedimentary environments, and contains both the concept of sedimentary environment and the concept of deposition. The depositional systems formed at the same time and occupying equal positions in the spatial framework of natural geography are linked together to form a depositional system tract. Different depositional system tracts have different occurrence locations relative to sea level. Based on this, the system tract can be divided into three types such as lowstand system tract (LST), transgressive system tract (TST), and highstand system tract (HST). Some people also list the depositional systems formed in the process of regression and call them the regressive system tract (Liu Zhaojun *et al.*, 2002). The system tract is a combination of depositional systems closely coexisting in geographical distribution. Any depositional system tract is isochronous. Biological fossils or isotope data are the main tools to determine isochronous surface, but in practice, it is generally determined by discontinuous surface.

"Facies" or "sedimentary facies" in sedimentology is a basic concept in geology, but it is also a long-term controversial concept. In the early development of geology, the term "facies" was introduced into geological literature by the Danish scholar Steno (1669). At that time, Steno only used "facies" to represent "period" and "stage" in the sense of stratigraphy. The first to give "facies" the meaning of sedimentology was the Swiss scholar A.Gressly (1838). When A.Gressly was studying the Jurassic strata in northwestern Switzerland, he found that the strata had great changes in lithology and paleontology. Therefore, A. Gressly used "facies" to describe such changes. He deemed that the various changes in the stratigraphic unit-"facies" have two main characteristics: (1) stratigraphic units with similar lithology must have the same paleontological combination; (2) stratigraphic units of different lithologies cannot have the biota of the same species (G. V. Miodleton, 1978). However, later geologists were confused when using the term "facies", and various understandings appeared, such as "turbidite facies", "biological reef facies", etc. In some cases, "facies" referred to a sedimentary environment, such as "fluvial facies" and "littoral facies" etc., In some cases, "facies" was related to a tectonic environment, such as "molasse facies" and "flysch facies" etc., but few people used "facies" as a stratigraphic unit in stratigraphy. The meaning of the term "facies" was confusing, so some people once argued that when using the term "facies", "as long as the meaning of the word was clearly pointed out, various uses of the term 'facies' were feasible" (HG Rending, 1986).

In recent years, with the rapid development of sedimentology, the understanding of "facies" has gradually become unified. At present, most people in geological circles at home and abroad regard "facies" or "sedimentary facies" as material manifestations of sedimentary environments. In a certain depositional environment, a certain deposition is carried out and a certain deposition combination is formed. The various characteristics of sedimentary environment and deposition will inevitably leave some records in these sedimentary products. These records are mainly manifested in differences in rock composition, geometry, texture, structure, and biological fossils. Therefore, "facies" should be the regular synthesis of lithological and paleontological characteristics that can indicate depositional conditions (Л. Б. Рухин, 1953, 1959; H.E. Peineck, I.B. Singh, 1980). According to this definition, "facies" and "environment" are not the same concept. "Environment" is the condition and cause, while "facies" is the product and result of various effect in the environment.

It can be seen that the sedimentary environment is the determining factor in the formation of sedimentary rock characteristics, and sedimentary rock characteristics are the material manifestations of the sedimentary environment. In other words, the former is the basic cause of the latter, and the latter is the inevitable result of the development of the former.

With the deepening of research, people are used to further dividing sedimentary facies into "subfacies" and "microfacies". "Subfacies" is the further subdivision of a certain sedimentary facies. The term "microfacies" today is different from its original meaning. Today's "microfacies" mostly refers to the further subdivision of "subfacies", and is used to indicate the small facies in a large facies environment. In the early days of sedimentary facies research, the term "microfacies" represented the meaning of "structural facies". Brown (1943) proposed the term "microfacies" to indicate the characteristics of the microenvironment displayed on a microscope. In other words, "microfacies" is the sum of all paleontological and sedimentological signs that can be classified in slices, peels, and polished sections. However, this definition requires the following conditions: the magnification factor of the microscope used in the research on slices, peels, and polished sections can reach nearly 200. Paleontological signs and sedimentological signs are considered equally. Data are classified (e.g., biological combination, limestone classification and morphological components) according to qualitative and quantitative signs. Microfacies classification is verified by field geological data, paleoecological interpretation, and comparison with possibly geochemical signs.

The division of depositional systems and sedimentary facies should be based on natural geographic conditions or geomorphic features and the comprehensive characteristics of sediments, and should be based on the principles of simplicity, ease of memory and understanding. Sedimentary facies are classified according to the three levels such as sedimentary background, depositional system and reservoir facies type (Table 3-1), and then the corresponding sedimentary subfacies and microfacies can be determined according to the sub-environment, microenvironment and sediment characteristics of each type of facies.

Table 3-1 Classification of sedimentary facies

Sedimentary background	Depositional system	Reservoir facies type
Continental	Alluvial fan	Mudflow fan
		Braided river fan
		Low bending meandering river fan
	River	Meandering river
		Braided river
		Anastomosing river
	Terminal fan	
	Aeolian desert	
	Glacier	
	Lacustrine	Fan delta
		Braided delta
		Delta
		Beach dam
		Turbidity current
		Windstorm
Transitional	Deltaic coast	Fan delta and braided delta
		River-dominated delta
		Wave-dominated delta
		Tide-dominated delta
	Non-deltaic coast	Tidal-flat
		Estuary
		Coastal plain and barrier island
Marine	Neritic	Tide-dominated shallow sea
		Storm wave-dominated shallow sea
		Ocean current-dominated shallow sea
	Bathyal—Pelagic deposit	Turbidite fan and non-fan turbidity facies
	Marine carbonate rock	Biogenic reef
		Carbonate shelf
		Platform slope, basin

Section 2　Alluvial Fan Glutenite Bodies

Ⅰ. Overview

Where mountain rivers or intermittent floods flow out of a mountain pass and enter an alluvial plain, a large amount of detrital materials (gravels, sands, mud) are quickly accumulated at the mountain pass due to sudden slowing of slope, reduction of river flow velocity, dispersion of water flow, and weakening of river transport capacity, so as to form a fan or cone tilted toward the plain, which is called an alluvial fan (cone) body or proluvial fan (cone) body.

Alluvial fans often appear in groups, are distributed along the foothills and connected laterally to form fan skirts, and become the prominent marginal facies in the margin of a sedimentary basin. Alluvial fans are mostly distributed along large faults in the boundary of the basin. When faults are active, mountains are uplifted quickly, and weathering and denudation are rapid, there are many coarse fragments generated and fans are also large. Alluvial fans are very developed in inland sedimentary basins, especially large inland depression basins or small rifted basins in arid regions.

Alluvial fans are adjacent to the eluvial facies and clinothem facies in the provenance direction; alluvial fans tend to be connected with alluvial plains or aeolian dry salt lakes in the direction of sedimentary areas, and can even directly prograde into lakes or coastal seas so as to form fan deltas. This depends on geographic location, structural conditions and regional climate conditions. According to climatic conditions, alluvial fans can be roughly divided into dry land fans and wet land fans (Figure 3-1), and their characteristic differences are shown in Table 3-2.

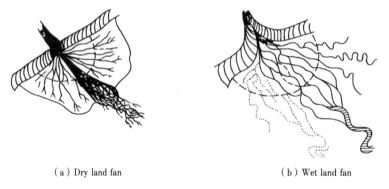

(a) Dry land fan   (b) Wet land fan

Figure 3-1   Dryland fan and wet land fan system (according to Galloway, 1983)

Table 3-2   Comparison of main characteristics of dry land fans and wet land fans

Type	Dry land fan	Wet land fan
River nature	Sudden and intermittent torrents	Perennial river
Fan radius	1.5-8km, maximum 25km	50-140km
Slope	Steep, 3°-10°	Gentle, mostly less than 1.5°
Sedimentary characteristics	Developed mudflow and sheetflood, relatively few braided river sedimentary facies	Dominated by braided river

The overall lithological characteristics of alluvial fans include coarse and messy composition, wide grain size distribution from mud and sands to boulders, high content of gravels and sands, and poor sorting property and roundness. In addition, the clastic components completely inherit the components of the parent rocks in the provenance area.

II. Alluvial fan sand body structure

(I) External geometrical morphology

An alluvial fan is fan-shaped or cone-shaped on the plane; the cross section of the fan body is a lenticle with a flat bottom and a convex top, and the vertical section is a wedge-shaped body with a flat bottom and a slight concave top (Figure 3-2). From the mountain pass outwards, both the thickness of the fan body and that of the single gravel layer become small, and the fragments change from coarse to fine, and the sorting property changes from extremely poor to good.

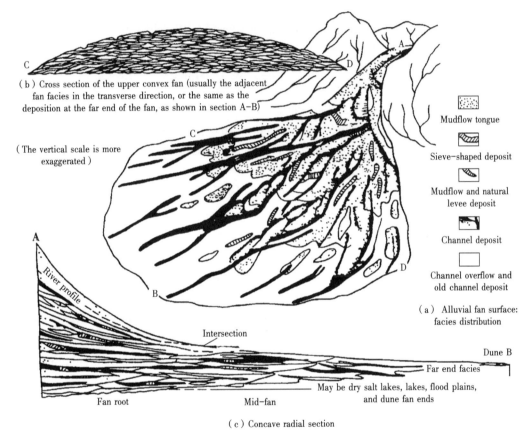

Figure 3-2 Sedimentary facies distribution and geomorphic features of an ideal alluvial fan
(according to Spearing, 1975)

(II) Facies belt distribution

The distribution of sand bodies inside an alluvial fan is very complicated and the regularity is poor. From the source area to the deposition area, the alluvial fan can be roughly divided into three subfacies.

(1) Fan root: also known as fan top, fan head, proximal end, and proximal base. The fan root is close to the mountain pass and has a large slope angle. The depositional types mainly include mudflow deposits (muddy, argillaceous glutenites) and/or sheetflood deposits (sandstones with relatively low mud content), and a small amount of riverbed filling deposits and/or sieve-

shaped deposits. The sand bodies are thick, coarse in size, and poorly sorted.

(2) Mid-fan: formed in a medium or low slope zone. The mid-fan is generally composed of alternate layers of braided channel deposits with mudflow and sheetflood deposits. Large-scale multi-layered cross-beddings can be developed in channel deposits, where the gravels are mostly arranged in an imbricate shape, and the flat surface is inclined to the mountain pass. Mudflow deposits mostly have a blocky structure and the gravel distribution is disorderly. Sheetflood deposits are often massive, and there may also be cross beddings or fine lamina. In general, the grain size of the deposits in the mid-fan is slightly smaller than that in the fan root, and the sorting property is better than that in the fan root.

(3) Fan end: also known as fan edge and distal facies, distributed at the tail of the alluvial fan, characterized by a low slope angle. the Fan-end deposition is dominated by flooding; sands, silty sands and argillaceous deposits are formed, wavy and horizontal beddings and massive structures are visible, and the thickness of the sand layers becomes small. However, braided channels can also be developed at the fan end, and well-sorted sandy deposits are formed, where cross beddings and parallel beddings are developed, and gravel fragments are arranged in an imbricate shape. In general, the sand bodies have small thicknesses and good sorting properties, but their shale content may be high.

Due to the rapid deposition of alluvial fans and complex changes in longitudinal and transverse lithology, the division of the microfacies is usually difficult. At present, there is no unified division scheme for the microfacies of alluvial fans.

(Ⅲ) Vertical sequences

Generally speaking, an ultra-thick alluvial fan is a complex from vertical superposition and lateral connection of multiple lobes formed by multiple torrents. For a single layer formed by a torrent, as the energy of the torrent decreases, the vertical section can show a positive rhythm that tapers upward slightly, and the bottom is the scouring surface. For a complex, if the subsequent lobes recede toward the mountain land, the complex shows a sequence that becomes thin upwards, which is a positive cycle; on the contrary, if the subsequent fan continues to advance into the basin, it is a reverse cycle.

The SP curve of the downhole profile of the alluvial fan appears as a blocky shape with inconspicuous stratification or a blocky shape with a sawtooth. If the shale content of the fan root is high, the permeability will be greatly reduced. Due to the decrease in shale content, the permeability of the mid-fan is high, and the SP amplitude increases accordingly, so that the lithologic section of the positive cycle shows funnel-shaped reverse cycle logging curve characteristics (Figure 3-3).

Ⅲ. Reservoir properties of alluvial fan sand bodies

Alluvial fan glutenite bodies are mainly coarse clastic rocks. Due to the particularity of alluvial fan deposition, not all alluvial fan coarse clastic rocks can form good reservoirs. The reservoir properties of alluvial fans are a comprehensive function of factors such as the nature of the parent rocks in the source area, climatic conditions, deposition types, facies belts, etc.

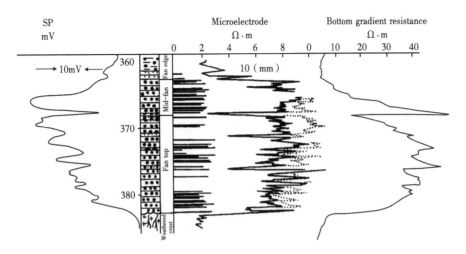

Figure 3-3 Vertical sequences and electrical characteristics of coarse clastic rocks in the alluvial fan (according to Zhang Jiyi, 1985)

When the parent rocks in the source area develop argillaceous rocks and less vegetation, under dry climate conditions, mudflows are very developed, and the shale content in sheetflood and channel deposits is high, forming extremely poorly sorted argillaceous glutenite mixed with gravels, sands and mud. The reservoir properties of such deposits are poor and they generally do not form reservoirs.

If the parent rocks in the source area have low shale content and the climate is not very dry or is even humid, mudflows are not very developed and are mainly distributed at the fan top, and alluvial fan deposits are characterized by a mixture of argillaceous gravels and sands. Such deposits can form oil and gas reservoirs.

From the perspective of the three subfacies of the alluvial fan, the reservoir properties of the fan-top glutenite are complicated. This is because the glutenite can contain poorly porous mudflow deposits, sheetflood deposits with variable reservoir properties (meaning that the reservoir properties change with changes in shale content), and channel filling deposits with relatively good reservoir properties, and even can develop sieve-shaped deposits with very good porosity and permeability. The quality of reservoirs and reservoir properties depends on the relative proportion of these deposit types. The reservoir properties of the mid-fan are relatively good, braided filling deposits are relatively developed, the deposits have been sorted to a certain extent (but in general, their sorting property is still poor), and the shale content is relatively low, so the mid-fan has reservoir properties to a certain degree. The fan end is dominated by sheetflood deposits, with relatively more suspended mud and relatively poor reservoir properties.

The rock structure of alluvial fan reservoirs has typical "complex modal" structure characteristics, showing that part or all of the large pores with gravels as the framework are filled with sand grains; in addition, sand grains are partly filled with clay. This intricate structure combination of gravels, sands and mud is called a "complex modal" structure. The pore distribution is also very complicated in this rock structure. The pore throats are distributed with multiple dispersed peaks, and the heterogeneity of the micro-pores is extremely strong.

In general, the reservoir properties of alluvial fan glutenite are very complicated. When exploring and developing such reservoirs, be very cautious and analyze specific conditions.

Section 3 River Sand Bodies

In areas with long-term slow subsidence, if the climate is semi-arid to humid, rivers will be the main geological agent on land. Rivers move about between alluvial fans and delta plains, forming vast alluvial plains. In a broad sense, rivers include braided rivers distributed radially on alluvial fans and distributary channels on delta plains. However, fluvial facies are usually limited to an alluvial plain, that is, the river deposits in the geomorphic unit from the alluvial fan end to the first diversion point on the delta plain.

There are various shapes and types of rivers. But to sum up, rivers can be divided into four types such as straight rivers, braided rivers, meandering rivers, and anastomosing rivers, which is also the general classification in current river sedimentology (Figure 3-4).

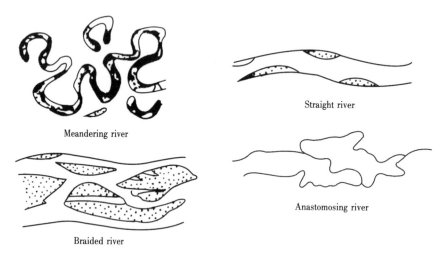

Figure 3-4 River classification (according to Miall, 1977)

This river classification is based on the braid index and bending of river channels. Rust (1978) proposed an index to divide four types of rivers (Table 3-3). But strictly speaking, the river pattern change is actually a continuous spectrogram. Different rivers have different sand body distribution patterns.

Table 3-3 River classification index

Type	Single channel (braided index<1)	Multi-channel (braided index>1)
Low bending (<1.5°)	Straight river	Braided river
High bending (<1.5°)	Meandering river	Anastomosing river

I. Straight river sand bodies

The straight river is a single-channel river with very small bending and relatively stable banks. The development of straight rivers often requires some special structures or geographical conditions, such as faulted troughs or strong river banks caused by vegetation development. Because river energy dominates in lake deltas, straight distributary rivers are often formed on delta plains. However, even for a straight river, its river bottom valley line (river bed) is often curved.

The deposition of such rivers is mainly aggradation and a small quantity of lateral accretion. Therefore, river sand bodies are dominated by channel aggradation microfacies, also including small marginal banks. The channel sand bodies are in the shape of a strip areally, and have an external shape with a concave bottom and a flat top on the section. Channel aggradation materials only undergo rough differentiation during the deposition process, so a rough positive rhythm thick in the lower part and thin in the upper part is formed on the vertical section.

II. Braided river sand bodies

A braided river is a wide and shallow river; the river channel is divided by many diaras or alluvial islands, and the water current continuously diverges and rejoins in multiple channels around many diaras. Both diaras and channels are unstable.

Braided rivers are formed in environments with large sloping, large flow changes, poor erosion resistance of river banks, easy migration, and a large ratio of river bed load to suspended load. Generally, in the upper reaches of rivers, due to the large sloping, large flow changes and coarse detrital materials transported, braided rivers are developed very easily. Therefore, braided rivers are mostly developed between alluvial fans and meandering rivers.

(I) **Sand body geometrical morphology**

Braided rivers have features such as large sloping, large flow changes, bottom load dominant, a large ratio of river bed load to suspended load, and coarse detrital materials transported. Depending on the natural geographical environment, there can be gravelly braided rivers and sandy braided rivers, where pebbly sandy deposits predominate.

Braided rivers are wide and shallow rivers with a large width-to-depth ratio, so the sand bodies have a large width-to-thickness ratio, their geometrical morphology shows a wide flat plate shape, the two sides are symmetrical, and the planar form shows continuous wide belt shape.

Due to loosely braided river banks, their poor erosion resistance, and very rapid lateral migration, so the sand bodies of multiple genetic units can easily migrate into large-area connected sand bodies in the lateral direction.

(II) **Types and internal structure features of sand bodies**

Braided river sand bodies are mainly diara bars and abandoned channel filling sand bodies, while the natural levees and splays etc. are not well developed.

Diara bars are the main type of sand bodies in braided rivers, including longitudinal sand bars, transverse sand bars and lateral sand bars (actually belonging to longitudinal sand bars) (Figure 3-5). Longitudinal sand bars are the most common, while transverse sand bars are mainly developed when transverse flow occurs as the flow rate of seasonal rivers decreases after flooding. Sand bars can be sandy bars or gravelly bars, depending on the materials supplied upstream.

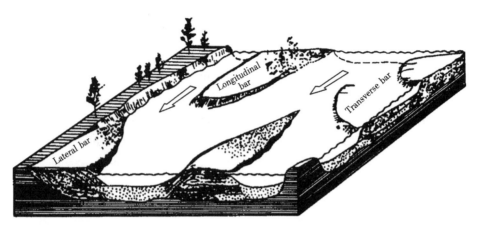

Figure 3-5　Braided river deposition model (according to Galloway, 1983)

The main part of the braided river diara bar is sand bodies formed by the vertical accretion of debris carried by multiple deposition events in a certain environment. The flooding energy of deposition events is different, and their changes are not regular; therefore, the formed clastic deposits are different in thickness vertically and do not have the typical upward-fining rhythm; moreover, there is also no certain sedimentary structure sequence, large-scale plate cross-beddings and parallel beddings are developed, and trough cross-beddings can also be developed. Such irregular sedimentary sand bodies show the characteristics such as high and low alternate changes on the permeability profile, and argillaceous interlayers are not easily deposited and preserved inside the sand bodies.

When a braided river is abandoned, it can form a waste-filled channel sand body, and the sand body is generally thick. This is due to the large slope of the braided river, the large flow energy and slow abandonment. However, when the river channel is abandoned, some suspended sediments can also be filled to form a thin argillaceous layer; its lateral distribution does not exceed the width of a channel, so the continuity is poor.

Braided rivers are relatively developed in continental sedimentary basins in China, which is related to the close source and large slope of continental basins. Especially in the short axis direction of a basin, the supply of coarse debris is sufficient, the slope is high, the provenance area is close to the center of the lake basin, the alluvial plain is narrow, and thus braided rivers are developed very easily. Many oilfields with braided river sand bodies as reservoirs have been discovered in China, e.g., the braided river sand bodies of the 2nd and 3rd sand groups of Sha-2 Member in Shengtuo Oilfield (Figure 3-6). The grain size in the sand bodies changes irregularly, the median grain diameter is 0.3-0.6mm, fine gravels are dispersed in the sand bodies, and the sorting property is poor; cross beddings are developed, and parallel beddings are also common; top layer subfacies are not developed; argillaceous interlayers in the sand bodies are almost absent; the geometry of the sand bodies is belt-shaped, with a width of 200-666m, locally connected to form a sand body with a width of about 2km.

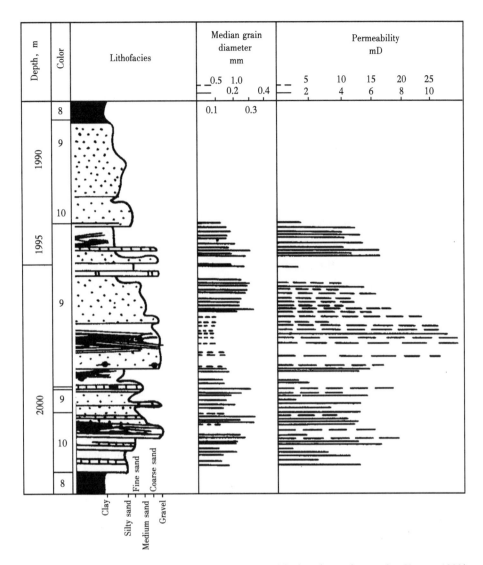

Figure 3-6 Lithofacies histogram of braided river sand bodies (according to Qiu Yinan, 1992)

III. Meandering river sand bodies

A meandering river is characterized by a single curved channel, and has a smaller slope, a larger depth, a smaller width-depth ratio, and a smaller ratio of bed load to suspended load in the debris carried than a braided river. Meandering rivers are generally developed in lower alluvial plains.

(I) **Sand body geometrical morphology**

Meandering rivers belong to mixed load rivers and suspended load rivers, where the sediments are mainly sands, silty sands and mud. There may be bottom conglomerates at the bottom of meandering rivers. The sediments of high bending meandering rivers are mainly silty sands and mud.

In general, a meandering river has a large depth and a small width-depth ratio, the two sides of the river channel are asymmetrical, and the bottom of the river is concave; therefore, the geometrical morphology of the sand bodies shows flat top, concave bottom and two asymmetrical sides. Due to the erosion of the concave bank of a meandering river and the lateral accretion of its convex bank, the planar shape of the sand bodies shows curved edges and beaded meandering belts. High-bending river sand bodies are shoelace-shaped.

Meandering river sediments have an obvious dual structure, the lower sequence has an upward-fining positive rhythm, and top layer subfacies and floodplain mudstones are developed. In the three-dimensional space, the volume of the sand bodies is smaller than that of the mudstones in the surrounding floodplain.

(Ⅱ) **Types and internal structure features of sand bodies**

There are plentiful types of meandering river sand body microfacies, including point bars, natural levees, splays, erosion ditch bars, etc. Point bars are the primary type of sand bodies in meandering rivers, and belong to channel sand bodies. Natural levees, splays, erosion ditch bars, etc. belong to the sand bodies of overbank genesis, and are the secondary types of sand bodies in meandering rivers.

1. Point bar sand bodies

Meandering river point bars are formed by lateral accretion on the convex bank of a meandering river. During the activities of a meandering river, the concave bank is constantly eroded, the convex bank continues to aggrade laterally, and the curvature of the river increases accordingly. The basic formation unit of a point bar sand body is a lateral accretion body, and a point bar is generally composed of multiple lateral accretion bodies. Lateral accretion bodies are sedimentary sand bodies formed by the periodic flooding of rivers. A lateral accretion body is deposited during a flooding event, and each lateral accretion body is an isochronous unit. The point bar sand body composed of multiple lateral accretion bodies is crescent-shaped on the plan view (Figure 3-7) and wedge-shaped on the profile (Figure 3-8), and is a regular imbricate sand body spatially. Based on this, Xue Peihua (1991) proposed the "point bar lateral accretion body deposition pattern" (Figure 3-8).

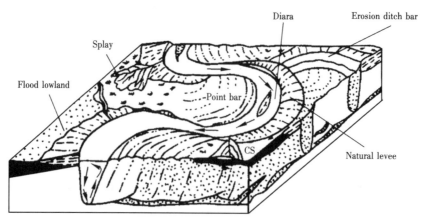

Figure 3-7 Deposition model of meandering river (according to LeBlanc, 1972, modified)

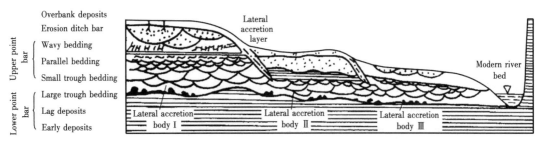

Figure 3-8　Point bar lateral accretion body deposition pattern (according to Xue Peihua, 1991)

The point bar sand body has a typical positive rhythm sequence where the grain size becomes small upwards vertically and the sedimentary structure scale becomes small upwards. When each lateral accretion body is deposited, from the bottom of the river bed to the bank, an upward-fining rhythm is formed due to the decrease in water flow energy; moreover, because of the different flow velocities and water depth everywhere, different bed shapes are formed. Generally, large trough cross-beddings gradually change to small trough cross-beddings and ripple cross-beddings, and sometimes can be sandwiched with parallel beddings of the upper flow regime. The reflection of the upward-fining positive rhythm sequence in the porosity and permeability of reservoirsis: the thickest riverbed lag deposition section at the bottom has the highest porosity and permeability, and it is even the ultra-high permeability section; the porosity and permeability become small gradually upwards; the porosity and permeability of the top overbank deposits are the smallest. The discontinuous thin argillaceous interlayers in the point bar sand body formed by lateral accretion can generally be divided into two types. (1) Suspended deposits during the flooding decline period occur at the end of each deposition event, and thus always appear in the upper part of the sand body. (2) The lateral mudstones deposited between two flooding deposition events are draped on the lateral accretion surface and have a certain intersection angle with the formation face. Such lateral accretion mudstones may be dry-cracked and washed away by the next period of flooding without being preserved. Under rapid accretion conditions, such lateral accretion mudstones may be completely preserved in sand bodies and thus become barriers and interlayers in oilfield development. In the meandering river lateral accretion point bars on the alluvial plain, such lateral accretion mudstones can extend from the upper point bar to the lower point bar up to 2/3 of the sand body thickness, while the lateral accretion mudstones below the perennial water level of rivers cannot be preserved due to the erosion of flowing water. Therefore, such lateral argillaceous interlayers generally appear in the mid-upper part of the vertical section of the point bar, so that the point bar sand body appears as a so-called "semi-connected body" (Qiu Yinan, 1985; Xue Peihua, 1991) where the lower half is connected and barriers and interlayers are complicatedly distributed in the upper half (Figure 3-8).

2. Overbank deposit sand bodies

Meandering rivers also develop natural levees, splays, erosion ditch bars and other overbank deposit sand bodies in addition to point bars. Such sand bodies occupy only a secondary position in meandering river sand bodies.

Natural levees are developed on both sides of a river channel, are generally thin alternate layers of fine sands, silty sands and mudstones, and can grow root systems. A splay is a fan-shaped sand body formed on a floodplain due to river breaching during the flooding period. The splay is generally composed of fine sands and silty sands, and has a high content of argillaceous matrixes. An erosion ditch bar is a strip-shaped sand body formed when a meandering river is straightened at the flood peak and then aggrades vertically during flood fall. The sand body is thick, but its shale content may be high. These sand bodies are generally small in scale, and form alternate layers of sandstones and mudstones with the widely developed floodplain mudstones on the profile. These sand bodies are located on the point bar sand bodies, forming the upper sequence of the typical "dual structure" of meandering rivers. Areally, these sand bodies can connect the point bars of various river sections into a meandering belt sand body. However, due to the small grain size, poor sorting property, high shale content, and generally low porosity and permeability, overbank deposit sand bodies tend to form low permeability sand bodies or tight sand bodies. Therefore, there is a large planar difference in permeability in a meandering belt sand body.

The lateral continuity of sand bodies in meandering belts is related to the width and bending of river channels. When a river is diverted by avulsion, the old meandering belt is abandoned and a new meandering belt begins to form. The degree of connectivity of sand bodies in different meandering belts is controlled by the relative magnitude of deposition rate, subsidence rate and river avulsion diversion frequency. When the deposition rate is relatively high and the subsidence rate is relatively low, sand bodies connected to each other are easily formed; otherwise, isolated sand bodies are easily formed.

In China's Mesozoic and Cenozoic petroliferous basins, high-bending meandering rivers are generally developed in the long-axis longitudinal depositional systems of basins and their shrinking period. $Pu-I_2$ sand layer in the northeast of Saertu Oilfield in Songliao Basin is a typical high-bending meandering river sand body.

$Pu-I_2$ sand body shows a typical point bar sequence (Figure 3-9). The bottom of the sand body is in abrupt contact with the underlying floodplain mudstones via the erosion surface; above the erosion surface are pebbly medium sandstones, and the gravels include parent rocks of the source area and intraformational boulder clay; the upward lithology is fine-grained to medium-grained sandstones with large trough or plate cross-beddings, sandstones with small cross-beddings, and siltstones and fine sandstones with ripple cross-lamina successively; the uppermost part is floodplain massive mudstones with root system and soil formation. A complete point-bar sequence is 5-7m thick and can be internally divided into 3-5 small rhythms. The upper part of each rhythm is sandwiched with thin silty and argillaceous interlayers. The sand body is continuous and stable, and has simple geometry; a meander belt has a width of 800-1000m, and a permeability of 1000-2000mD. This sand layer is the first main reservoir in Saertu Oilfield.

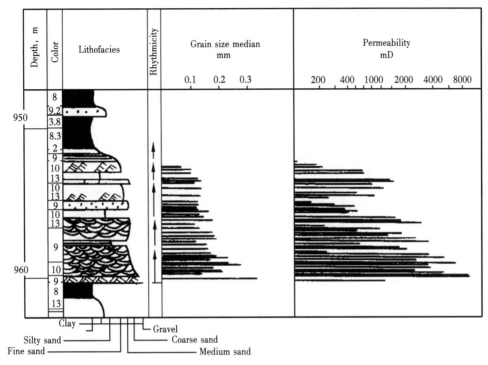

Figure 3-9 Lithofacies histogram of braided river sand body
(Pu-I$_{1+2}$ layer in Well ZJ3-23 in Daqing Oilfield) (according to Qiu Yinan, 1992)

IV. Anastomosed river sand bodies

The anastomosed river, also known as netted river, is a multi-channel river flowing around a fixed diara. The river channel is stable due to the strong diara.

The 2D style of the anastomosing river is similar to that of the braided river, and the main difference between them is that the anastomosed river channel is stable, has a low width-depth ratio, and thus a well-arranged 3D space combination (Figure 3-10). The braided river has a high channel width-depth ratio and is characterized by rapid migration, so it lacks an orderly arrangement in 3D space. In addition, the anastomosed river is also characterized by large-scale natural levees, while the braided river hardly develops natural levees or only has very low natural levees.

The formation of an anastomosed river requires some special conditions, that is, the sedimentary basin must continuously subside or the rise of the local datum plane of the basin is controlled to ensure rapid and continuous channel aggradation; the sediments injected into the basin must be sufficient to maintain the alluvial plain environment; the banks are strong against impact, so that the banks and natural levees can be stabilized; when vegetation is developed under humid climate conditions, it is more conducive to the formation of an anastomosed river. The most suitable tectonic environment for an anastomosed river is structurally active inter-mountain basins and molasses inter-mountain plains. The anastomosed river is a fixed anastomosed multi-channel river, and the deposition rate and subsidence rate maintain a balanced compensation for a long time, so the sand body geometry shows a typical narrow and thick anastomosed strip-shaped sand body (Figure 3-10).

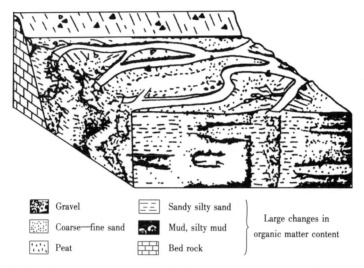

Figure 3-10 Ideal sedimentary environment model of anastomosing river (according to G.Klein, 1985)

The main type of sand bodies in the anastomosed river is channel aggradation sand bodies, which have a large grain size and often contain gravels. The aggradation belongs to rapid filling-type deposition. The clastics are only roughly differentiated by gravity action during the deposition process, and are deposited coarsely and then finely, so that the formed sand bodies have a rough positive rhythm sequence. Due to the continuous subsidence and aggradation of the river channel, a sand body filled and superimposed multiple times is formed, that is, multiple positive rhythm units are superimposed into a thick sand body, and the sand body has large-scale cross beddings. When the river channel is finally abandoned, it may evolve into a small meandering river and small point sand bars are deposited. The thin argillaceous and silty layers filled in the river channel during short-term abandonment may be preserved between positive rhythms, and become barriers with relatively good continuity in thee sand body.

Other types of sand bodies in the anastomosed river are natural levee and small splay sand bodies, but they are not dominant.

There have been many reports on anastomosed river sand bodies as reservoirs abroad. The reservoir of Yan-10 Formation in Maling Oilfield in Ordos Basin in China is considered to be restricted valley filling deposits similar to anastomosed rivers. This set of deposits is the main reservoir of Maling Oilfield. The lithology is dominated by glutenite, which account for 70% of the thickness on the profile. The mineral maturity is low, and mudflow deposits can be seen locally. The lithofacies sequence is an overall upward-fining sequence composed of multiple small positive rhythms. There are centimeter-thick waste fillings (laminated silty sands or mudstones) on each small rhythm. Large cross-beddings with a thickness of tens of centimeters are developed in the sand body. The sand body shows typical shoelace shape, and is only a few hundred meters wide, and the inheritance of each layer of river valley is quite strong (Figure 3-11). Due to strong diagenesis, the reservoir has become a low permeability reservoir.

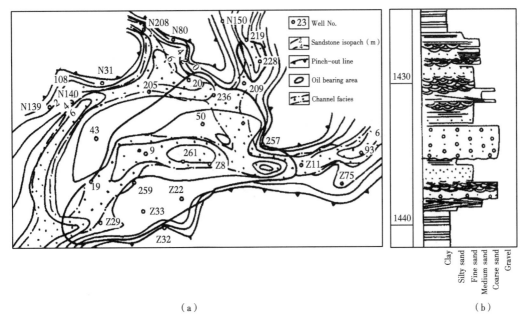

Figure 3-11 Geometry (a) and lithofacies profile (b) of the sand body of the anastomosed river
(restricted valley filling) (Yan-10 Formation in Maling Oilfield in Ordos Basin)
(According to Qiu Yinan, 1992)

Section 4 Lacustrine Sand Bodies

A lake is a relatively stable water body surrounded by land. There are not many modern lakes in the world, with a total area of only $250 \times 10^4 km^2$, accounting for only 1.8% of the global land area. The area of modern lakes in China is only $8 \times 10^4 km^2$, which is less than 1% of the national land area. However, lakes were very developed in China during the Mesozoic and Cenozoic. Since the Mesozoic, many large lakes have developed in China due to the large-scale withdrawal of sea water and the expansion of land. For example, the area of lakes in Ordos Basin in the Late Triassic was $9 \times 10^4 km^2$, equivalent to the area of 20 Qinghai Lakes; the lake area of Songliao Basin during the deposition of Member 1 of Qingshankou Formation and Member 1 of Nenjiang Formation in the Cretaceous was up to $8.7 \times 10^4 km^2$ and $15 \times 10^4 km^2$, respectively. More than 90% of China's proven oil reserves and oil production are from Mesozoic and Cenozoic lacustrine deposit. Therefore, in-depth research on the distribution and reservoir properties of lake sand bodies is of great significance to China's petroleum exploration and development.

Lakes are surrounded by land, and the supply of terrestrial materials is very ample; therefore, lacustrine sand bodies (including sandstone bodies and conglomerate bodies) are very abundant and there are many types of lacustrine sand bodies. At present, the classification and naming of sand body types are not uniform. Comprehensively considering the location of sand bodies in a lake and their geneses, lacustrine sand bodies are roughly divided into the following types: delta sand bodies, shallow beach bar sand bodies and deepwater turbidite sand bodies. Their distribution in the lake basin is shown in Figure 3-12 and Figure 3-13.

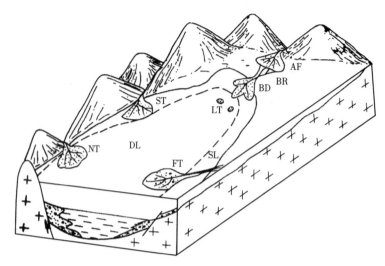

Figure 3-12 Schematic diagram of sedimentary facies during the deep depression and expansion period of the rifted lake basin (according to Wu Chongyun, 1992, modified)
AF—Alluvial fan; BR—Braided river; BD—Braided river delta; NT—Near-shore turbidite fan;
FT—Far-shore turbidite fan; LT—Lenticle of turbidity; SL—Shallow lake area; DL—Deep lake area;
SF—Subaqueous alluvial fan (transgressive fan delta)

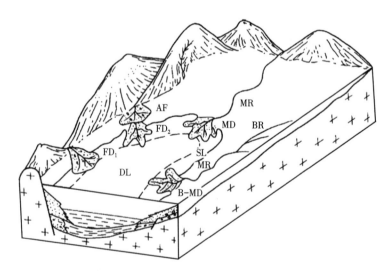

Figure 3-13 Schematic diagram of sedimentary facies during the shrinkage period of the rifted lake basin (according to Wu Chongyun, 1992, modified)
AF—Alluvial fan; MR—Meandering river; MD—Normal delta; B—MD—Braided river-normal delta;
FD—Fan delta; BD—Braided river delta; SL—Shallow lake area; DL—Deep lake area

I. Delta sand bodies

In the past, the concept of delta usually referred to a normal delta formed by a meandering river entering a lake. In fact, deltas in a broad sense include normal deltas, braided river deltas, fan deltas, etc.

A normal delta refers to the triangular sandy and argillaceous sedimentary body formed at the shallow gentle slope of a meandering river entering the lake (sea) in an onshore plain region. At present, the commonly said delta refers to this normal delta, and some people call it long river delta (Wu Chongyun, 1992). A braided river delta is a delta rich in sands and gravels and formed by the progradation of a braided river system (including river-dominated humid climate alluvial fans and glacial water alluvial fans) into stable water bodies (lake, sea). A fan delta is an alluvial fan that directly prograDes from an adjacent highland into stable water bodies (A. Holmes, 1995). The classification signs of various deltas are shown in Figure 3-14.

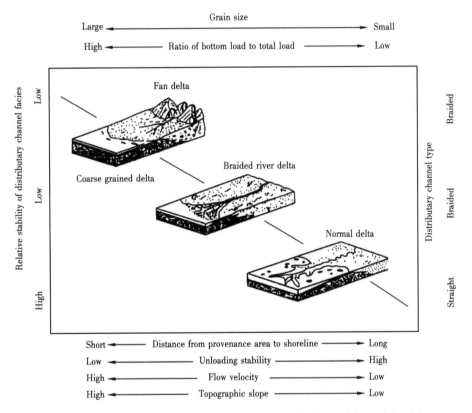

Figure 3-14 Classification signs of normal delta, braided river delta and fan delta
(according to McPherson, 1987)

All types of deltas in a lake have a characteristic three-belt (layer) structure. Delta plain belts, delta front belts and prodelta clay belts appear successively from shore to lake areally; three layers are superimposed on the vertical section, and their sequence changes depending on transgression and regression of deltas. However, the sedimentary characteristics of different types of deltas have certain differences.

(Ⅰ) **Normal delta**

A normal delta is what is commonly referred to as a delta. The normal delta is located far away from the provenance area, and the flow path of a river before it enters a lake is long (Figure 3-15), so the normal delta is also called long river delta (Wu Chongyun, 1992). The delta has a three-belt (layer) structure, and the sand bodies are mainly distributed in the delta plain facies and front facies.

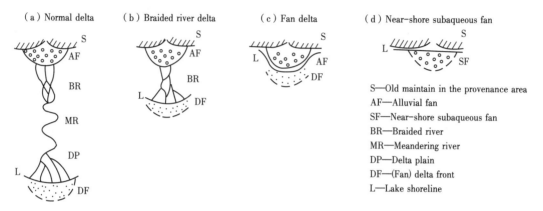

Figure 3-15 Schematic diagram of the distribution locations of various deltas in the lake basin

1. Sand bodies of delta plain facies

Delta plain facies is located in the upper part of the triangular shore between the first diversion point of the river's downstream reach and the lakeshore and are characterized by distributary channel deposits and interchannel floodplain deposits. The favorable reservoir sand bodies of the plain facies are mainly above-water branch channel sand bodies. In addition, there are also natural levee sand bodies, splay sand bodies or abandoned channel sand bodies, but they are small in scale and thin, and occupy a secondary position in the sand bodies of plain belts. Floodplain regions are close to the lakes, so the groundwater level is high and plants are easily grown in floodplain regions. In addition to silty sands and argillaceous sediments, carbonaceous shales and even coal seams tend to be developed. This is an important feature of the delta plain facies and also one of the important identification marks.

Distributary channel sand bodies are the main type of sand bodies in the delta plain facies. After the main channel diverges, the bending of the channel is reduced. Generally, there are two types of distributary channel sand bodies, namely, curved distributary channel sand bodies and straight distributary channel sand bodies. Curved distributary channels are developed in the upper reaches of a bird's foot delta; with the continuous progradation to the lake area, straight distributary channels predominate all the more.

A single distributary channel sand body is strip-shaped along the water flow direction, shows a geometric shape with a flat top and concave bottom on the section, and is embedded in interdistributary argillaceous deposits. The internal structure of the curved distributary channel sand body is similar to that of a meandering river, the deposition is dominated by lateral accretion, and the sand body has an upward-fining positive rhythmic sequence, but the scale of the sand body is much smaller than that of a meandering river. The deposition of straight distributary channel sand bodies is dominated by aggradation. The active period of such distributary channels is very short after the formation of them. The formation of a river channel, deposit filling, and drying-up and transfer is completed in a short time unit. The deposition and filling of clastics are also often the main reason for river channel abandonment. Such rapid-filling type deposits are roughly differentiated by gravity action, forming a rough positive rhythm sequence. The range of the formed permeability rhythm is smaller than that of the lateral accretion point bar sand body. Sometimes, multiple distributary channel sand bodies in different time units are alternately superimposed in the vertical direction into a thick sand layer, thereby forming multiple composite positive rhythm sequences.

Distributary channels are mostly branch-shaped areally, and some are parallel-strip-shaped. In the same time unit, the main distributary channel and multiple secondary distributary channels move and are deposited at the same time, and divergent streams are frequently abandoned and diverted. Therefore, in the 3D space, multiple distributary channel sand bodies are distributed in a "maze" shape, showing a very complicated spatial combination of multiple small sand bodies (Figure 3-16). Along the water flow direction, the sand bodies are in a strip shape and have good continuity. However, in the direction vertical to water current, the sand bodies are lenticular with a flat top and a concave bottom, and have poor transverse continuity.

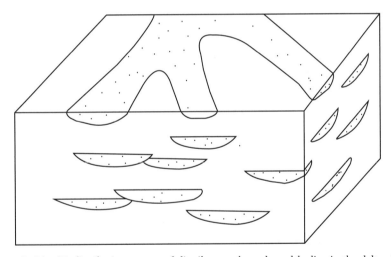

Figure 3-16　3D distribution pattern of distributary channel sand bodies in the delta plain

2. Sand bodies of delta front facies

A delta front belt is a concentrated development belt of sand bodies in the delta, lies in the littoral-shallow lake gentle slope belt below an estuary, and is the most characteristic belt where rivers and lakes work together. The main types of sand bodies are underwater distributary channel sand bodies, interdistributary splay sand bodies, estuary sand bars and sheet sands. Sheet sands are actually the sand bodies formed by the former types of sand bodies through the action of lakes and waves.

1) Underwater distributary channel

An underwater distributary channel is the extension of a distributary channel on a delta plain to the water bodies of a lake, and the degree of development of the underwater distributary channel is related to the intensity of river action. The stronger the river action, the more developed the underwater distributary channel. The sedimentary characteristics of underwater distributary channels are similar to those of above-water distributary channels of plain facies, that is, the sand body is in the shape of a narrow strip, the cross section is in the shape of a concave bottom and a flat top, and there is an upward-fining positive rhythm in the vertical direction. However, its sedimentary scale is smaller than that of an underwater distributary channel. In fact, for normal deltas, front facies sand bodies are dominated by estuary sand bars, and underwater distributary channels generally play a secondary role.

2) Interdistributary splay sand bodies

During the flooding period, underwater distributary channels may also be breached, forming a splay between the distributary channels. Generally, splay sediments are thinner and finer than distributary channel sediments, and have high shale content. However, for some large deltas, splay sand bodies can also form considerable reservoirs, and they often form isolated lithological traps.

3) Mouth bar

Mouth bars are the most characteristic sand bodies in deltas and often become the most important reservoirs in a delta system. They are sedimentary sand bodies formed at the estuary as distributary channels carry detrital materials into a lake for unloading, and are distributed in the front edge of a delta.

The geometry of mouth bars is leaf-like or finger-like areally, which depends on the relative magnitude of river energy and lake water energy. On the longitudinal section vertical to the lake bank, a mouth bar is an asymmetric lenticle with a flat bottom and a convex top; it is thick at the estuary end and has a large grain size, which is the main body of the mouth bar (Figure 3-17). The sand bodies have good sorting property due to the screening by lake waves. There may be multi-directional inclined medium cross-beddings and parallel beddings. Towards the lake center, the thickness and grain size of sand layers gradually decrease, and they are mostly thin alternate layers of silty and fine sand layers and argillaceous layers, with small cross beddings, wavy beddings, lenticular beddings, and flaser beddings. As the water depth increases, the energy decreases. Biological disturbance structures are developed. This belt is the tail end of the mouth bar, that is, the distal bar. In fact, the main body and tail end (distal bar) of the mouth bar is a whole (Figure 3-17). On the cross section of the mouth bar, it shows a typical lenticle shape, and is thick in the center and thin on both sides. An underwater distributary channel can aggrade on the mouth bar and cut its top, thereby superimposing a narrow and thin straight strip-shaped sand body on the mouth bar.

Due to the progradation of deltas, most of mouth bars show a typical upward-coarsening reverse rhythm sequence (Figure 3-18), the porosity and permeability of the sand body also shows a reverse rhythm pattern (i.e. low porosity and permeability in the lower part and high ones in the upper part), and the section with the highest porosity and permeability is located in the sand body top. Argillaceous interlayers can appear at the bottom of the sand body, and their transverse continuity is good, while argillaceous interlayers are rarely seen in the upper part of the sand body.

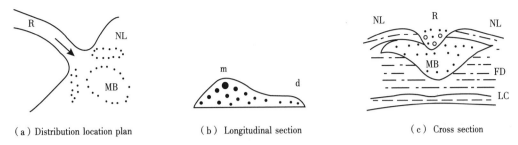

(a) Distribution location plan (b) Longitudinal section (c) Cross section

Figure 3-17 Schematic diagram of the location and shape of the mouth bar
(according to Wu Chongyun, 1992)

R—Distributary channel; NL—Subaqueous natural levee; MB—Mouth bar; M—The main body of
the mouth bar; d—The tail end of the mouth bar (distal bar); FD—Front delta clay; LC—Lacustrine clay

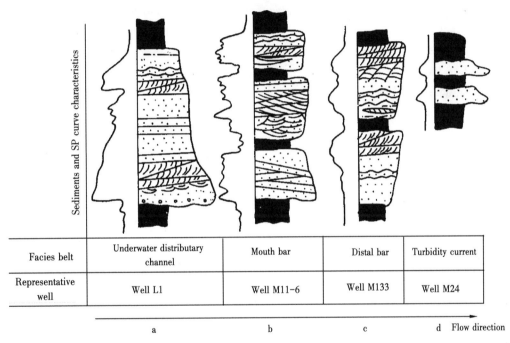

Figure 3-18 Sedimentary sequence characteristics of various microfacies sand bodies in the delta
(according to Zhao Chenglin, 1992)

4) Front sheet sands

Front sheet sand bodies are developed in areas where the actions of lake waves and lake currents are strong. Mouth bars migrate laterally, so that mouth bars are connected to form wide stripped sand bodies parallel to the shoreline, or called sheet sand bodies. The sand quality of sheet sand bodies is pure and they have good sorting property, so their reservoir properties are good. The sand bodies become thick toward the shore and thin toward the lake. Of course, not every delta develops sheet sands. Sheet sands are relatively developed only in destructive deltas where waves and lake currents (coastal currents) are strong, while sheet sands are not developed in constructive deltas.

3. Prodelta clay belt and slump turbidite sand bodies

A prodelta clay belt is located in front of a front belt, and its lithology is mainly dark mudstones sandwiched with thin silty sand layers. Sometimes the front belt collapses, and fluxoturbidites appear in prodelta mudstones, forming turbidite lenticle sand bodies. Prodelta facies and shallow lake—semi-deep lake facies are actually difficult to distinguish, and they are of gradual transition.

(Ⅱ) Braided river delta

The delta rich in sands and gravels and formed by the progradation of a braided river system into a lake (sea) is called braided river delta. In the narrow part of the slope in the direction of the short axis or the long axis of a lake basin, the gradient of the onshore slope and the underwater slope is large, the lake bank is close to the mountain foot, the river is short, and it enters the lake only when it develops to the braided river stage, thus forming a braided river delta. The flow path of the river before it enters the lake is short, so the delta is also called short river delta (Wu Chongyun, 1992). In a rifted lake, braided river deltas are very common, and are mainly distributed on the side of the short-axis gentle slope (or narrow slope in the long-axis direction). On the steep side of the short axis of the lake basin, due to the steep slope and being close to the mountain, the alluvial fan directly enters the lake to form a fan delta. However, as the fan delta continues to prograde, the slope becomes long and its gradient becomes small, which will gradually transform into a braided river delta.

A braided river delta also has a similar three-layer (belt) structure to a normal delta, but its sedimentary characteristics are different from those of the normal delta.

(1) The distributary channels of the braided river delta plain have the characteristics of braided rivers, that is, the channel sediments have a large width-to-thickness ratio and are in the shape of a wide flat plate. The geometry of the sand bodies is similar to that of a braided river (Figure 3-5); the clastics are coarse, and the content of sands and gravels is high (while the normal delta is dominated by sands and silty sands); channel sand bodies have no typical "dual structure", that is, top subfacies or overbank deposits are few; the river channels are unstable and easy to migrate, so the coarse clastic sand bodies are often distributed continuously areally.

(2) Underwater channels are developed in the front of braided river deltas. The water flow intensity of braided rivers is relatively strong, detrital materials are abundant, and the ratio of bed load to suspended load is large. Therefore, after entering water bodies, the channel sediments are relatively developed. Mouth bars are in a secondary position, and quite different from normal deltas. Near-source underwater distributary channels are braided and can be crossed and merged, but far-source distributary channels are often stripped.

(3) Braided river deltas are often small, but they are often distributed in groups and can be connected continuously.

In fact, the sedimentary characteristics of braided river deltas are between those of normal deltas and fan deltas.

(Ⅲ) Fan delta

The term "fan-delta or fan delta" was first proposed by A. Holmes (1965). A fan delta is defined as an alluvial fan that directly progrades from an adjacent highland into stable water bodies (sea, lake). Therefore, it is a genetic term, rather than simply referring to fan-like delta. The important conditions for the development of a fan delta include large height difference of coast (lakeshore) topography, steep slopes on coast (lakeshore), short distance to the provenance, and sufficient detrital material supply.

A fan delta also has a three-layer (belt) structure, namely fan delta plain, fan delta front and pro-fan delta. The fan delta has the following characteristics:

(1) The fan delta is generally located near the mountain foot, and is often associated with boundary faults in a lake basin. In a rifted lake basin, the fan delta is mainly located on the short-axis steep slope side of the lake basin, while the braided river delta of the same period is often distributed on the short-axis gentle slope side of the lake basin.

(2) The fan delta plain facies are actually alluvial fans. That is, the onshore part is composed of alternate layers of debris flow (mudflow), sheetflood, sieve tongue and transient braided channel sediments. A braided river delta is mainly composed of numerous braided river channels or braided river plain facies, generally without debris flow and sieve deposits. Normal delta plains appear as distributary channels with low undulation and meandering river characteristics and mosaic deposits (sand bodies embedded in mudstones) composed of interchannel swamps.

(3) The mouth bars in the front of the fan delta are poorly developed or even there is a lack of mouth bars. This is mainly determined by estuary dynamics. That is, the supply of sediments is fast and short in time and the distributary channels are unstable. Even if a mouth bar is formed at the temporary distributary channel mouth, it is easily destroyed by later deposition (including strong alluviation). In addition, the fan delta front is usually rich in underwater gravity flow deposits, and in some cases, gravity flow deposits even predominate (Gloppen and Steel, 1981). In the Lengdong-Leijia region (steep slope in the middle section of the western depression) in Liaohe western depression in the Bohai Bay Basin, the fan delta front of Sha-3 Member is characterized by the development of a large set of granular flow-derived massive glutenites sandwiched with traction flow-derived glutenites with plate cross-beddings. Both braided river deltas and normal deltas have limited mouth bars, and the mouth bars of meandering river deltas are more developed. Generally, gravity flow deposits are not developed in the front of the two types of deltas. The gravity flow deposition of sediments generated by front slump can occur in a prodelta belt, but this can be distinguished by facies combination research. In fact, fan delta, braided river delta and normal delta are components of a continuous pedigree, and fan delta and normal delta are the two end members of the delta pedigree.

II. Shallow beach, bar sand bodies

Shallow beach and bar sand bodies are mainly found in the gentle slope littoral-shallow lake areas in the margin of lakes, local uplifts in lakes, or lake bays etc. In the micro-depression expansion period of a lake basin, the lake area is large, the lake bottom is flat, the shallow lake area is large, and beach bars are the most developed. On the contrary, in the deep depression period of the lake basin, the water is deep and the slope is steep, and beach bars are few and also small. Another condition for the development of beaches and bars is that the action of lake waves is strong and not affected by river injection. Sandy beach bars are an important type of lake sand bodies.

The materials of sandy beach bars mainly come from the nearby large near-shore sand bodies such as deltas etc. and are transported, washed and deposited by lake waves and lake currents. The typical characteristics of beach bar sand bodies include high sandstone composition maturity (significantly higher than that of other types of sand bodies, that is, high quartz content), and high structural maturity (low shale content, good sorting property, and high psephicity). The most common grains are medium, fine sand and silty sands, and there are also a small amount of gravels, which often contain oolites, biological shells and fragments; in addition, the heavy mineral content is high. Wavy beddings, wavy cross-beddings, cross beddings tilted in multiple directions (wave-built cross-beddings), etc. are common. The sand body reservoirs have high quality, good porosity and good permeability.

(Ⅰ) **Beach sands**

Beach sands are sand bodies formed by lake waves and currents on a flat topography. Beach sands are generally distributed parallel to the shoreline, show a wide strip shape (sheet shape), and cover a large distribution area (Figure 3-19). The vertical profile characteristics of beach sands include frequent alternate layers of sandstones and mudstones, many sandstone layers but small single layer thickness (mostly less than 2m), unobvious grain sequence or anti-rhythm, good sorting property of a single sand body, generally uniform in-layer permeability, and rare argillaceous interlayers.

(Ⅱ) **Bar sands**

Bar sand bodies include near-shore sand bars, sand spits and underwater paleo-uplift beach bars.

Near-shore sand bars are long-stripped, and most of them are parallel to the shoreline (Figure 3-19). The formation mechanism of nearshore sand bars is similar to that of littoral barrier islands. They are mainly coarse debris deposits formed by shoreward lake waves (vertical to the shoreline or obliquely intersected with the shoreline) in breaker zones. Sand spits are formed at the turning position of the lake shoreline. When lots of detrital materials eroded and transported by lake waves and alongshore currents pass through the turning position of the lake shoreline, the detrital materials are deposited due to the attenuation of the water current energy, thus become long-stripped sand spits at a certain angle with the shoreline (Figure 3-19), and can be gradually developed and extended. Paleo-uplifts can be formed in the center of some rifted lake basins due to factors such as tectonic movement, volcanic eruption, etc. Beach bar deposits can also be formed above these underwater paleo-uplifts due to the strong water body energy. Such beach bars are usually parallel to the long axis of a lake basin (Figure 3-19).

The cross-section of sand bars tends to show a geometric shape with a flat bottom and a convex top. The thickness of a single layer of bar sands is large, generally a few meters or even larger, and they appear as alternate layers of thick sandstones and thick mudstones on the section.

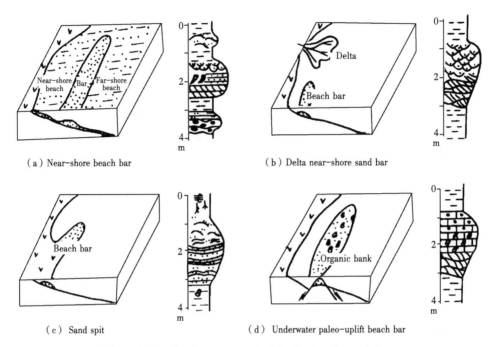

(a) Near-shore beach bar (b) Delta near-shore sand bar

(c) Sand spit (d) Underwater paleo-uplift beach bar

Figure 3-19 Sedimentary model of shallow beaches and bars
(according to Zhu Xiaomin, 1994, modified)

Sand bars can be formed during the transgression stage or the regression stage, but most of the sand bars preserved in ancient times are of regression type, while the sand bars formed during transgression are easily transformed by lake water into sheet beach sands.

Most of bar sands show a reverse rhythm sequence, fine in the lower part and thick in the upper part. During deposition, bar sands are screened by lake waves repeatedly, and fine-grained materials are screened to the bar edge, so that bar top sands become thick and clean, thus promoting the formation of reserve rhythms of bar sands. The porosity and permeability of the sand bodies also show a rhythm becoming good upwards, and the section with the highest porosity and permeability is in the bar top. Argillaceous interlayers are rare in the sand bodies.

Beach and bar sand bodies do not occupy a major position in a lake basin. However, sand bodies are adjacent to source areas and are often surrounded by source rocks, and tend to have high porosity and permeability, so they can become good reservoirs. Beach – bar sand body reservoirs are highly productive, and their waterflood recovery ratio is among the highest among sandstone reservoirs in lake basins.

III. Deep water turbidite sand bodies

Turbidite sand bodies can be used to summarize all sand bodies formed by gravity flow of sediments in lake deepwater environment (semi-deep lake – deep lake). In terms of sediment transport mechanism, sediment gravity flow includes debris flow, grain flow, liquefaction flow and turbidity flow. Turbidity flow is the most common and important one of sediment gravity flows. Other gravity flows are easily transformed into turbidity flow in the process of transporting to deep water.

In China's Mesozoic and Cenozoic continental lakes, turbidite sand bodies are very developed, and there are many types of turbidite sand bodies, which are very important reservoirs. Famous scholars such as Wu Chongyun (1985, 1992), Zhao Chenglin (1986, 1992), *et al.* systematically studied and classified the turbidite sand bodies in China's continental lake basins. According to the distribution position of sand bodies, Wu Chongyun (1992) divided turbidite sand bodies into six types, namely, near-shore turbidite fan sand bodies, far-shore turbidite fan sand bodies with water supply channels, turbidite sand bodies in front of near-shore shallow water sand bodies, trough fault turbidite sand bodies, underwater local uplift turbidite sand bodies, and turbidite sand bodies in the central lake bottom plain. Zhao Chenglin (1992) divided the turbidite sand body deposition system into two major types mainly based on sand body shape, namely sublacustrine fan deposition system and axial gravity flow channel deposition system.

There are two main formation mechanisms of deep-water turbidite sand bodies. (1) The gravity flow of flood genesis, that is, ashore floods enter the deep water environment of a lake basin through the water supply channel so as to form a turbidite fan, (2) The gravity flow of slump genesis, that is, near-shore shallow water sand bodies (such as various delta sand bodies) slump and then are transported so as to form a turbidite fan in deep water. There are many reasons for the slump of shallow water sand bodies: earthquake, storm (mainly storm backflow), excessive accumulation of shallow water sand bodies themselves, etc. Gravity flow is blocked by faults or other underwater elevations during the transportation process and can make a bend, thereby forming axial gravity flow channel turbidite sand bodies.

Comprehensively considering the distribution pattern and genetic mechanism of turbidite sand bodies, deep-water turbidite sand bodies are divided into two major types, namely turbidite fan sand bodies (including near-shore turbidite fans, far-shore turbidite fans, and slump turbidite fans) and axial gravity flow channel sand bodies.

(Ⅰ) **Turbidite fan sand bodies**

Turbidite fan sand bodies are fan-shaped turbidite sand bodies. The deep-water turbidite fans in a lake basin are also called sublacustrine fans.

Turbidite fans are composed of sediments (sedimentary rocks) of various gravity flow geneses, that is, coarse clastic facies having obvious sediment gravity flow characteristics and formed by event deposition in gray, grayish black and black mudstone facies formed by normal deposition in a deep water environment, including massive sandstone (conglomerate) facies, graded bedding sandstone (conglomerate) facies, pebbly sandstone facies, stacked graded sandstone (conglomerate) facies, typical turbidite facies with Bouma sequence, etc. A turbidite fan is shaped like a fan, so it is easy to delineate its range by using seismic data or dense well pattern data.

The division of subfacies and microfacies types of lacustrine turbidite fan sand bodies can be compared with the ancient submarine fan models summarized by Walker (1978). A turbidite fan can be divided into feeder channel, inner fan, mid-fan and outer fan subfacies, which can be further divided into several microfacies. The inner fan is dominated by the main channel, and the side edge of the channel is the main levee and terrace overflow microfacies; the mid-fan is the main part of the fan body, dominated by braided trench microfacies, the side edge of the trench is the levee and inter-trench, and the outer side of the trench is the central microfacies; the outer fan is mainly tip microfacies.

According to the formation mechanism and distribution position of turbidite fans, lacustrine turbidite fan sand bodies are divided into three types.

1. Steep slope near-shore flood turbidite fan sand bodies

Steep slope near-shore flood turbidite fan sand bodies are often distributed on the steep banks of rifted lake basins. Mountain torrents flow straight into a lake along steep slopes (usually fault planes), and lots of detrital materials are quickly accumulated where the slope suddenly slows down in deep water areas. Moreover, because of the very high flow velocity of torrents, supported by the lake water, torrents still have the ability to move forward and undercut after entering the lake. Therefore, a braided channel is formed, and some detrital materials continue to be transported and deposited, forming a large-scale turbidite fan (Figure 3-20a). The formation mechanism of this kind of turbidite fan is mountain flood, and the fan body develops in the steep slope near-shore zone, so it is called steep slope near-shore flood turbidite fan, referred to as near-shore turbidite fan. The fan body is wedge-shaped on the inclined section, and the root is close to the section of the bedrock. The sand layer stretches towards the center of the lake and can be subdivided into three subfacies zones in the plane such as inner fan, mid-fan and outer fan.

The inner fan is a single main channel, and the lithology is mainly hybrid conglomerates supported by matrix and positively graded conglomerates supported by grains. The mid-fan is a braided channel area and the main part of the fan body, and the lithology is multiple superimpositions of positively graded pebbly sandstones and massive sandstones. The main lithology of the outer fan is dark mudstones sandwiched with thin silty and fine sandstones, showing the characteristics of the CDE section of Bouma sequence.

Such turbidite sand bodies are close to the edge of a lake basin, and the supply of detrital materials is sufficient, so the lithology is coarse, the fan area and fan thickness are both large, the buried depth is relatively small, and the sand bodies are drilled easily. Therefore, they are the sand bodies with the most abundant hydrocarbons and the highest exploration success rate among turbidite sand bodies. The superimposed sandstones of the mid-fan braided channel have the best physical properties (low shale content), and are the most important reservoir facies belts of such fans. The thin silty and fine sandstones in the outer fan can form pinch-out lithologic reservoirs in sand bodies. The sand bodies in the inner fan area tend to have poor reservoir properties because of very low structural maturity.

2. Gentle slope far-shore flood turbidite fan sand bodies

If there is a fault perpendicular to the lakeshore on the short-axis gentle slope side of a lake basin, valleys and trenches that reach the lake bottom will be developed in general, so that torrents entering the lake are difficult to accumulate on the edge of the lake basin, but continue to be transported along the trenches in the form of gravity flow; until torrents reach the deep water area, lots of detrital materials are accumulated, forming a turbidite fan distant from the lakeshore

(Figure 3-20b).

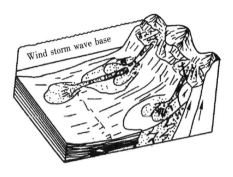
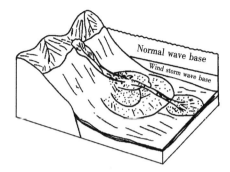

(a) Steep slope near-shore flood turbidite fan sand body

(b) Gentle slope far-shore flood turbidite fan sand body

Figure 3-20 Turbidite fan facies model diagram (according to Zhao Chenglin, 1992)

The gentle slope far-shore flood turbidite fan sand body is very similar to the steep slope near-shore flood turbidite fan sand body in morphology, zoning, and lithology. They are both coarse clastic turbidites, and their material sources are ashore torrents.

3. Fluxoturbidite fan and lenticular sand bodies

The delta and beach bar sand bodies in the shallow water zone in the margin of a lake basin will form a certain slope and be in an unstable state when they accumulate thickly. If there are external dynamic influences (such as earthquakes, storms), the sand bodies easily slump and are transported to deep water areas so as to form slump turbidite fan sand bodies or small lenticular turbidite sand bodies.

Fluxoturbidite fan and lenticular sand bodies have the following features:

(1) The lithology is restricted by the lithology of the shallow water sand bodies (i.e. the sand bodies in the source area), and is generally fine;

(2) Containing a large amount of debris in the basin, such as mudstone tearing debris, boulder clay, sandy debris or gravel debris of siltstones or fine sandstones, and limestone or dolomite debris;

(3) Inverted deformation structures (wrinkles), contemporaneous small faults, liquefaction and drainage structures, etc. are common;

(4) The sand bodies are small, but there are many sand bodies, which are distributed in groups around the source area;

(5) Sand bodies have various shapes, and are often fan-shaped, lenticular, etc.

Fluxoturbidite fan and lenticular sand bodies can also become good reservoirs. Although sand bodies are small, there are many sand bodies and they often appear in groups. Because these small sand bodies are surrounded by mudstones, they generally form abnormally high pressure reservoirs with high and stable production. For small lenticular turbidite sand bodies, it is difficult to explore them because of their small size. At present, the seismic resolution does not fully meet the identification accuracy of such sand bodies.

(Ⅱ) Axial gravity flow channel sand bodies

In a deep-water gravity flow deposition system, there are non-fan-shaped turbidite bodies in addition to fan-shaped turbidite bodies. Among the non-fan-shaped turbidite bodies, the primary ones are axial gravity flow channel sand bodies.

On the slope of a lake basin, if there is a fault trough parallel to the lakeshore (long and narrow low-lying terrain formed by two faults that are in the same direction and inclined to each other), when turbidity currents flow down the slope to the fault trough, they will be blocked to a certain extent and then flow along the trough (i.e. turning gravity flow), thus forming a stripped turbidite sand body parallel to the lakeshore. There can be multiple turbidity current supply points on the slope, so the section of any point in the fault trough may be a random superposition of turbidity sediments from multiple sources (Figure 3-21).

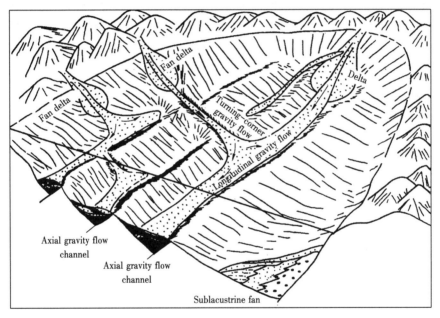

Figure 3-21   Genesis model diagram of axial gravity flow channel sand body
(according to Zhao Chenglin, 1992)

The axial gravity flow channel can be distributed not only along a slope fault trough, but also in a deep lake fault trough. The turbidity current supply can come from both the short axis direction and the long axis direction (Figure 3-21).

The petrological characteristics of axial gravity flow channel sand bodies are similar to those of turbidite fans in rock types, grain size distribution characteristics, sedimentary structures, low component maturity, and low structural maturity that reflect deep-water gravity flow deposits. However, the distribution patterns of the two types of sand bodies are quite different, so that they can be distinguished.

The turbidite sand bodies in lake basins are most developed during the maximum deep depression expansion period of rifted lake basins, e.g. the deposition period of Member 3 of Oligocene Shahejie Formation in the Bohai Bay Basin. During this deposition period, the lake water depth, the deep lake area, and the lakeshore slope were large, and turbidite sand bodies were developed from the lakeshore edge to the lake center, forming a very important type of lake basin sand bodies in this period. In other periods of lake basin development, as long as there are deep lake areas, there will be small turbidite sand bodies, but no large near-shore turbidite fans. During the deep depression expansion period of a rifted lake basin, the deep lake area is very large. Although the water depth is not as large as that in the deep depression period of the rifted lake basin, turbidite sand bodies are also developed.

In terms of deep-water turbidite sand body size, oil storage properties, and exploration difficulty, the near-shore turbidite fan sand bodies and far-shore turbidite fan sand bodies are the best. With the advancement of exploration technology and the development and perfection of sedimentology theory, the ability to identify turbidite sand bodies will be greatly improved, and lenticular sand bodies that are small but appear in groups can be accurately delineated. Deep-water turbidite sand bodies are important targets for finding deep reservoirs and subtle (non-structural) reservoirs, as well as one of the important fields for tapping the potential of old oil regions.

Chapter 4 Fine Stratigraphic Structure Description

Stratigraphic Structure description is to establish a fine and reasonable stratigraphic framework on the basis of fine stratigraphic division and correlation. Stratigraphic division and correlation are mainly intended to subdivide the stratigraphic section into different levels of layer groups according to the characteristics such as lithological associations, sedimentary cycles, stratigraphic contact relationships, etc., establish the isochronous comparison relationships at all levels among the wells of an oilfield, and achieve unified layering within the oilfield. Stratigraphic division and correlation are the means to study the characteristics of reservoir morphology and the spatial distribution of parameters. Large stratigraphic boundaries within the scope of oil and gas fields are easily divided and correlated, but it is difficult to divide and correlate stratigraphic units below sand group level (or smaller levels); therefore, fine stratigraphic division and correlation are the primary research content of reservoir description. A reasonable reservoir framework can be established only by dividing strata and carrying out correct isochronous correlation according to geological laws, so as to further reveal the heterogeneous characteristics of reservoirs and provide guidance to the reasonable development of oil and gas fields. The degree of understanding of reservoirs depends on the fineness of layer group division. In the middle and late stages of oilfield development, the division and correlation of layer groups are required to be accurate to genetic units (or single sand bodies). At present, there are mainly two ideas and methods for the fine division and correlation of reservoirs in the middle and late stages of development. (1) the "cycle correlation, hierarchical control" method established in response to the sedimentary characteristics of China's continental reservoirs, including many layers, thin layers, alternate layers of sandstones with mudstones, narrow planar facies belts, fast facies change, and poor lateral continuity. (2) the high-resolution sequence stratigraphy method of base level cycles proposed by T. A. Cross et al. A fine and reasonable stratigraphic framework can be established combining the identification of marker horizons, the distribution characteristics of effective sand bodies and the response of dynamic data.

Section 1 Division and Correlation of Reservoirs by the "Cycle Correlation and Hierarchical Control" Method

"Cycle correlation, hierarchical control" refers to the correlation of cycles at all levels under the control of the marker horizon taking advantage of the cyclicity of sedimentary rocks and the characteristics of hierarchy from large cycles to small cycles, that is, correlation of Level 2 cycles in Level 1 cycles, and correlation of Level 3 cycles in Level 2 cycles until Level 4 cycles. Reservoir groups, sandstone groups, and layer boundaries are correlated in the corresponding cycles at all levels, and the correlation accuracy is controlled level by level. This is the commonly used layer correlation method "cycle correlation, hierarchical control".

Ⅰ. Division of reservoir groups

(Ⅰ) Division and correlation of sedimentary cycles

1. The meaning of sedimentary cycle

Sedimentary cycle refers to the phenomenon of regularly repeated appearance of some similar lithologies vertically on the stratigraphic section. This phenomenon is mainly manifested in the color, lithology, texture, structure, etc. of rocks. The most obvious manifestation of this phenomenon is the granularity of rocks, which is called rhythmicity.

The formation of a sedimentary cycle is mainly due to the periodic up-and-down movement of the crust. As the crust descends, ingression occurs, changing water bodies from shallow to deep and forming an upward-fining ingression sequence on the section, which is called a positive cycle. As the crust rises, transgression occurs, changing water bodies from deep to shallow and forming an upward-coarsening transgression sequence on the section, which is a reverse cycle. A complete cycle refers to the upward-fining and then upward-coarsening ingression-transgression sequence formed in the section as the crust descends and then rises and water bodies change from shallow to deep and then from deep to shallow.

The up-and-down movement of the crust is regional, and the sedimentary cycle characteristics shown by the same up-and-down movement are the same or similar. This is the theoretical basis for stratigraphic division and correlation using sedimentary cycles.

The up-and-down movement of the crust is unbalanced, which is manifested in different up-and-down scales (time, amplitude, range); in addition, there are small-scale up-and-down movements under the overall up-and-down background. Therefore, the cycles on the stratigraphic section hierarchy, that is, there are small cycles in a large cycle. Hierarchical correlation of reservoirs can be performed using cycles from large to small levels. This is called the theoretical basis for "cycle correlation, hierarchical control".

2. Sedimentary cycle division method

Hydrocarbon-bearing intervals are divided into cyclic sedimentary intervals with different stable distribution ranges on the basis of regional stratigraphic division and oil-bearing strata division fully considering the contact relationship between layers combined with the vertical evolution law of sedimentary facies according to core data and logging curve shape characteristics. This is called the sedimentary cycle division method.

3. Analysis of the sedimentary facies of reservoirs

(1) Collect regional sedimentary facies results and determine the regional sedimentary background of hydrocarbon-bearing intervals.

(2) Subdivide the sedimentary microfacies of each interval of a single well making full use of various facies identification markers based on core data, and then determine the vertical evolution law of the sedimentary facies of hydrocarbon-bearing intervals in the single well.

(3) Determine the facies belts changes of hydrocarbon-bearing intervals areally based on the

division of sedimentary microfacies of each interval in the single well.

(4) Determine the specific correlation methods to be used for reservoirs of different sedimentary geneses according to the depositional environment of reservoirs.

4. Study of the relationship between lithology and electrical properties

(1) Select coring wells with complete core data and logging data to study the litho-electric relationship and analyze the display of various lithologies, various sedimentary cycles and various lithological marker horizons on various logging curves, thereby providing a basis for the division and correlation of reservoirs using logging curves.

(2) According to the characteristics of response of various logging curves to the reservoir characteristics, cyclic characteristics and marker horizons of the region, the logging curves sensitive to reservoirs are selected as the logging curve series for the division and correlation of reservoirs. It shall be able to identify lithology, physical properties, and hydrocarbon-bearing characteristics, the cycle characteristics of lithological associations of reservoirs, and boundaries of various rock formations.

5. Single-well sedimentary cycle division

(1) Select wells or well sections with complete core data, and divide sedimentary cycles such as positive cycles, reverse cycles, and complex cycles according to the vertical combination type of lithology and the contact relationship between layers. The deposits must be continuous in the same cycle.

(2) Divide the sedimentary cycles of different levels. Sedimentary cycles are generally divided into four levels from large to small.

① Level 1 sedimentary cycle. Level 1 sedimentary cycle is composed of a set of cyclic deposits including some reservoir groups, including the entire oil-bearing series of strata, and can be correlated in sedimentary basins. It is equivalent to the combination of source beds with reservoirs, or the combination of reservoirs with caprocks. Each set of oil-bearing series of strata generally has paleontological or micro-paleontological marker horizons to control cycle boundaries. The boundary of a sedimentary cycle is generally on the erosion surface or at the boundary where the depositional environment has changed significantly.

② Level 2 sedimentary cycle. Level 2 sedimentary cycle refers to the cyclic deposits composed of different lithofacies sections, and can be tracked and correlated in a secondary structural unit. It contains several reservoir groups composed of some sandstone groups. The distribution of reservoirs is similar to their characteristics, and it is a set of reservoir combinations that can form a development unit. There are mudstones of appropriate thickness above and below it to completely separate it from adjacent reservoirs. Generally, there is a standard horizon or a marker horizon to control the cycle boundary, or the boundary of obvious regressive or transgressive deposits can also be used as the cycle boundary.

③ Level 3 sedimentary cycle. Level 3 sedimentary cycle refers to the cyclic deposits composed of several different types of small (single) layers or Level 4 cycles in the same lithofacies section. Level 3 sedimentary cycle can be correlated in oilfields. It is roughly equivalent to a sandstone group. Concentratedly developed oil-bearing sandstones have certain connectivity, and the upper and lower mudstone barriers are distributed relatively stably. According to lithological association type, evolution law, thickness change and logging curve shape combination characteristics, the upper and lower mudstone layers can be used as the basis for determining the cycle boundary during correlation.

④ Level 4 sedimentary cycle (or called rhythm). It is a sedimentary cycle composed of single (small) layers of different lithologies, and can be correlated within a block.

The division of cycle levels in reservoir correlation is the development and deepening based on regional stratigraphic correlation. The corresponding relationship between the cycle levels of regional stratigraphic correlation and those of reservoir correlation is shown in Table 4-1.

Table 4-1 Comparison of sedimentary cycles

Regional stratigraphic correlation		Reservoir correlation	
Sedimentary cycle level	Stratigraphic unit	Sedimentary cycle level	Reservoir unit
Level 1	System	Level 1	Oil-bearing series of strata
Level 2	Formation	Level 2	Some reservoir groups
Level 3	Member	Level 3	Sand group
Level 4	Sand group	Level 4	Some single reservoirs

6. Correlation of sedimentary cycle boundaries

(1) Carry out the correlation of sedimentary cycles based on paleontological characteristics, lithological characteristics and logging curve shape characteristics, find out the common cycle boundaries of most wells, modify the inconsistent single-well cycle boundaries, and make the sedimentary cycle boundaries of single wells unified.

(2) Analyze the planar changes in the lithology and thickness of various sedimentary cycles, and ascertain the relationship between the sedimentary cycles in different regions so as to provide a basis for reservoir correlation.

(Ⅱ) **Division of reservoir groups**

1. Basis for dividing reservoir groups

The following factors shall be considered in the division of reservoir groups:

(1) Reservoir characteristics such as the depositional environment, distribution status, rock properties, etc. of reservoirs;

(2) Separating conditions such as the thickness, distribution range, and lithological characteristics of the barrier between reservoirs;

(3) The properties of fluids and pressure systems in reservoirs.

2. Division of reservoir groups

(1) In a secondary sedimentary cycle, the hydrocarbon-bearing intervals with relatively similar depositional environment, distribution, rock properties, physical features and reservoir properties are classified as an oil (gas) reservoir group. An oil (gas) reservoir group can be composed of one or several sand groups.

(2) Oil (gas) reservoir groups shall be separated by relatively thick and stably distributed barriers, and the boundary of oil (gas) reservoir groups shall be as consistent as possible with that

of the sedimentary cycle.

(3) The divided oil (gas) reservoir group can be used as the basic units of combined development series of strata in the initial stage of development.

3. Division of sandstone groups

(1) The adjacent sections with concentrated oil (gas) reservoirs in an oil (gas) reservoir group are classified as a sandstone group. The divided sandstone group shall be as consistent as possible with the horizon of a Level 3 sedimentary cycle.

(2) A sandstone group can contain several layers. There are stable barriers which separate sandstone groups.

(3) The number and boundaries of sandstone groups within the same oil and gas field shall be unified.

4. Division of layers

(1) The top and bottom of the oil (gas) reservoir are separated by impermeable rock formations is classified as a layer. Several single layers can be included in the same layer. The number of well points separated between two layers in a block shall be greater than the number of well points that are combined.

(2) The boundary of the divided layer shall be as consistent as that of a Level 4 sedimentary cycle. The number of layers in each block can be different, but the layering boundaries shall be consistent.

II. Reservoir correlation

(I) Correlation principle

Reservoir correlation level by level is performed using logging curve shapes and combination features based on ancient organisms and lithologic features as well as sedimentary cycles under the control of the correlation marker horizon. Different correlation methods shall be used for different regions and different facies belts according to the sedimentary geneses of reservoirs.

(II) Correlation methods and steps

1. Marker horizon selection

(1) The interval with stable lithology, prominent features, wide distribution, and easy-to-identify logging curve shape characteristics or the bedding surface with obvious upper and lower differences can be selected as the correlation marker horizon. Typical marker horizons include special rock formations such as fossil layers, oil shales, carbonaceous shales, limestones, dolomites, pure mudstones, etc.

(2) The interval with obvious lithological association characteristics and easy-to-identify logging curve shape characteristics or the bedding surface with obvious upper and lower differences can be selected as the correlation marker horizon.

(3) The correlation marker horizon shall be selected near the boundary of sedimentary cycles and the boundary of different lithofacies sections.

(4) The auxiliary marker horizons distributed in local regions shall be identified.

2. Establishment of standard correlation profiles

(1) One or several standard profiles for reservoir correlation in different directions shall be established according to the area the correlation region and the planar distribution stability of reservoirs.

(2) The standard correlation profile shall run through the entire correlation region, and coring wells shall be fully selected. The sedimentary sequence of the well or well section where the standard correlation profile is selected shall not have repeated strata or lack some strata.

(3) Determine the layering boundary of each well on the standard profile by correlation.

3. Correlation of the layer boundary of each well

(1) Based on the layer group division result of the standard well and through inter-well correlation, the boundaries of reservoir groups, sandstone groups and layers of other wells are divided, and are verified using adjacent wells.

(2) Through the correlation of reservoir groups, determine the faulted point depth, the faulted thickness, the faulted horizon, etc. of the well point where a fault is drilled, and mark the number of the well.

4. Uniform layering for a block

(1) Uniform layering shall be performed for single wells in a block as per a certain direction and sequence, so that the layer group boundaries of each well in the block are consistent.

(2) In a complex fault block region, the corresponding strong reflection phase shall be calibrated on the basis of seismograms using 3D seismic data, and then transverse tracking is performed areally so as to ensure the consistency of the layer group division boundaries of different well points in the oilfield or the block.

5. Correlation of oil (gas) reservoir groups

The boundaries of reservoir groups in a Level 2 sedimentary cycle are correlated under the control of the correlation marker horizon or the auxiliary marker horizon according to the shape characteristics of lithological association logging curves and the planar change law of the thickness of reservoir groups.

6. Correlation of sandstone groups

Under the control of reservoir group boundaries, the boundaries of sandstone groups are correlated according to the nature of Level 3 sedimentary cycles, the characteristics of lithological associations, logging curve shape and thickness change law.

7. Correlation of layers

(1) Under the control of sandstone group boundaries, the boundaries of layers are divided based on the correlation of Level 4 cycles. According to the different geneses of sedimentary cycles, different specific correlation methods are used.

(2) For reservoirs deposited in lacustrine facies, the boundaries of layers are divided using the method of correlation of the same layers with similar cyclicity and lithology, similar curve shape, and roughly equal thickness in adjacent wells according to the gradual change characteristics of lithology and thickness areally. If the cyclicity of layers of individual well points is not obvious, the boundaries of layers shall be determined according to the thickness ratio of

layers in the sandstone group.

(3) The lithology and thickness of reservoirs dominated by fluviation have sudden changes laterally; therefore, the boundaries of layers shall be divided using the unequal thickness correlation method based on the sedimentary characteristics of fluvial sedimentation cycles such as undulating scouring of bottom boundaries etc. and the principle that the cycle top boundary of the same layers is roughly horizontal.

8. Correlation of the connectivity of single layers

(1) The connectivity of single layers between two adjacent wells is correlated according to a certain direction and sequence on the basis of dividing the layering boundaries of single wells (Figure 4-1).

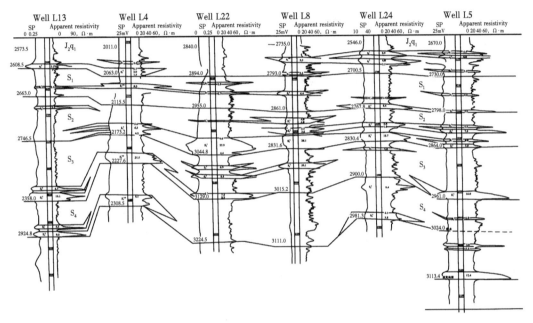

Figure 4-1 Correlation of sand layers in Qiketai-Sanjianfang Formation of Well L4—L5 in Qiuling Oilfield

(2) The connectivity of single layers between well points is determined under the control of layering boundaries according to the different sedimentary geneses of reservoirs as well as the sedimentary mechanism of each microfacies and the connectivity between different sand bodies of sedimentary microfacies.

For example, the reservoirs deposited in fluvial facies have abrupt changes in the sedimentary genesis of sand bodies areally; therefore, the connection status of single layers between well points shall be determined based on the contact relationship between sand bodies. For thick layers superimposed by multi-stage channel sand bodies, the connection relationship of single-stage channel sand bodies shall be correlated (Figure 4-2).

For reservoirs formed in other depositional environments, the connectivity of single layers between wells can be determined using the corresponding correlation method according to the specific sedimentary characteristics of the reservoirs and sand body distribution law.

(3) When the connectivity of single layers between well points is determined, the reservoirs

in the two walls of a fault cannot be connected.

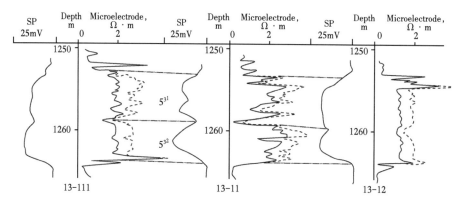

Figure 4-2 Correlation mode of superimposed sand bodies of fluvial facies reservoirs

Section 2 High-Resolution Sequence Stratigraphy Analysis Technology

Sequence stratigraphy analysis technology is a branch of sedimentary stratigraphy developed on the basis of stratigraphy in the 1980s, and is a new technology and new method for dividing and correlating sedimentary strata.

Sequence stratigraphy is to study the strata with genetic connections and cyclic lithology sequence correlations in the chronostratigraphic framework which boundary is the unconformity surface or its corresponding conformity surface. Sequence stratigraphy can also be defined as a discipline to study genetic correlations in the chronostratigraphic framework (Vail Wagoner, 1988, 1990). A sedimentary sequence is a combination of various sediments deposited between sea level fluctuation cycles. The geometry and lithology of the stratigraphic units in a sequence are controlled by four basic factors, including tectonic subsidence, global sea level fluctuation, sediment supply rate and climate change. Tectonic subsidence provides accommodation space for deposition of sediments; global sea level change controls the distribution patterns of strata and lithofacies; sediment supply rate controls the sediment filling process and the change of the paleo-water depth of basins; the climate controls the type and quantity of sediments. Generally speaking, the three parameters such as tectonic subsidence rate, sea level fluctuation rate, and sediment supply rate control the geometry of a sedimentary basin, and the subsidence rate and sea level fluctuation comprehensively control the change in the accommodation space of sediments. Vail (1987) once deemed that the global sea level fluctuation is the most basic factor controlling stratigraphic superposition patterns.

The core of the sequence stratigraphy analysis technology is to establish an isochronous stratigraphic framework throughout a basin. On this basis, basin filling sequences are interpreted as sequence stratigraphy units of different levels, and the further research is conducted on the division and transverse isochronous correlation of sequence stratigraphy units at all levels. With the deep exploration and development and the needs of fine description of modern reservoirs, the high-resolution sequence stratigraphy school represented by the Genetic Stratigraphy Research Group of the Colorado School of Mines led by Cross has rapidly emerged. Based on core data, outcrop data, logging data and high-resolution seismic profile data, it predicts 3D stratigraphic relationships, establishes a reservoir sequence stratigraphy framework at region level, oilfield level and even reservoir level, and evaluates the distribution of reservoirs, barriers, and source beds using the fine sequence division and correlation technology. The advent of high-resolution sequence stratigraphy has not only expanded the research scope of sequence stratigraphy, but also enriched and perfected the theoretical basis of sequence stratigraphy, so that sequence stratigraphy has taken an important step towards high precision and quantification.

I. Theoretical basis of high-resolution sequence stratigraphy

The theoretical basis of high-resolution sequence stratigraphy advocated by Cross is the stratigraphic process – response sedimentation dynamics. Due to the change in the ratio of accommodation space of sediments to sediment supply (A/S) in a base level cycle process, the volume distribution of sediments in the same depositional system tract or facies tract occurs so as to result in changes in the preservation degree of sediments, the accumulation pattern of strata, facies sequence, facies type, and rock structure. These changes are a function of the location in the base level cycle process and accommodation space. The stratigraphic distribution form of isochronous stratigraphic units controlled by the base level cycle is regular and predictable. Therefore, discussing the base level cycle process and the accommodation space change is the key to understanding the formation mechanism of continental sequences.

The basic principles and methods of high-resolution sequence stratigraphy include stratigraphic base level principle, volume division principle, facies differentiation principle and cycle isochronous correlation rule. The stratigraphic base level principle is the theoretical basis, the temporal and spatial evolution process of strata, and the "cause"; the volume division principle and facies differentiation principle are sedimentary responses, and the "effects"; the cycle isochronous correlation rule is the application method.

(I) **Basic terms**

1. Sedimentary facies

Sedimentary facies objectively describes a combination of identifiable physical, biological, and/or chemical characteristics used to distinguish different types of sedimentary rocks. The marker features involved in defining facies include mineral features, biological features, physical properties, structural features, sedimentary structure features, etc.

2. Relief elements

Relief elements refer to the 3D topography composed of sediments. The number and scale of relief elements are different in different depositional systems. For example, anastomosing channel belts, flood plains, and lakes in a delta plain depositional system have a Level 1 scale; the sub-level relief elements include point bars, diaras, splays, etc.; the more secondary relief elements include sand streaks, sand dunes, and argillaceous drapes of beaches and channels.

3. Depositional system

Depositional system is a 3D combination of relief elements with genetic connections in the adjacent depositional environments within the continuous time and space. During the entire stratigraphic time period, the migration of a relief element or geographic environment is accompanied by the complete preservation, complete replacement or incomplete preservation of the former relief element or environment (Cross et al., 1993).

4. Facies sequence

Facies sequence is the 3D combination of facies. There are three mechanisms for the generation of a facies sequence: the lateral migration of relief elements (such as the lateral migration of channels in a floodplain environment), the lateral migration of a connected environment (such as the progradation from a coastal plain to shallow sea), and the change of hydrodynamic force at a specific geographic location (such as the upward-fining channel facies sequence deposited under the condition of gradual weakening of hydrodynamic force).

5. Facies tract

Facies tract is the stratigraphic record of depositional systems (Cross et al., 1993). The facies tract is the 3D combination of facies sequences in the same time period. The facies sequences comprising a facies tract shall have the same depositional environment and be controlled by the same deposition, biological action and chemical action. Due to being controlled by the strata associated with accommodated space changes, the physical and geometric characteristics of the facies tract in different depositional systems change with time and space.

6. Superimposition pattern

It refers to a recognizable stratigraphic accumulation mode due to the regular changes of the accommodation space with geographical location and time. According to geometry, it can be divided into three forms: Seaward Stepping (SS), Landward Stepping (LS), and Vertical Stepping (VS). Superimposition pattern is the product of the change in the ratio (A/S) of accommodation space (A) to sediment flux (S) fluctuation cycle, that is, the base level cycle. The sediment flux is theoretically equivalent to the sediment supply. When the A/S ratio increases, the accommodation space increases toward the land direction, forming an LS-type stratigraphic superimposition pattern; when the A/S value decreases, the accommodation space decreases toward the land direction, and the migration distance of sediments towards the basin direction increases, forming an SS-type superimposition pattern; when the A/S value is stable, a VS-type superimposition pattern is formed. These superimposition patterns and the parts they represent in long-term base-level cycles can be associated with Vail's sedimentary sequence bordered by unconformity surfaces in the following manner (Vail, 1977). SS units (roughly equivalent to HST) are deposited during the long-term base level fall period. VS units (roughly equivalent to LST) after a series of SS are deposited at the beginning of long-term base level rise. LS units (roughly equivalent to TST) are deposited during the long-term base level rise period. After a series of LS, another series of VS units (roughly equivalent to the early HST) are deposited at the end of the long-term base level rise period and at the beginning of the long-term base level fall period. In the lithologic section, the superimposition pattern can be identified based on the correlation of the beginning and end facies of each stepping unit.

7. Accommodation space

Accommodation space refers to the cumulative space for accumulation of sediments that is generated or disappears in the course of time. The accommodation space defines the volume of sediments that may be deposited in all geographic locations. The concept of accommodation space in high-resolution sequence stratigraphy is more precisely an effective accommodation space. The A/S value change can be analyzed by directly linking depositional characteristics with the increase or decrease in accommodation space. But in fact, sediment supply is discontinuous and changes continuously as sediment flux increases or decreases. As the A/S value decreases, the sediment supply at a specific location on the sedimentary profile is relatively larger than the accommodation space, and the stratigraphic system tends to reduce the volume of sediments preserved per unit time at a specific geographic location, unless there is excess space for sediment accumulation. As the A/S value increases, the space available for sediment accumulation at a specific location on the sedimentary profile is relatively larger than the sediment supply volume, and the volume of sediments preserved per unit time at the specific location increases, unless sediments are not enough to fill the new space. These changes are manifested in the stratigraphic record as rich facies types and stratigraphic boundaries that indicate the integrity and preservation degree of strata. By observing the increasing or decreasing trend of the A/S value in a certain geographical environment, the changes in the A/S value in other places can also be predicted. In general, for a specific environment, the larger the A/S ratio, the better the preservation of relief elements; as the A/S value decreases, the types and proportions of preserved releif elements also decrease.

(Ⅱ) Basic principle

1. Base level principle

In 1964, Wheeler proposed a base level concept suitable for stratigraphic analysis on the basis of the previous base level concept. The genetic stratigraphic research group quoted and developed Wheeler's base level concept, and analyzed the process-response principle of base-level cycles and genetic sequence formation. As shown in Figure 4-3, the stratigraphic base level

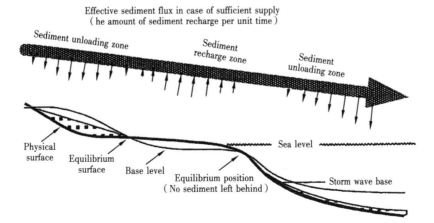

Figure 4-3　Schematic diagram of stratigraphic base level principle (according to T.A.Croos, 1994, modified)

is neither the sea level nor the horizontal plane extending to the land, nor a surface equilibrium profile, but an abstract equipotential surface that vibrates up and down and swings laterally relative to the geophysical surface. The stratigraphic base level describes the interaction between the generation and disappearance of accommodation space. In the process of its change, the base level always has a unidirectional movement to the maximum or minimum of its amplitude, forming a complete ascending and descending cycle. An ascending and descending cycle of the base level is called a base level cycle. The base level can swing completely above or below the ground surface, or can swing through the ground surface to below the ground surface and then back to the ground surface, which is called the base level crossing cycle. In a base level cycle, the base level can pass through the ground surface once or twice. A base level cycle is isochronous. The rocks preserved in a base-level cyclic change process are a genetic stratigraphic unit, that is, genetic sequence, which takes time as the boundary and is a temporal stratigraphic unit.

The undulation of the base level relative to the ground surface is accompanied by changes in the accommodation space of sediments. When the base level is above the ground surface, it provides space for sediment accumulation, deposition occurs, and any erosion is local or temporary. When the base level is below the ground surface, the accommodating space disappears, erosion occurs, and any deposition is local or temporary. When the base level is consistent with the ground surface, neither deposition nor erosion occurs. Therefore, in the time domain of base level changes, there are four geological action states in different geographical locations on the surface, namely, deposition, erosion, non-deposition caused by passing of sediments, and starvation deposition or even non-deposition caused by non-recharge of sediments. The time-space event representing the change of the base level cycle in the stratigraphic record appears as a rock record and a sedimentary boundary. Therefore, a genetic sequence can be composed of rocks formed by the base level ascending hemicycle and the base level descending hemicycle, or by rocks and boundaries. In fact, the base level can be regarded as a potential energy surface, which describes the position where the gradient, sediment supply and accommodation space are mutually balanced through moving of the ground surface up and down due to energy requirements (Figure 4-3). The stratigraphic base level describes the balance between the generation of accommodation space and the distribution of sediments on the ground surface; therefore, base level changes can be identified from a large number of sedimentological and stratigraphic features formed by changes in the A/S ratio in the stratigraphic record.

In the case of low accommodation space, the relief elements in the depositional system move up and down on the potential energy surface for sediment accumulation, and thus are less preserved. When a relief element moves to a place, the original relief element of this place disappears or is replaced. In the case of low accommodation space, the mixing, erosion and cutting of relief elements are serious, and the types of relief elements originally existing in the environment are incomplete. Due to the low A/S value, the possibility of sediment accumulation is small. As the accommodation space decreases and the sediment flux increases at a specific location, more sediments are transported to other locations downstream.

On the contrary, in the case of high accommodation space, there is large space for the preservation of relief elements on the sedimentary potential energy surface; moreover, when relief elements migrate along the sedimentary profile, the original relief elements are less likely to be eroded and cut, resulting in an increase in the accommodation space, and also an increase in the volume of sediments preserved in a specific environment as well as the diversity and proportion of the original relief elements. When the A/S value is high, the amount of sediments deposited is large and the amount of sediments transported to other places decreases.

2. Volume distribution principle

The base level cycle and the dynamic system of its accompanying accommodation space changes control the structure and sedimentary characteristics of strata. For this reason, the genetic stratigraphic research group put forward the concept of sediment volume distribution. Sediment volume distribution refers to the process in which sediments are distributed to different facies tracts in the genetic strata. It is the product of the 4D dynamic change of the accommodation space in different depositional environments during the change of the base level. Sediment volume distribution is directly accompanied by many sedimentological and stratigraphic responses such as the preservation degree of the original topographic form, the thickness of sediments, the internal structure, etc. During the fall of the base level, the position of the effective accommodation space moves toward the basin, and the accommodation space increases toward the basin and decreases toward the land. Therefore, the volume of sediments in the environment near the basin gradually increases, and the volume of sediments in the environment near the provenance decreases. During the rise of the base level, the position of the effective accommodation space moves toward the land, the accommodation space increases toward the land, and the volume of the sediments in the environment near the provenance increases. In the genetic sequence formed by the long-term base level crossing cycle, the accumulation mode of strata and the migration of their geographic location are also related to their position in the base level cycle. The progradation and accumulation mode of migration toward the basin is formed during the fall period of the long-term base level cycle, and the resulting vertical accretion strata are formed at the beginning of the base level cycle rise. The retrogradation and accumulation mode of migration toward the land appears in the rise period of the base level, and the resulting accretion strata appear at the end of base level rise and the early stage of base level fall (Figure 4-4).

Volume distribution changes the symmetry of stratigraphic cycles in time and space. Cycle symmetry records the time ratio of the rise and fall of the base level preserved in the form of rocks. Three extreme forms of symmetry can be identified: asymmetric base level descending cycles, symmetric base level ascending cycles, and asymmetric base level ascending cycles. The change of cycle symmetry is accompanied by the change of the strata thickness in various facies tracts at different positions with the accommodation space. Strata thickness (sediment deposition rate) per unit time is inversely proportional to the frequency of strata discontinuities and directly proportional to facies heterogeneity.

3. Principle of facies differentiation

Along with the change of accommodation space and the volume distribution of sediments, the facies sequences, facies combinations, facies types and facies diversities preserved in the same

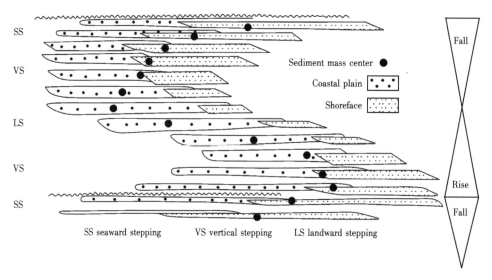

Figure 4-4 Stratigraphic superimposition pattern and volume distribution (according to T.A. Cross, 1994)

depositional environment also differ significantly. This phenomenon is called facies differentiation. Facies differentiation directly affects the physical characteristics of reservoirs, such as the continuity of reservoirs in 3D space, geometry, lithology, lithofacies type and even petrophysical properties. The channel sand bodies formed in high accommodation space are obviously different from those formed in low accommodation space in geometry, sand body continuity, lateral continuity, mutual cutting degree, bottom shape type, preservation degree, and the thickness and type of bottom lag sediments. These facies differentiation characteristics also directly affect the physical properties of reservoirs, as well as the channels of oil, gas, and water and the entire oil displacement system. Figure 4-5 shows the characteristics of channel sediments under different accommodation space conditions.

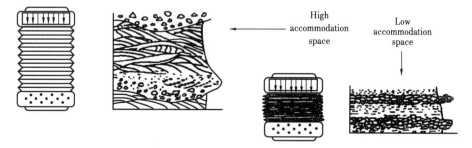

Figure 4-5 Facies differentiation in channel sediments (according to T.A. Cross, 1994)

II. Application method of high-resolution sequence stratigraphy

(I) Identification of base level cycles

The base level cycles of the corresponding levels are recorded by the stratigraphic cycles of different levels in the stratigraphic record. The key to high-resolution isochronous stratigraphic correlation is to identify these stratigraphic cycles that represent multi-level base level cycles in the stratigraphic record. According to the principle of base level cycle and accommodation space changes, the cyclicity of strata is the response of strata to the spatial migration of the deposition caused by changes in the position of the base level relative to the surface, erosion, non-deposition formed by no sediment left behind, and sediment under-compensation over time. Therefore, there must be "traces" in each level of stratigraphic cycle that can reflect the changes in the A/S value during the time experienced by the corresponding level of base level cycle. Base level cycles are identified based on outcrop data, drilling data, logging data, and seismic data as well as these "traces". This is the foundation for high-resolution sequence division and correlation.

1. Identification method of base level cycles

Base level cycles can move across the surface during their change. The time elapsed for the base level cycle across the surface is composed of the rock record formed when the base level is above the surface and the unconformity interface generated when the base level drops below the surface. The base level can also only move above the surface. In this case, the sediments during the rise period and fall period of the base level are preserved. Depositional discontinuities, especially unconformities, are not developed in the stratigraphic record, but the rise and fall of the base level relative to the surface can still be reflected in the stratigraphic record. Unlike the concept of "sequence" in EXXON's classic sequence stratigraphy, the interface of high-resolution sequence stratigraphy units (time units) is not necessarily an unconformity. It can be an unconformity or a deposition discontinuity, or a transition surface of deposition. In regions where unconformities or deposition discontinuities are not developed, the interface of the base level cycle is usually identified by the transformation of deposition.

The sequence stratigraphy analysis of one-dimensional profiles is realized through the identification and division of different levels of base level cycles. The division of multi-level base level cycles shall begin with identifying the most basic genetic stratigraphic units that constitute stratigraphic cycles, then the vertical arrangement or superimposition pattern of continuous genetic stratigraphic units is analyzed, and short-term cycles are gradually merged into long-term stratigraphic cycles.

The identification of either short-term stratigraphic cycles or long-term stratigraphic cycles is carried out through trend analysis of A/S value changes. The change trend of the A/S value in a short-term cycle can be identified by the facies sequence, facies combination and facies differentiation that can indicate the water depth and the preservation degree of sediments during their formation. The change trend of the A/S value in longer-term base level cycles can be identified by the superimposition pattern of short-term cycles, the change of the symmetry of cycles, the trend of thickening or thinning of cycles, the nature and occurrence frequency of stratigraphic discontinuities, the occurrence position and proportion of rocks and boundaries, etc.

In summary, the sedimentological and stratigraphic characteristics used to identify different levels of base level cycles include the following (Figure 4-6):

(1) The vertical change in the physical properties of a single phase.

(2) Changes in facies sequence and facies combination.

(3) Changes in cycle symmetry.

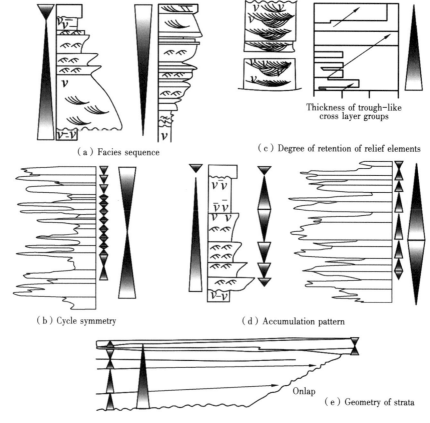

Figure 4-6 Identification marks of base level cycles

(4) Changes in cycle superimposition pattern.

(5) Changes in strata geometry and contact relationships.

Outcrop core data are usually the basis for identifying short-term base level cycles. Logging curve analysis is the best way to identify long-term base level cycles based on the superposition pattern analysis of short-term cycles. Seismic data can be used to identify the boundary of a third-order sequence through the analysis of the properties of reflection terminators; in addition, the seismic profile after fine well-seismic calibration can be used to further identify high-level base level cycles within the third-order sequence. The identification accuracy and reliability of base level cycles determined mainly using any type of data can be improved only through mutual lithologic-electrical-seismic calibration and verification (Figure 4-7).

2. Identification marks of base level cycles

1) Identification marks on lithological profile

Core data and drilling data, especially 3D outcrop profiles, have a higher resolution than logging-seismic reflection profiles, and thus are the basis for the identification of base level cycles, and especially short-term base level cycles (genetic sequences).

The shortest-term stratigraphic cycle (genetic sequence) on the stratigraphic profile is identified on the basis of facies sequence analysis. This is because facies sequence characteristics

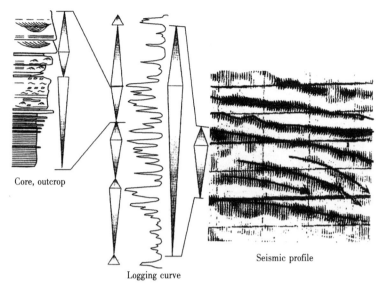

Figure 4-7 Sequence stratigraphy analysis method by combining multiple data

and longitudinal facies differentiation are directly related to changes in accommodation space during the short-term base level cycle change. To identify a base level cycle on the lithological profile, the depositional system type and facies composition of the profile and the relative relationship of facies and facies sequence changes with water depth changes shall be firstly ascertained, and then the change trend of the A/S value is identified according to facies sequence and facies combination features. The identification marks of cycle boundaries on the lithological profile are as follows.

(1) The scouring phenomenon in the stratigraphic profile and the overlying retained sediments, or the erosion surface meaning that the base level falls below the ground surface, or the transgressive scouring surface during base level rise. The difference between the latter and the former is that the amplitude of the scouring surface is smaller, and there are intrabasinal clasts above it in most cases.

(2) Downward migration of the shoreland onlap as the sequence boundary is often manifested as movement of sedimentary facies toward the basin in the drilling profile. For example, shallow-water sediments directly overlay deep-water sediments, and fluvial glutenites and turbidity glutenites directly overlay deep-water mudstones. There is a lack of transitional environmental sediments between the two types of sediments in general.

(3) The transition position of lithofacies type or facies combination on the vertical profile, e.g. the transition position from the facies sequence or facies combination of upward-shoaling water bodies to the facies sequence or facies combination of gradually upward-deepening water bodies.

(4) Cyclic changes in the thickness of sandstones and mudstones; for example, below the sequence boundary, the grain size of sandstones increases upwards, and the sand-mud ratio increases upwards; above the sequence boundary, the opposite is true. Such change characteristics of cycles are often manifested by the change of superimposition patterns.

According to the above characteristics, short-term base level cycles can be identified in different depositional environments, as shown in Figure 4-8.

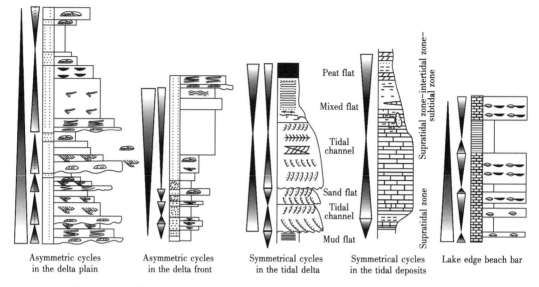

Figure 4-8 Identification of short-term cycles in different depositional environments

2) Logging curve identification marks

The high-resolution characteristics of logging curves provide a good data basis for the identification and division of various levels of base level cycles. The determination of base level cycles based on logging curves, especially the determination of the cycle boundaries, is carried out on the basis of the analysis of coring sections. In other words, a logging response model of short-term cycles and cycle boundaries shall be firstly established using coring sections so as to provide guidance to the division of cycles based on the logging curves of regional non-coring wells.

In order to avoid the multiplicity of the geological significance represented by logging curves, a reasonable logging combination series shall be selected based on the geological characteristics of the oilfield in the identification and division of base level cycles using logging data.

Long-term base level cycles can be determined by analyzing the superposition patterns of short-term cycles, and logging curves are particularly effective for this analysis. This is because the specific superimposition patterns of the short-term cycles that make up long-term cycles move unidirectionally to the maximum (maximum accommodation space) or minimum (minimum accommodation space) of the amplitude during the ascending and descending of the long-term base level cycles. These superimposition patterns often have distinct logging responses (Figure 4-9). The superimposition pattern of progradation toward the sea (lake) basin is formed in the fall period of the long-term base level cycle, during which the A/S value is less than 1; in addition, compared with the adjacent underlying cycle, the overlying short-term cycle shows the characteristics of reduced accommodation space in sedimentology and petrology. The regressive superimposition pattern of advancing toward the land is formed in the rise period of the long-term base level cycle, during which the A/S value is more than 1; in addition, compared with the adjacent underlying cycle, the overlying short-term cycle shows the characteristics of increased accommodation space in sedimentology and petrology. The aggradational superimposition pattern of the short-term base level cycle appears in the transition period of the long-term base level cycle from ascending to descending or from descending to ascending, during which the A/S value is equal to 1. There is little change in the accommodation space during the formation of adjacent short-term cycles. Figure 4-10 illustrates how to determine a mid-term base level cycle using the superimposition patterns of short-term cycles.

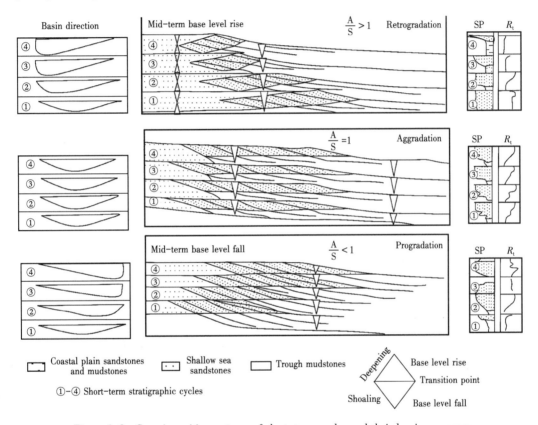

Figure 4-9　Superimposition patterns of short-term cycles and their logging response
(according to M.H. Gardner, 1964)

3) Identification marks on seismic profile

The seismic reflection interface tracks the time surface, so seismic reflection profiles can be used for sequence stratigraphy analysis. However, limited by the vertical resolution of seismic information, the division accuracy of seismic base level cycles is closely related to the quality and resolution of seismic data. Generally speaking, seismic reflection profiles can only be used to identify long-term base level cycles. The marks used to identify seismic sequence boundaries in seismic stratigraphy are also suitable for the analysis of cycle boundaries, e.g. the unconformities distributed regionally or the termination types of seismic reflection events reflecting the discordant relationship of strata, i.e. toplap, truncation, onlap, etc.

There are four sedimentary processes during the movement of the base level relative to the

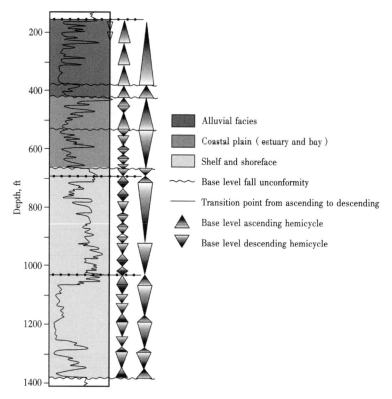

Figure 4-10 Combination of short-term cycles into a mid-term cycle using logging data
(according to Gross, 1996)

ground surface, i. e., deposition, erosion, sediment passing erosion and deposition non-compensation. The erosion of the base level below the ground surface appears as truncation on the seismic profile, which is the seismic sequence boundary and also the long-term base level cycle boundary. When the base level coincides with the ground surface, the passing erosion of the later sediments on the surface of the earlier sediments often appears as a toplap phenomenon on the seismic profile. This kind of sedimentary discontinuity is developed in the delta and fan delta environment with progradation. When the base level is above the ground surface, the non-compensation caused by the relatively insufficient supply of sediments appears as an onlap phenomenon on the seismic profile. Therefore, the important boundary in the base level cycle can be identified according to the properties of seismic reflection terminators.

4) High-resolution sequence division and correlation combined with well-to-seismic

The identification and division of multi-level base levels are the basis for the establishment of a high-resolution stratigraphic framework, for the purpose of transforming the one-dimensional information obtained in the wellbore into the prediction of three-dimensional stratigraphic relations. Although the vertical resolution of logging information is high, its transverse detection range is very small. Seismic information is the opposite, and has good transverse continuity and poor vertical resolution. Giving full play to their respective advantages is the key to precise division and correlation of high-resolution sequences. The accuracy of sequence division and correlation is further improved generally by means of vertical seismic profile (VSP) data, fine calibration of synthetic seismograms, and seismic inversion technology under well constraints.

(II) Base level cycle correlation—isochronous correlation rule

High-resolution stratigraphic correlation is the correlation of the contemporaneous strata and boundaries, not the correlation of the rock types of cycle amplitude. A complete base-level crossing cycle and the increase and decrease in its associated accommodation space are composed of a complete stratigraphic cycle representing a bisected time unit (each part represents the base level rise and fall) in the stratigraphic record, and in some cases, are only composed of asymmetric hemicycles and the boundaries representing erosion and non-deposition.

Genetic sequence correlation is performed based on the identification of base level cycles by analyzing accumulation modes. The analysis of accumulation modes is a practical method developed by the genetic stratigraphic research group. The principle of this method: the position and boundary of genetic sequences as well as the spatial distribution, accumulation patterns, etc. of genetic sequences are identified using facies sequences and stratigraphic boundaries, and then the fluctuation trend of A/S value is inferred using facies sequences, stratigraphic boundaries and preservation degree. The identification of base level cycles on the same time scale is the basis of stratigraphic correlation.

The turning point of the base level cycle in the correlation of genetic sequences, i.e., the transition position of the base level from descending to ascending or from ascending to descending, can be used as the preferred position for time-stratigraphic correlation. This is because the turning point is the limit position of the unidirectional change where the accommodation space increases to the maximum or decreases to the minimum. There can be rock-to-rock correlation, rock-to-boundary correlation, and boundary-to-boundary correlation in the correlation of base level cycles (Figure 4-11). Base level cycles can reflect the entire time of the stratigraphic record by the combination of rocks with boundaries, and thus the location of the accommodation space for geographic migration is identified. The temporal and spatial distribution characteristics of base level changes, deposition, non-deposition, sediment passing and erosion are successfully reflected on the time-space diagram (Figure 4-12). The space-time representation method shows the space-time regions in which various actions operate, regardless of whether the product is a stratum or a sedimentary layer. Accordingly, a standard stratigraphic cross-section is equivalent to a stratigraphic response map, and a time-space cross-section is equivalent to a stratigraphic action map. Figure 4-12 shows the cross-section of the rock formations of the base level cycle and the

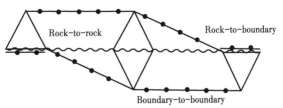

Figure 4-11 Correlation of genetic sequences

matching Wheeler diagram. The time-space diagram is the most effective method for time-space inversion of stratigraphic profiles, and helps to understand the stratigraphic response (rock + boundary) of geological processes (time + space), and to determine the time of rock-to-rock correlation, rock-to-boundary correlation, and boundary-to-boundary correlation.

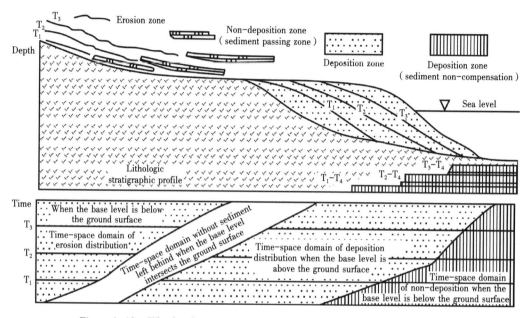

Figure 4-12 Wheeler diagram of geological processes (according to Wheeler)

The genetic stratigraphic research team of T.A. Cross summarized a correlation scheme for genetic strata in a shallow sea environment based on the analysis of the coastal plain-shallow sea depositional system. As shown in Figure 4-13, the accommodation space decreases during the fall period of the long-term base level; with the decrease of long-term accommodation space, short-term stratigraphic cycles in shallow seas and coastal plains are getting thinner and thinner, and most of them are base level descending asymmetric cycles, and are top-truncated by base level descending unconformities or sediment passing surfaces. With the reduction of the accommodation space, the sediments deposited during the rise of the base level are eroded during the fall of the base level. The transition point of the base level from fall to rise marks the beginning of an event. In the coastal plain, more sediments are deposited and preserved in the base level ascending and descending cycles, and the stratigraphic cycles are more symmetrical and thicker.

In the more landward position, as the long-term accommodation space decreases, there is a tendency of deposition only during the rise of the base level. When the base level drops, the accommodation space is reduced, and sediments pass by or are eroded. These parts appear as stratigraphic boundaries in the base-level crossing cycle. If the sediments deposited during base level rise are not eroded away during base level fall, the stratigraphic cycle is more likely to be an asymmetrical base level rise cycle at the turning point of long-term base level rise. When the long-term base level rises, stratigraphic cycles are more symmetrical and thicker. This reflects that sediments are deposited and preserved during both the rise period and the fall period of the base

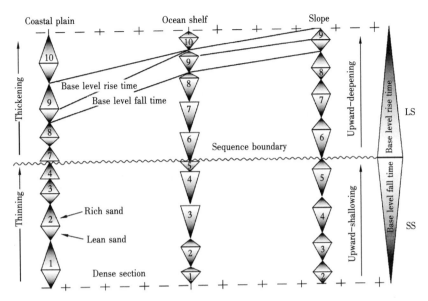

Figure 4-13 Correlation scheme for genetic sequence of shallow sea environment
(according to T.A.Cross, 1994)

level. Similarly, at the turning point of the long-term base level from fall to rise, the shorter the distance to the provenance, the higher the possibility of the asymmetrical base level ascending cycles; the longer the distance to the provenance, the higher the possibility of descending cycles.

The sediment flux tends to increase in the more seaward position; moreover, deposition occurs when the long-term base level drops, so that the stratigraphic cycle appears as a base level descending asymmetric cycle at the turning point and is thick. When the long-term base level rises, the symmetry of the stratigraphic cycle is enhanced and it becomes thinner, reflecting that deposition occurs in both base level hemicycles, but the supply of sediments is constantly decreasing.

Once the high-resolution time surface is placed in a rock framework for consideration and correlation, the information about the characteristics of facies and facies sequences is also stored in this framework. Facies characteristics and facies sequence are also combined with the high-resolution time framework on control points. However, what is very helpful to the knowledge about stratigraphy and has not been seen before is that the description of the petrophysical characteristics, geometry characteristics and continuity characteristics of the strata on each control point gives the content of accommodation space and datum level changes. The characteristics of the strata deposited outside and between control points can also be predicted by the time information of mass conservation, volume division and facies differentiation. Thus, if what kind of facies deposited at a specific location and within a specific time, base level and accommodation space changes are known, the relative volume and facies characteristics of the strata between and outside control points can be predicted.

From the dynamic point of view of base level and accommodation space, the volume distribution, sediment preservation degree, stratigraphic accumulation pattern, facies sequence characteristics and facies types in the same depositional system tract or facies tract are not fixed, but they are a function of their position in the base level cycle and accommodation space, that is, a function of time and space. Therefore, from the perspective of sedimentary dynamics, the method of analyzing the changes in the accommodation space caused by the changes in the base level during sediment accumulation and interpreting stratigraphic configuration and sedimentological characteristics is fundamentally different from the traditional facies model analogy method, and has powerful advantages.

Section 3   Fine Stratigraphic Structure Establishment Example

The purpose of fine stratigraphic division and correlation is to establish fine stratigraphic structure through subdivision of layers and reveal the distribution law of genetic unit sand bodies. Taking Block Xingerzhong of Daqing Oilfield as an example, this section describes the establishment of fine stratigraphic configuration.

I. Stratigraphic characteristics

The target horizons of research in Block Xingerzhong are Saertu reservoir group and Putaohua reservoir group, where the strata experienced strong regression from the late stage of Qingshankou Formation to the early stage of Yaojia Formation and the gradual transgression and stable deposition from the middle stage of Yaojia Formation to the Nenjiang Period, the river energy is strong, and the material supply is abundant, forming a set of large-scale river delta sedimentary bodies with long extension. The strata appear as a regressive and transgressive sedimentary cycle from bottom to top. The strata are characterized by alternate layers of sandstone and mudstone deposits. Their lithology is mainly alternate layers of gray and dark gray mudstones and sandy mudstones with gray, brown and white-gray fine sandstones and siltstones, and the sandstones generally contain oil (Figure 4-14). The overall thickness of sandstones varies from 1-8m, and the thick sand bodies are mainly distributed in the middle of Sa-2 Member and the bottom of Pu-1 Member, and have a thickness of about 4-8m. Relatively, thick sandstones account for a relatively low proportion and 10% of the total sand bodies, but due to their good physical properties and wide distribution, they constitute the main reservoirs in Daqing Oilfield.

The burial depth of Saertu and Putaohua reservoir groups is 800-1200m. The thickness of Saertu reservoir group is about 140m, and the thickness of Putaohua reservoir group is about 100m. Saertu reservoir group is divided into three members, and Putaohua reservoir group is divided into two members. According to the traditional division method of Daqing Oilfield, the lithology of tabulated reservoirs is mainly fine sandstones and siltstones, with an average air permeability of 278mD, an average porosity of 25.6%, and an oil saturation of 63.5%. Compared with tabulated reservoirs, untabulated reservoirs have fine sandstone grains, high shale content, poor sorting property of sandstones and mudstones, low porosity and permeability, and low oil saturation; in addition, their lithology is mainly argillaceous siltstones; the silty sand content is 65.5%, the shale content is 14.2%, the average air permeability is 20mD, the average porosity is

21.6%, and the average oil saturation is 42.2%.

System	Series	Formation	Member	Lithologic profile	Relative lake level 40 20 0	Thickness m	Lithology	Sedimentary facies evolution	Oil-bearing assemblage
	Paleogene–Quaternary						Loess, yellowish-gray sand layers, sandy gravel layers. Yellowish-green and gray mudstones and glutenites sandwiched with lignite	Fluvial diluvium	
Cretaceous	Upper Series	Mingshui Fm.	2			0~381	Alternate layers of brownish red and grayish-green mudstones with sandstones	Alternate fluvial facies and lacustrine facies	
			1			0~243	Dark gray and grayish-green mudstones and sandstones, glutenites		
		Sifangtai Fm.				0~413	Alternate layers of brownish red and grayish-green mudstones with light brownish red sandstones and glutenites		
		Nenjiang Fm.	5			0~355	Grayish-green and brownish-red mudstones sandwiched with off-white sandstones	Alternate fluvial facies (dominat) and lacustrine facies	
			4			0~300	Grayish-green and gray mudstones sandwiched with grayish-white sandstones		Heidimiao reservoir
			3			47~118	Alternate layers of off-white sandstones with grayish-black mudstones		
			2			50~252	Grayish-black mudstones and basal oil shales	Deep lacustrine facies	Saertu reservoir
			1			27~222	Grayish-black mudstones sandwiched with grayish-white sandstones		
	Middle Series	Yaojia Fm.	3			0~150	Alternate layers of grayish-green mudstones with grayish-white sandstones	Delta	Putaohua reservoir
			2						
			1			0~100	Grayish-white and grayish-green sandstones sandwiched with grayish-green and brownish-red mudstones		
		Qingshankou Formation	3			53~552	Grayish-green and grayish-black mudstones sandwiched with grayish-white sandstones and ostracods layers		Gaotaizi reservoir
			2						
			1			25~112	Grayish-black mudstones and basal oil shales	Deep lacustrine facies	
		Quantou Fm.	4			0~128	Alternate layers of grayish-white sandstones with brownish-red and grayish-green mudstones	Delta	Fuyu reservoir
			3			0~529	Alternate layers of brownish-red and dark purplish-red mudstones with brownish-gray and light grayish-green sandstones		Yangdachengzi reservoir
			2			0~479	Purplish-brown mudstones sandwiched with light purplish-red mudstones	River	
			1			0~885	Brownish-gray, grayish-white and light grayish-green sandstones sandwiched with purplish-brown and dark brownish-red mudstones		
	Lower Series	Denglouku Fm.	4			0~212	Alternate layers of brownish-gray and grayish-white sandstones with purplish-brown and brownish-gray mudstones	Fluvial–delta lake facies	
			3			0~562	Grayish-white and brownish-gray sandstones sandwiched with grayish-black and purplish-brown mudstones		
			2			0~700	Grayish-black, greenish-gray and purplish-brown mudstones sandwiched with grayish-white and brownish-gray sandstones		
			1			0~119	Varicolored conglomerates sandwiched with purplish-brown and grayish-black mudstones	Diluvium	

Figure 4-14 Comprehensive stratigraphic column of the research area (modified from the "Continental Hydrocarbon Generation, Migration and Accumulation in Songliao Basin")

II. Fine stratigraphic structure establishment idea

Block Xingerzhong is a development block, and the division and correlation of sequences must be combined with regional research. Therefore, firstly a macroscopic sequence framework is established by learning from previous research results. Then with the guidance of high-resolution sequence stratigraphy theory, stratigraphic division and correlation are further performed in the macroscopic sequence framework to subdivide the single sand layers of the research area. The flooding surface is selected as the sequence boundary in the process of sequence division according to the sedimentary characteristics of this area such as stable distribution of flooding mudstones and easy tracking and correlation of them. In addition, the four-level division scheme including long-term cycles, medium-term cycles, short-term cycles and ultra-short-term cycles is selected.

The basic data of the research area include drilling and coring data, logging data and production performance data. The scale and accuracy of these data are different, and their roles in sequence division are also different. Therefore, according to the data of the research area and the research purpose, the principle "marker horizon control, base level cycle correlation, and step-by-step refinement" is used in the process of fine stratigraphic correlation, specifically including the following content.

(1) The division of core base level cycles, which can achieve the identification of ultra-short-term base level cycles;

(2) Core calibration of logging curves to determine the high-resolution stratigraphic cycle characteristics of logging curves. Mid-term base level cycles are divided based on the division and combination modes of ultra-short-term and short-term base level cycles;

(3) High-resolution stratigraphic cycle correlation using logging curves to establish a high-resolution isochronous stratigraphic framework;

(4) Cycle correlation, hierarchical control and splitting of single sand layers.

III. Identification and correlation of base level cycles

Base level cycles are divided through the identification of base level cycle boundary marks. The resolution of different data is quite different, so the levels of base level cycles that can be divided are different. Base level cycles are divided this time mainly based on core data and logging data.

(I) **Identification of base level cycles using core data**

Core data have very high resolution and can be used to identify short-term and ultra-short-term base level cycles. The contact surface or erosion surface between different rock types is often an important boundary for identification of base level cycles using core data.

1. Identification of ultra-short-term base-level cycle boundaries

The following depositional boundaries can be identified on cores. (1) The bedding surface inside rocks (lamina surface, series of strata surface, series of strata group boundary): formed during the deposition of clastic sediments due to changes in hydrodynamic force or flow regime. (2) The contact surface between different rock types: the rock type conversion surface reflects the jump in the correlation between accommodation space and sediment supply. There are many boundaries between different rock types, such as mudstone-sandstone boundary, mudstone-salt rock boundary, mudstone-gypsum salt boundary, etc. Mudstone-sandstone boundaries widely exist in the research area, including two types such as gradual change and sudden change. The conversion from sandstones to mudstones indicates a rise in the base level, while the conversion from mudstones to sandstones is the opposite. (3) Erosion surface: reflects the beginning of the ascending process after the base level drops to the lowest position. (4) The base-level cycle transition surface represented by deep lake mudstones: the appearance of deep lake mudstones indicates that the base level rises to the highest position and that there is a transition surface where the base level rises and falls. A set of typical deep lake mudstone deposits are developed in the top of Saertu reservoir in the research area, forming a regional marker horizon distributed throughout the area.

Among the various depositional boundaries mentioned above, the appearance of deep lake mudstones can be generally used to determine the transition process of one-stage base level rise and fall, while the erosion surface represents a transition point of the base level from fall to rise, and other natural boundaries need to be analyzed specifically.

2. Determination of ultra-short-term base level cycles

An ultra-short-term base-level cycle is equivalent to the stratigraphic record of a single deposition event, and consists of a single lithology formed in a single deposition event or superposition of related lithologies. In each hemicycle of the ultra-short-term cycle, the change in water depth is unidirectional, which is generally manifested as a set of continuous lithofacies assemblage. For example, the positive rhythmic deposition of distributary channels, the upward-fining grain sequence of delta front sheet sands, etc. in the area reflect the increment of a single layer, and there is evidence that the water depth increases or decreases across the boundary. The identification basis for the ultra-short-term base level cycle is the base level change information recorded in the lithofacies assemblage. The change direction of the base level is determined by looking for the change of water depth in the rock sequence or the preservation degree of sedimentary landforms or the denudation trend of sediments. The specific identification method is generally to first determine the transition point of the base level cycle, and then the change direction of the base level in the sequence formation process within the sequence.

Symmetrical cycles, asymmetrical ascending hemicycles and asymmetrical descending hemicycles can be identified using the cores from the research intervals. Symmetrical cycles and asymmetric ascending hemicycles are the main types of ultra-short-term cycles in the area (Figure 4-15). The

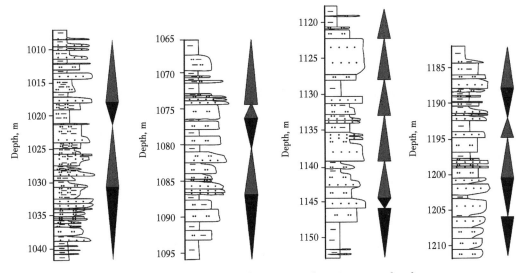

Figure 4-15   Identification of short-term cycles using core onlap data

symmetrical cycle formed by accommodation space has both base level ascending hemicycle deposits and base level descending hemicycle deposits, but is not completely symmetrical. Usually the ascending hemicycle is thicker than the descending hemicycle. Under the condition of sufficient provenance supply, the rise of the base level is conducive to the preservation of sediments, which is characterized by the extensive development of sheet sands. The asymmetric ascending hemicycle is mainly developed in the delta plain subfacies (the upper part of Pu-1 Member), and is characterized by thick or superimposed distributary channel sandstone deposits. The sandstones have good homogeneity and high oil saturation, where trough cross-beddings are developed, and parallel beddings and wavy beddings can be seen in the top fine-grained part. The asymmetric descending hemicycle is mainly due to the fact that during the fall of the base level, sediments advance toward the lake to form transformed underwater distributary channel sands covering the lacustrine mudstones, where wavy beddings and parallel beddings are developed.

(II) **Identification of base level cycles using well data**

The oilfield in the middle and late stages of development has very rich logging data. On the basis of litho-electric calibration of coring wells, the identification of base level cycles using logging curves is the main method of stratigraphic division and correlation. According to the lithological characteristics of the research intervals, the logging series curves mainly including GR and SP and secondarily resistivity and interval transit time are selected to identify cycle boundaries and divide base level cycles.

1. Identification of cycle boundaries using logging curves

Two types of cycle boundaries can be identified on logging curves: erosion surface and flooding surface. The erosion surface is generally at the bottom of sandstones, and its existence is judged from the abrupt contact relationship between sandstones and the underlying strata, while the flooding surface is marked by the appearance of flooding mudstones, which is stably distributed and easily tracked and correlated.

The erosion surface is the transition boundary between base level fall and base level rise, and is the boundary within the sequence. The flooding surface is the transition position between base level rise and base level fall, and is a sequence boundary. The cycle boundary can be determined according to the change of GR value or SP value. The GR value increases upwards, indicating that the water bodies are deepened as a whole, and forming a retrograding stratigraphic superposition pattern. On the contrary, the upward reduction of the GR value indicates that the water bodies become shallower, forming a prograding stratigraphic superimposition pattern. The position of GR value conversion represents the cycle boundary.

The conversion surface within the short-term base level cycle is directly determined by the change in the GR value or SP value. The transition surface inside the mid-term base level is located at the transition position of the stratigraphic superimposition pattern composed of short-term cycles. Generally, the change position from the prograding stratigraphic superimposition pattern to the retrograding superposition pattern is the transition surface inside the mid-term base level cycle.

2. Determination of base level cycles using logging data

Using logging curves to divide and identify base level cycles is mainly based on the shape of curves and their combination relationship. The GR and SP curves are sensitive to lithological changes in the research area, and are the main basis for dividing base level cycles; in addition, other curves such as resistivity curve etc. are referred to. The specific method: firstly calibrate the logging data according to the core characteristics of the cored intervals, establish logging response models for different ultra-short-term base level cycles, analyze their superimposition patterns, and identify short-term cycles. The top and bottom of a short-term cycle are the transition surface from the rise to the fall of the base level, which is equivalent to the flooding surface. The cores are marked by thick mudstones, and the change direction of the base level inside the cycle is judged according to the superimposition pattern of ultra-short-term cycles. Symmetric and ascending asymmetric cycles are mainly developed, and descending asymmetric cycles are rare (Figure 4-16). Then, medium-term base level cycles are determined on the basis of the division of short-term cycles according to their superimposition pattern and boundary identification marks. The target interval in the research area has been comprehensively analyzed and medium-term and short-term cycles have been divided using this method of identifying boundaries and cycles with logging curves.

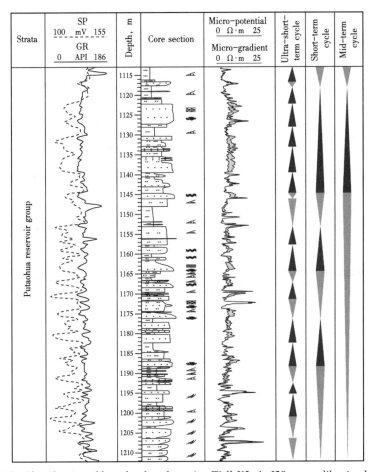

Figure 4-16 Identification of base level cycles using Well X2-1-J29 core calibration logging data

Ⅳ. Establishment of fine stratigraphic framework

Inter-well cycles are correlated and a sequence stratigraphy framework is established under the control of the regional marker horizon based on the division of single-well base level cycles (Figure 4-17). The correlation is based on the following principle: the deep lake mudstones widely distributed, stably deposited and easily identified provide constraints for the correlation of mid-term and long-term base level cycles. The correlation of mid-term cycles and their internal strata is achieved based on the stratigraphic superimposition pattern composed of mid-term cycles. The correspondence between short-term cycles is determined, and the correlation of short-term cycles is completed within different superposition pattern frameworks according to the position of short-term cycles. The correlation method uses the closed method from point to line and from line to surface. The stratigraphic framework of the target area has been established level by level according to the steps of single-well cycle division, skeleton profile correlation, and closed correlation of multi-well profiles.

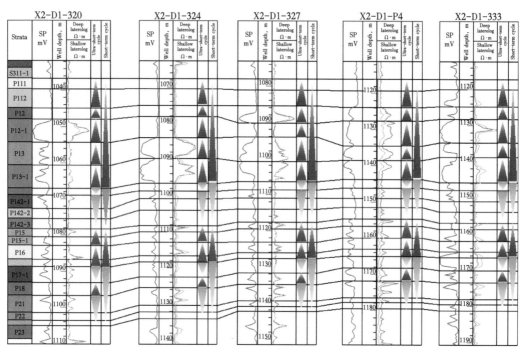

Figure 4-17 Skeleton correlation profile of Putaohua reservoir group

Saertu reservoir group is further divided into 7 short-term cycles, and Putaohua reservoir group is divided into 3 short-term cycles. A short-term cycle is divided into several ultra-short-term cycles so as to form a fine stratigraphic framework with hierarchical control and step-by-step refinement. The correlation results show that the stratigraphic deposition in the area is stable and the thickness of the strata does not change much.

V. Correlation of single sand layers

The sand bodies of about 1–2m in the research area account for a large proportion, and the sand bodies with large thickness are mostly the superposition of multi-stage single sand bodies. Therefore, it is necessary to split and correlate single sand layers in a sequence based on the establishment of the sequence framework in order to further reflect the distribution of single sand bodies. The following single sand layer correlation modes shall be considered in the splitting and correlation process.

(I) **Marker horizon flattening correlation mode**

The position of the single sand bodies deposited in the same period shall be roughly parallel to the marker horizon, that is, the single sand bodies of the same period shall have an approximately equal distance from the marker horizon (or an isochronous surface). For example, the top surfaces of the same river channel shall be an isochronous surface, and shall have an approximately equal "elevation" from the marker horizon. Sheet sands are the sand bodies carried by rivers into lakes and formed from selection and transformation by lake water, and have characteristics such as stable distribution etc. and the same "elevation" distribution with the marker horizon. Therefore, flattening with the marker horizon stably distributed throughout the area is helpful for splitting and correlating single sand layers. Figure 4-18 shows the flattening correlation data using the mudstones stably distributed in the top of Saertu as the marker horizon. The shorter the distance from the marker horizon, the higher the correlation accuracy.

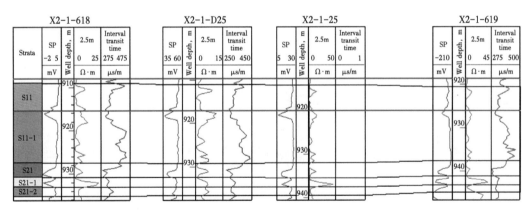

Figure 4-18 Flattening correlation mode of the top marker horizon of Saertu reservoir group in the research area

(II) **Superimposed channel sand body correlation mode**

Due to the erosion of distributary channel sand bodies, sedimentary sand bodies in different periods are easily superimposed. In regions with multi-stage channel sand bodies, the late channel sand body scouring causes the upper or all of the channel units deposited in the early period to be eroded, and new channel sandstones are deposited, forming thick channel sand bodies inter-superimposed vertically. Therefore, single sand layers can be split and correlated according to core data, the residual mudstones inside sandstone bodies, logging curves and the distribution characteristics of sand bodies in adjacent wells (Figure 4-19).

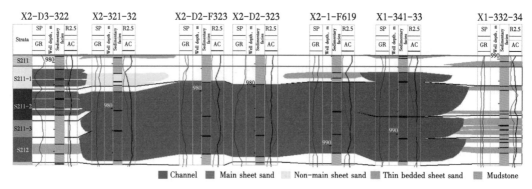

Figure 4-19 Correlation model of superimposition and splitting of channel sand bodies in the anatomy area

(Ⅲ) Channel undercut sand body correlation mode

Channel sand body deposits usually have strong undercut erosion. Due to the uneven erosion of the different parts of a river channel in the late period, in some parts, such as those near the main stream of the channel, especially on the side of the meandering river near the concave bank, the river bottom is strongly eroded, and the channel sediments directly cover the early channel sands, forming thick undercut sand bodies. For such sand bodies with undercutting phenomenon, the change law of sand body thickness distribution shall be considered, and the sand body undercut correlation mode shall be adopted (Figure 4-20).

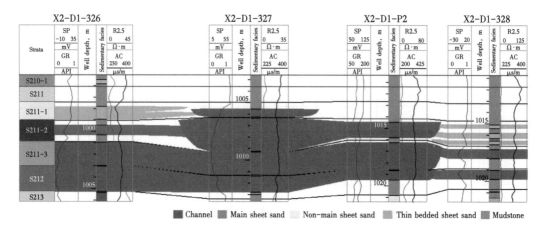

Figure 4-20 Correlation model of undercutting of channel sand bodies in the anatomy area

(Ⅳ) Lateral facies change correlation mode

Facies change is an universal geological law in nature. Different sedimentary facies types have different facies change characteristics. In the depositional environment of a river channel, the facies change is rapid, and the thickness of the sandstones changes greatly laterally. Within the same deposition time unit, even adjacent locations may belong to different sedimentary microfacies. For example, in case of changing from river channel deposits to overbank deposits, both the lithology and logging curve characteristics are quite different. Sheet sands are the product of lake water transformation effect, so the facies change is slow, most contact relationships are transitional contact, and eroded contact facies change is rare. When the facies change correlation mode is used, it is necessary to make full use of the law of facies sequence gradual change, and consider the rationality of the spatial combination of various sedimentary microfacies (Figure 4-21).

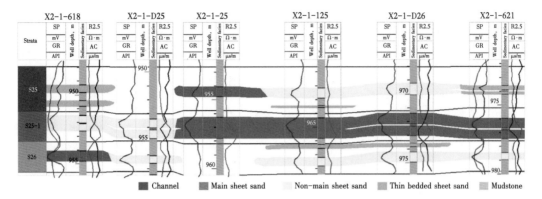

Figure 4-21 Correlation model of lateral facies change of sheet sands in the anatomy area

(Ⅴ) Analysis of correlation results

The single sand layers in the whole area have been split and correlated and the entire target interval has been divided into 84 single sand layers using the single sand layer splitting correlation mode on the basis of isochronous stratigraphic framework establishment (Table 4-2). In addition, the rationality of splitting correlation results has been verified in the correlation process by continuously compiling top surface structure maps and strata thickness maps, and the well points causing singular changes have been checked. On the one hand, the wrong splitting correlation of single sand layers can lead to a sudden change in structural singular points and strata thickness, and thus there is a need for splitting correlation again. On the other hand, not all singular points are unreasonable, and there are reasonable mutation points locally. This is mainly due to the rapid local deposition rate, resulting in relatively large strata thickness. For example, the rapid accumulation and undercutting erosion of channel sand bodies cause large lateral changes in strata thickness. Therefore, uncoordinated singular change points need to be reasonably analyzed based on the development characteristics of stratigraphic sand bodies.

Table 4-2 Division scheme for Xingerzhong strata

Series	Formation	Reservoir group	Sand group	Number of layer(s)	Number of single sand layer(s)
Lower Cretaceous Series	Nenjiang Fm.	Saertu	Sa-1 Member	1	2
			Sa-2 Member	16	29
			Sa-3 Member	11	20
	Yaojia Fm.	Putaohua	Pu-1 Member	8	16
			Pu-2 Member	12	17

The final correlation results show that the distribution of most layers in the area is stable, and the thickness of the strata is not much different; in addition, the thickness of most single sand layers is less than 3m, the thickness of some layers is 3-4m, and the thickness of a few layers is greater than 4m, which is basically equivalent to the thickness of the genetic unit sand bodies. This can reflect the distribution characteristics of the genetic unit sand bodies and has laid the foundation for the subsequent sedimentary microfacies analysis and single sand body description.

Chapter 5 Description of Sedimentary Microfacies of Reservoirs

Sedimentary facies description is one of the key points of geological work, generally including the determination of sedimentary facies types, the identification of sedimentary subfacies and microfacies, the research on the development characteristics, spatial distribution characteristics and evolution laws of sedimentary microfacies, etc. Sedimentary facies research anatomizes the distribution law and sedimentary characteristics of geological bodies from the perspective of genesis, and provides a basis for the analysis of the geometry, size, distribution and connectivity of reservoirs. In the development stage, the division and depiction of sedimentary microfacies are of guidance significance to predicting the distribution characteristics of sand bodies, revealing the heterogeneity of reservoirs and understanding the movement law of oil and water, and also provides constraints for further description of single sand bodies.

Section 1 Basic Methods of Sedimentary Microfacies Description

The main method of sedimentary microfacies research: starting from single-well analysis of cores, establish microfacies identification marks, and divide microfacies types based on the previous understanding of sedimentary background and depositional systems; then obtain the spatial distribution characteristics of sedimentary microfacies through well-tie correlation analysis and the study of planar distribution law.

I. Basic data for sedimentary microfacies research

The basic data required in the sedimentary microfacies analysis for fine reservoir description work in the development stage are quite different from those in the exploration stage. The analysis of sedimentary microfacies in the development stage is based on the understanding of sedimentary facies in the exploration stage; therefore, it is necessary to fully understand the types and characteristics of sedimentary facies. In addition, with the deployment of development wells, logging data are more abundant, and the production performance data of each well can also provide references for microfacies analysis. The basic data for sedimentary microfacies research can be divided into the following types.

(1) Macroscopic research data. They include regional sedimentary background data, and the recognized data of division of depositional systems, sedimentary facies and subfacies.

(2) Coring well data. Various data on lithology, sedimentary texture, sedimentary structure, paleontology, porosity, permeability, shale content, calcareous content, geochemistry, etc. are obtained from the observation, analysis and testing of cores; moreover, the scale and development

frequency of genetic unit sand bodies can be quantitatively analyzed using cores.

(3) Logging data. There are many logging data in the development stage relative to the exploration stage, which are the main basis for sedimentary microfacies analysis.

(4) Seismic data. They are mainly high-resolution 3D seismic data, and can improve the basis for lateral prediction of sedimentary microfacies.

(5) Production test data, test production data and production data. They indirectly reflect the connectivity between sedimentary microfacies.

II. Basic steps of sedimentary microfacies research

Sedimentary microfacies research in the development stage: the types of sedimentary facies are firstly determined and then subfacies are accurately divided. Sedimentary microfacies research is carried out step by step on the premise of subfacies identification. Sedimentary microfacies research is generally performed in accordance with the steps of the analysis of single-well facies, profile facies and planar facies.

(I) Single-well facies analysis

1. Lithofacies analysis

The main basis for the division of lithofacies is core observation and description. Lithofacies analysis is the main basis for identifying sedimentary microfacies in fine reservoir description. Through the observation and description of color, lithology, sedimentary structure and grain size etc., the hydrodynamic conditions, geochemical actions and biochemical actions during deposition, distinguish the depositional environment and deposition characteristics of different microfacies are analyzed, and then the microfacies are divided.

For core analysis of lithofacies in the development stage, quantitative description shall be carried out as much as possible in addition to qualitative description. For example, the quantitative analysis of the vertical thickness and distribution frequency of various lithofacies can provide references for fine reservoir modeling and cross-well prediction. Lithofacies description results can be named by the combination of sedimentary microfacies, grain size, sedimentary structure and lithology. The names are not required to be comprehensive, but shall reflect the most representative identification features, such as braided river channel glutenite facies, braided river channel coarse sandstone, and other similar names.

2. Analysis of lithologic sequences and sedimentary cycles

A lithologic sequence is a combination of lithology, grain size, color, and sedimentary structure of rocks. Lithologic sequence analysis is an important basis for sedimentary facies analysis. Generally, a certain sedimentary microfacies has a corresponding vertical lithologic sequence, but the same lithologic sequence can be formed in several depositional environments, so there is no absolute correspondence between the lithologic sequence and the sedimentary microfacies. Therefore, lithologic sequence analysis must be combined with sedimentary facies types and depositional environments, and the research on lithologic sequences and sedimentary cycles related to specific facies types is more representative and can be used as an important basis for distinguishing microfacies types.

The study of lithologic sequences mainly starts from cores. The representative cored well sections are selected to work out lithologic sequence column diagrams, which cover not only sedimentological description, but also parameters reflecting the physical characteristics of reservoirs, as well as typical logging curves. Single-well sedimentary cycles can be determined through lithologic sequence analysis. And the analysis of sedimentary cycles at all levels is performed level by level taking the smallest sedimentary cycle as the unit. The purpose of sedimentary cycle analysis is to ascertain the evolution of microfacies vertically, further to confirm subfacies (macrofacies), and to check microfacies based on facies combination. Sedimentary cycle analysis is a comprehensive analysis of all facies markers. Sedimentary cycles at all levels reflect basin's tectonic activities, climate change, clast supply change, transgression and regression, the abandonment and transfer of sedimentary bodies, the energy difference between sedimentary events, and the energy change process of each sedimentary event.

3. Logging facies identification

Logging facies is a synthesis of logging responses that characterize stratigraphic characteristics. In a research area, after all, there are only a few coring wells, and drilling information is an inevitable method for logging facies identification. Sedimentary facies is represented by specific facies markers, while logging facies is represented by specific logging responses. Logging facies is equivalent to sedimentary facies. Different sedimentary facies belts result in different logging responses due to different composition, texture, and structure of rocks. However, due to the multiplicity of logging curves, the two are not all in one-to-one correspondence. Therefore, it is necessary to calibrate logging facies using known sedimentary facies. Firstly, the logging curves or parameters of cored wells are divided into several logging facies, and these logging facies are correlated with the sedimentary facies from core analysis to establish the correlation between the two; then in turn, a sedimentary facies analysis is made using the logging data of non-cored wells.

To identify logging facies, the sensitivity of logging curves is firstly analyzed, and a combination of logging curves representative of the research area is selected. Different logging curves have different logging responses to different lithologies. Logging series shall be selected mainly considering the ability of logging curves to distinguish lithology, thin layers, and the physical properties and oil-bearing properties of reservoirs. In general, typical logging data mainly include SP, GR, resistivity, interval transit time, density, neutron, diplog, etc. These logging data reflect the characteristics of lithology, physical properties, fluid properties, etc. from different aspects to varying degrees, and have universal practical significance to most formations. In addition, diplog and imaging logging also have unique effects in sedimentary structure, contact relationship and fracture analysis. Logging curve information includes not only the shape and combination features of logging curves, but also the size distribution of logging curve values and the relative changes of curve values.

The general process of logging facies analysis: firstly establish the corresponding relationship between cores and logging responses, and a logging facies library through the observation and analysis of cores; then conduct logging facies analysis for each well and each interval and divide sedimentary facies according to the logging facies library data; finally carry out comprehensive adjustment and establish the sedimentary microfacies division profile of each single well comprehensively considering the sedimentary characteristics of the whole area (Figure 5-1).

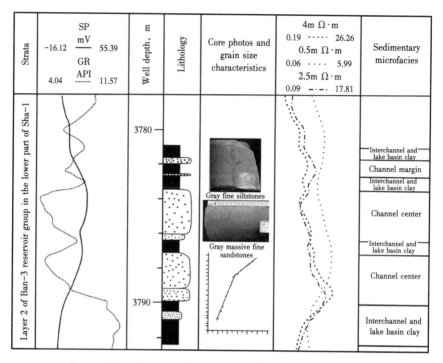

Figure 5-1 Single well facies analysis diagram of well lithology

(Ⅱ) **Analysis of profile facies correlation**

The analysis of profile facies correlation is carried out on the basis of single-well profile facies analysis. The relationship between the sedimentary facies of each single well is established and the distribution characteristics of the sedimentary facies in 2D space according to the isochronous stratigraphic division and correlation results are determined; moreover, the 3D spatial distribution characteristics of sedimentary facies can be obtained through correlation and analysis of multiple profiles.

Attention shall be paid to the three issues in the analysis of profile facies correlation.

(1) Homochronous correlation issue. The same deposition type can be developed in different periods, and similar logging curve characteristics do not mean that transverse correlation can be performed. Therefore, the analysis of sedimentary facies profile correlation must be carried out in an isochronous stratigraphic framework, which reflects the changes of sedimentary facies in different regions during the same period, so that the correlation of profiles is meaningful.

(2) The skeleton correlation profile must have two types: along the provenance direction and perpendicular to the provenance direction. Research on sedimentary microfacies has nothing to do with structural morphology. Therefore, it is not necessary to consider the axial derection problem of structures to establish correlation profiles.

(3) Facies change issue. The sedimentary microfacies in the same deposition period change transversely and are developed according to a certain combination law. Therefore, the development

mode of sedimentary microfacies shall be fully considered in the correlation of profile facies.

The correlation of sedimentary facies in the development stage is different from that in the exploration stage. In the exploration stage, the research scope of the survey area is large, the types of subfacies developed are complete, and even the types of facies deposited during the same period in the same research area will change. Therefore, the obtained facies correlation profile reflects the combination law of subfacies or facies in the macroscopic view, such as the distribution of subfacies from delta plain, delta front to prodelta. The development stage emphasizes the change law of microfacies, the research area is small and is often concentrated in a certain subfacies areally; therefore, the correlation profile obtained is the change of microfacies in the subfacies. In order to highlight the distribution of effective reservoirs, barriers and interlayers, the background facies are not deliberately depicted, and the formed correlation profile is similar to the sand body correlation profile (Figure 5-2).

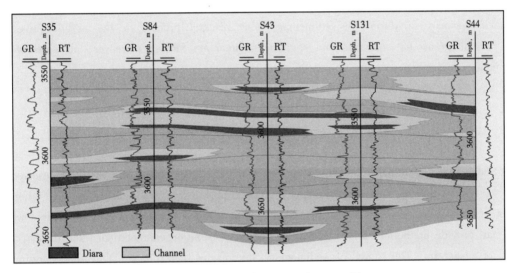

Figure 5-2 Braided river correlation profile

(Ⅲ) Planar facies analysis

Planar facies analysis is the comprehensive mapping work based on the correlation analysis of single wells and profiles combined with plan views such as sand body thickness maps etc. The types and distribution of sedimentary facies in the whole area are analyzed by drawing a series of basic maps such as profiles and plan views. These basic maps include comprehensive histograms, single-well profile facies analysis maps, facies profile correlation maps, stratigraphic isopach maps, sand layer isopach maps, sand-gross ratio maps, and distribution maps of rock types and mudstone types.

The planar facies analysis steps are as follow.

(1) Determine the provenance direction. The judgment indicators include heavy minerals, ancient flow direction, cathode luminescence, sand-gross ratio, grain size distribution characteristics, etc..

(2) Draw the profile facies map of each single well in the research area and the facies profile correlation diagram of the entire research area.

(3) Make a statistical analysis of sand-gross ratio and draw sand-gross ratio maps and stratigraphic isopach maps; in addition, use seismic data to constrain them, such as amplitude and RMS attribute maps that reflect lithology and strata thickness.

(4) Plot the sedimentary facies plan view according to the combination law of sedimentary microfacies.

(5) Check the consistency of the planar facies map with the single-well profile facies map and facies profile comparison diagram.

Sedimentary facies research is one of the important contents of petroleum exploration and development. Depositional environments and depositional conditions control the development degree, spatial distribution and internal structural characteristics of sand bodies, and the physical properties of sand bodies formed in different depositional environments are obviously different, which has different impacts on both accumulation and development of oil and gas.

It can be seen that the single-well profile facies map, facies profile correlation diagram, and planar facies map etc. during the sedimentary facies research process are the most intuitive and important diagrams for analyzing and studying sedimentary facies; moreover, all parts of the sedimentary facies research are an organic whole.

Section 2 Sedimentary Facies Research Examples in Dense Well Pattern Areas

Sedimentary microfacies analysis is the most abundant content of fine reservoir description. Due to regional differences and sedimentary type differences, description results are also different. Through the division of sedimentary microfacies, genetic types are assigned to reservoirs, which can thus provide predictions and explanations for the distribution characteristics and development laws of reservoirs. The study of sedimentary microfacies can deepen the understanding of reservoirs, and is the foundation for establishing reservoir geological models, and also an important constraint condition for reservoir parameter prediction. Based on core data and logging data as well as regional sedimentary facies research, the fine sedimentary microfacies study in the research area is performed in the fine isochronous stratigraphic framework, which mainly reflects the distribution law of combined microfacies.

I. Research idea and method

The study of sedimentary microfacies in a dense well pattern area has its particularity. Abundant drilling data increase the accuracy and credibility of microfacies research, as well as the complexity of microfacies distribution research. Therefore, the compilation of microfacies maps is a process of continuous revision. The research process flow chart is shown in Figure 5-3. The general idea: on the premise of understanding the regional sedimentary characteristics, based on the facies analysis of coring wells, a logging facies model is established through litho-electric calibration, and the single-well facies analysis of all wells is performed; then carry out reasonable microfacies combination analysis and establish a planar distribution map of sedimentary microfacies controlled by the division of logging facies of single wells and sand body thickness distribution characteristics and guided by the classic depositional model and the law of facies sequence.

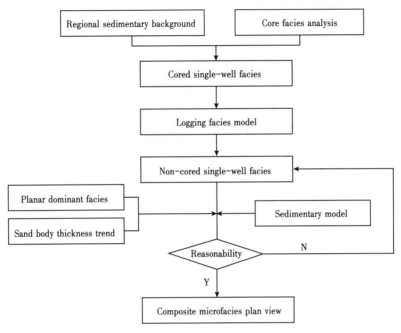

Figure 5-3 Composite microfacies hierarchy research flow chart

Take the S221 single layer in Sa-II Member in Xingerzhong as an example. First, perform a single-well facies analysis of the cored well according to the litho-electric calibration of it (Figure 5-4), and obtain the logging facies maps of various facies. Make a logging facies analysis of non-cored wells according to the logging response feature templates of various facies. Then complete the planar distribution analysis of composite microfacies of the S221 single layer under the guidance of the classic river delta deposition model and facies sequence law according to the distribution characteristics of sand body thickness (Figure 5-5).

The following principles shall be followed in the process of planar facies combination.

(1) The provenance direction in the research area is north by east, so the sandy ribbons shall be distributed along the provenance direction, and the sand body's lateral boundary is determined between the sand-bearing facies well point and the non-sand facies well point, and in principle, at 1/2 of the well spacing. But in the actual treatment process, it can be adjusted appropriately according to the thickness of the sand bodies. For crevasse splays and crevasse channels, no backflow phenomenon can occur.

(2) The average well spacing in the research area is about 150m. If two adjacent wells have the same facies attributes and the logging curves are similar, the inter-well treatment shall be performed in accordance with the facies attributes. If the facies attributes are the same and the logging curves are quite different, it is necessary to consider the facies attributes and curve shapes of the surrounding wells, and fitting prediction shall be performed based on the planar distribution pattern of sedimentary facies.

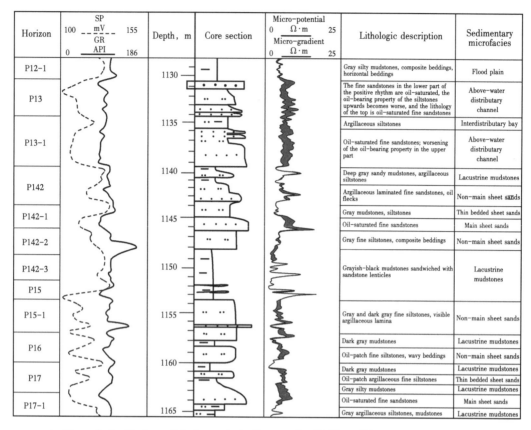

Figure 5-4　Single-well facies analysis diagram of well X2-1-J29 in Xingerzhong area

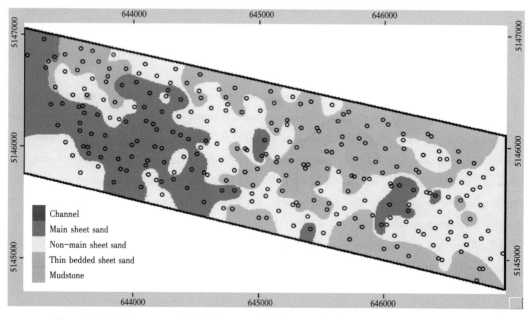

Figure 5-5　Sedimentary microfacies map of S211-1 single sand layer in Xingerzhong area

(3) For thin-layer massive sand bodies that exist in isolation, they are mostly regarded as thin-layer sheet sand or overbank sand deposits.

(4) The particularity of deposits in the area determines that mouth bar sand bodies are not developed. A small amount of developed mouth bar sand bodies are treated as main or non-main sheet sands.

(5) Sheet sands are widely distributed, and especially thin-layer sheet sands are continuously developed in general.

(6) Combining planar facies according to the classic depositional model and the law of facies sequence makes the overall planar facies distribution reasonable in sedimentary genesis.

II. Main sedimentary microfacies types and characteristics

(I) Regional sedimentary background

The analysis of the regional sedimentary background of the research area shows that the research area is in a large shallow river delta depositional system developed under a gentle slope background, and the provenance mainly comes from the north-by-east direction (Figure 5-6). The topographic slope of the basin bottom is about 1°, and the water bodies are shallow. This

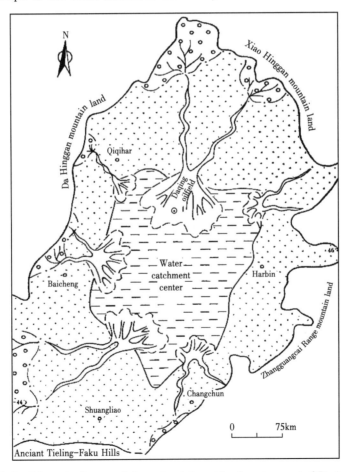

Figure 5-6　Schematic diagram of the regional depositional environment of Daqing Oilfield

special sedimentary background determines that a river still maintains high energy after entering a lake. in addition, the advancing distance to the lake basin water bodies is large, forming a constructive delta; moreover, the thickness of the sedimentary sand bodies is small, mouth bars are not developed, and there are obvious alternate layers of sandstones with mudstones.

(Ⅱ) Sedimentary microfacies division

The research area develops two subfacies environments—delta plain and delta front. The thickness of the strata in the delta front subfacies environment is large and about 200m, which mainly occur in Saertu reservoir group, the lower part of Pu-I Member, and Pu-II Member. The thickness of the strata in the delta plain subfacies environment is only around 30m, which are developed in the upper part of Pu-I Member.

1. Delta plain subfacies

The delta plain subfacies occupies a small proportion in the research area, only 13%, but it is mainly composed of superaqueous distributary channel deposits, develops sand bodies, has good physical properties, and generally contains oil, so it constitutes the main pay beds in the research area. The delta plain subfacies can be further divided into several types of microfacies, such as superaqueous distributary channel, sub-channel, abandoned channel, overbank deposit, interdistributary bay, flood plain, etc.

1) Superaqueous distributary channel

Superaqueous distributary channel deposits are equivalent to fluvial facies deposits, which can be of different river types, and are mainly high-curved meandering river deposits in the research area. Superaqueous distributary channel deposits generally have positive rhythm or massive deposit characteristics. They develop large-scale cross beddings and parallel beddings. Their bottom has a scouring surface, and they can also form scouring and superimposition of multi-stage sand bodies. The shape of the logging curves include bell shape, box shape or a combined shape, reflecting the natural attenuation process of river energy from strong to weak (Figure 5-7).

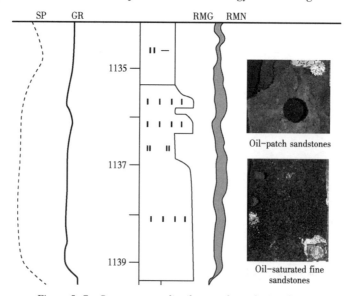

Figure 5-7 Superaqueous distributary channel microfacies

The thickness of the superaqueous distributary channel sand bodies is 4-6m, most of which are developed continuously.

2) Sub-channel

Sub-channel is mainly for the distributary channel sand bodies with small thickness and relatively poor physical properties. Sub-channel is generally fine siltstone and argillaceous fine siltstone deposits, the thickness of the sand bodies is about 4m, and the logging curves show relatively low-amplitude bell shape, box shape or a combined shape.

3) Abandoned channel

Abandoned channel microfacies is an important basis for the division of single channel sand bodies, while the research level of combined microfacies reflects the distribution characteristics of dominant facies, and does not reflect the vital significance of the identification of abandoned channel microfacies. Therefore, the characteristics and identification methods of abandoned river channel microfacies will be detailed in the description of typical single sand bodies in the next chapter.

4) Overbank deposit

Overbank sand body deposits mainly include crevasse splays, crevasse channels and infralittoral sand deposits. Natural levees are not developed in the research area (Figure 5-8). The crevasse splay deposits are mainly alternate layers of siltstones with silty mudstones, have smaller grain size than channel sand bodies and are not too thick (generally less than 2m). The electrical logging curve shows a toothed funnel shape and a low-amplitude bell shape, as well as a positive rhythm or a reverse rhythm in the vertical direction. The crevasse channels are in the shape of a narrow strip, and often coexist with crevasse splays, thin bedded overbank sands, etc.

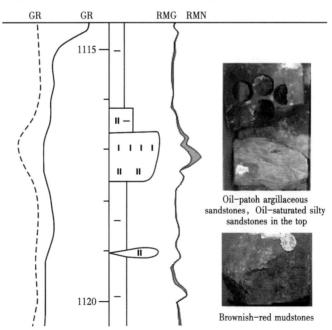

Figure 5-8 Overbank deposit

The transverse section is thin lenticular, and the longitudinal section is mostly wedge-shaped from the crevasse end to the tip. The deposits far away from the main channel will become thinner. Infralittoral sedimentary sands mainly arise from low-lying topography and seasonal flood events, the deposits overflow to low-lying places with the flood, and sands and mud are deposited according to gravity differentiation. Most of the sand bodies show an isolated potato shape, and a finger shape or a toothed bell shaped on the SP curve.

5) Interdistributary bay

Interdistributary bay is mainly composed of mudstone and silty mudstone deposits, sandstone ribbons are often visible, and deformed laminas are developed locally. On the electrical logging curve, the GR and SP are close to the baseline; the microelectrode curve has low amplitude and basically no amplitude difference.

6) Flood plain

Flood plain is mainly mudstone deposits. Compared with interdistributary bay, flood plain has purer argillaceous substances. Fossils of plant fragments are visible. The GR and SP curves are close to the baseline; the microelectrode curve has low amplitude and no amplitude difference.

2. Delta front subfacies

Delta front subfacies accounts for a large proportion in the research area, and is mainly sheet sand deposits, but the physical properties of the reservoirs are relatively poor and they do not constitute the main pay beds. The delta front subfacies in the area can be further divided into sedimentary microfacies types, such as underwater distributary channel, mouth bar, front sheet sand, interdistributary bay, lake basin clay, etc.

1) Underwater distributary channel

Underwater distributary channels generally have sedimentary features such as coarse bottom lithology and fine top lithology. Underwater distributary channels are massive sand bodies on bottom scouring — abrupt change surfaces and can also form scouring and superimposition of multi-stage sand bodies. Argillaceous laminas are visible in underwater distributary channels. The logging curve shows positively gradual change shapes (bell shape, wine bottle shape, tower pine shape, umbrella shape) and uniform massive shape - box shape. On the whole, the sand bodies of positively gradual change shapes and massive sand bodies predominate, reflecting the natural attenuation process of river energy from strong to weak (Figure 5-9). According to the development characteristics of the sand bodies in the two sections of the channel in the area, the sand bodies are large in scale, and have a thickness of about 2-4m. The reservoirs have good physical properties, second only to superaqueous distributary channel sand bodies.

2) Mouth bar

Mouth bars are developed at the mouths of underwater distributary channels. Screened by lake water, argillaceous deposits are taken away. Usually the sands are pure and have good physical properties. The curve shows funnel shape and toothed funnel shape, and has medium to high amplitude. The thickness of the sand bodies is about 2m (Figure 5-10). Typical mouth bar sand bodies are not developed in the target area. The identified mouth bars are characterized by thin-layer and sheet-like distribution. The curve slightly shows reverse cycle features. In the actual

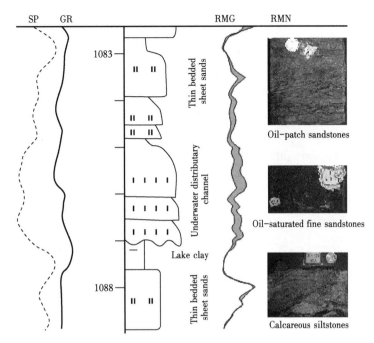

Figure 5-9 Underwater distributary channel microfacies

division, mouth bars are determined as main or non-main sheet sands according to specific characteristics.

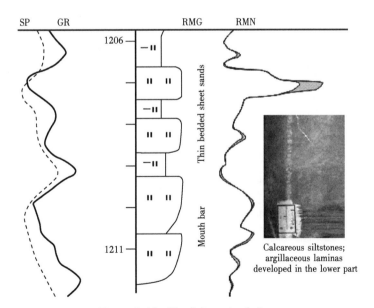

Figure 5-10 Mouth bar microfacies

3) Sheet sand

Sheet sand is the main microfacies type in the target area. After water enters a channel, the sediments carried by the channel are radially deposited due to the decrease in the river energy, and are modified by the lake water. The thickness of the sand bodies is not large, but their distribution is stable. Wavy beddings and composite beddings are developed. Logging curves mostly show thin or alternating spike shapes. From the perspective of internal structure, these sand bodies are composed of many single sand bodies formed by one-time or one-stage continuous flooding events. Each single sand body is separated by obvious interlayers from other sand bodies in space (Figure 5-11). The thickness of this type of sand bodies varies from 0.2-4m, and the physical properties of the reservoirs vary greatly. Therefore, according to the thickness, development stability and physical properties of sand bodies, they are further divided into main sheet sands, non-main sheet sands and untabulated reservoirs. The physical properties of the reservoirs gradually become poorer (Table 5-1).

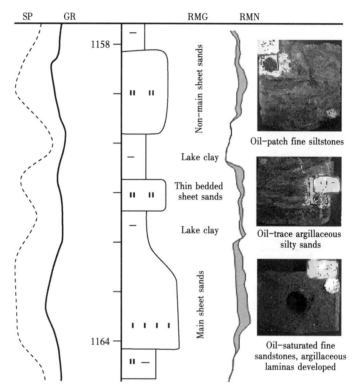

Figure 5-11 Sheet sand microfacies

Table 5-1 Subdivision of sheet sand microfacies

Microfacies type	Lithologic features	Sand body thickness	Effective thickness	Curve features
Main sheet sands	Mainly fine sandstones	Around 1-2m	≥0.5m	Relatively large curve amplitude, smooth and full
Non-main sheet sands	Mainly thin bedded fine sandstones and siltstones	Around 1m	<0.5m	Slightly small curve amplitude, toothed box shape, bell shape, funnel shape
Thin bedded sheet sands	Mainly argillaceous siltstones	<1m	0m	Low amplitude, spike shape

4) Interdistributary bay

Interdistributary bay is mainly composed of silty mudstone and argillaceous siltstone deposits, where deformed laminas and composite beddings are developed, and biological disturbance structures and charcoal debris are visible locally. The SP curve is a straight section close to the baseline; the microelectrode curve has low amplitude and basically no amplitude difference.

5) Lake basin clay

Compared with interdistributary bay, lake basin clay has purer argillaceous substances. Lake basin clay is mainly gray and dark gray mudstones, where horizontal laminas are developed, and ostracode and shell fossils are common.

III. Distribution features of sedimentary microfacies

(I) Planar distribution of composite microfacies

The distribution maps of the sedimentary microfacies of each layer have been compiled and they have been comprehensively compared and analyzed guided by sedimentology theory and comprehensively considering multiple factors according to previous research results.

1. Delta plain subfacies

The delta plain subfacies in the research area is developed in the upper part of Pu-I Member, and is characterized by the extensive distribution of distributary channel sand bodies. According to the shape of sand bodies, they can be divided into three types: continuous sheet, stripped, and isolated.

1) Continuous sheet channel sand bodies

Continuously distributed sheet channel sand bodies are actually formed by the lateral combination of multi-stage channel sand bodies with overbank sand bodies due to the lateral migration and diversion of high-curved rivers, and mainly appear in P13-1 and P13 layers. The thickness of continuous sand bodies is generally large. The thickness of the sand bodies in P13-1 layer is up to 9.8m, with an average of about 5.0m. The thickness of the sand bodies in P13 layer is up to 8.1m, with an average of about 4.5m. The lateral extension of continuous sand bodies is large. The width of the sand bodies in the two single layers is above 800m. The channel sand bodies in the P13-1 layer are distributed almost in the whole area, and have a width of up to about 2000m. The overbank deposits of this type of sand bodies are not developed, and only scattered overbank sand bodies appear in the marginal regions. This is related to the strong erosion effect in the process of river channel migration and diversion, which makes it difficult to preserve overbank sand bodies and reflect them in dominant facies maps. Such continuous sand body deposits are developed in the early stage of base level rise, and the A/S value is small. Sufficient materials supplied are quickly accumulated in the area, resulting in the continuous distribution of sand bodies with large thickness.

2) Stripped channel sand bodies

Channel sand bodies are embedded in flood plains and distributed in stripped shape, have a large width change, and can be combined with continuous sand bodies. The width of sand bodies distributed in stripped shape in the research area ranges from 150-600m. The greater the width of the sand bodies, the greater the thickness of the sand bodies. As the width of the sand bodies is larger and larger, they are distributed gradually continuously. Overbank sand bodies are relatively developed. They can be distributed on the edge of river channels, and can also be far away from river channels so as to form thin bedded infralittoral sands. Usually their scale is small, no more than 1-2 wells spacings. Overbank sand bodies can be developed continuously in local regions. Such channel sand bodies are formed when the river's lateral swing and migration capacity gradually decreases in the middle and late stages of base level cycle rise. Channels are mainly small channels, and generally several channels are distributed areally in stripped shape. Several sand body thickness centers are developed along the channels.

3) Isolated channel sand bodies

Isolated channel sand bodies are based on widely distributed flood plains. Channel sand bodies are scattered in potato shape and pod shape in flood plains, and overbank sand bodies are relatively developed. Such channel sand bodies are small in scale, the river energy is weak, and the erosion effect is not strong, which is conducive to the preservation of thin bedded sand bodies deposited in overbank microfacies. Thin bedded sands of less than 1m in thickness are often distributed continuously. Such channel sand bodies are formed at the end of base level cycle rise, where the A/S value increases, the sediment supply is relatively insufficient, and the rivers intermittently develop, and their lateral swing and migration capacity is low. The width of the channel sand bodies is about 160m, which is basically a well spacing. Overbank sand bodies are distributed around the channel sand bodies, and there are isolated thin-layer infralittoral sand deposits far away from the channels.

2. Delta front subfacies

Delta front is the main subfacies environment of the research area, and is characterized by the wide distribution of sheet sands. Underwater distributary channel sand bodies form skeletons interspersed in delta front subfacies. According to the distribution form and combination characteristics of different microfacies, the delta front subfacies is divided into three types: distributary channel—sheet sand deposit, main—non-main sheet sand deposit, and thin-layer sheet sand deposit.

1) Distributary channel—sheet sand deposit

Such microfacies combination is less developed, and is mainly in the Layer S211. The microfacies type is still dominated by the widespread distribution of sheet sands. Underwater distributary channels are distributed in strip shape, pod shape or potato shape in the sheet sands, and are in erosion contact with the sheet sands. The proportion of underwater distributary channels generally reaches more than 20%, and they have an obvious river channel shape. The thickness of underwater distributary channel sand bodies is about 2-3m, and their width is generally less than 300m. The sheet sands are widely distributed, and are mainly composed of main sheet sands and non-main sheet sands, where thin-layer sheet sands play a role of edging and bridging. Such microfacies combination appears in the early stage of base level rise, where the A/S value is small, the provenance is sufficient, and the river still has high energy after entering a lake, so the advancing distance to the lake basin is large and a good channel shape is maintained.

2) Main—non-main sheet sand deposit

Such microfacies combination is dominated by main sheet sands and non-main sheet sands. Main sheet sands are distributed continuously in strip shape. Non-main sheet sands are embedded around main sheet sands, while thin-layer sheet sands are alternately filled in stripped shape in the edges of the first two. The thickness of the sand bodies is mainly about 1.5-2m, and they are widely distributed in the whole area. Such microfacies combination is developed in the middle and late stages of base level rise. As the water bodies become deeper, the energy declines rapidly after the river enters the lake, forming wide and shallow channel sand bodies, as well as widely distributed main sheet sands and non-main sheet sands under the transformation effect of lake water. Therefore, main sheet sands still retain the shadow of channel sand body deposits. This type of microfacies occupies a large proportion in the research area.

3) Thin-layer sheet sand deposit

Thin-layer sheet sands are siltstone and argillaceous siltstone deposits that are formed by carrying of suspended fine-grained deposits by lake water and have long transportation distance and wide distribution range. Most of thin-layer sheet sands are distributed continuously and have good continuity. The thickness of the sand bodies is around 0.5-1m. Main sheet sands and non-main sheet sands are underdeveloped and distributed sporadically. Large sandstone pinch-out areas appear in some layers.

(Ⅱ) **Vertical evolution law**

Lacustrine deposits use stable mudstones as the sequence boundary. A complete one-stage cycle has experienced the process of changing of water bodies from deep to shallow and then to deep, showing a transition from a prograding superimposition model to a retrograding superimposition model. Saertu reservoir group and Putaohua reservoir group in the research area correspond to a mid-term cycle respectively. The base level of the mid-term cycle of Putaohua reservoir group has a relatively large change, which has experienced a process of deposition from underwater to above-water. The descending hemicycle has a high A/S ratio, showing developed thick mudstones, thin sandstones and relatively poor physical properties of sandstones. The ascending hemicycle has a low A/S ratio and sufficient provenance, which is manifested by continuously developed channel sandstone deposits, forming high-quality reservoirs. The overall A/S ratio of Saertu reservoir group is relatively high, all of which is underwater delta front deposits. Underwater distributary channel sandstones are developed near the transition surface of the base level cycle.

There are multiple short-term cycles in a mid-term cycle. Putaohua reservoir group is divided into 3 short-term cycles, and Saertu reservoir group is divided into 7 short-term cycles. The deposition thickness of a single short-term cycle varies from 10-20m. Vertically, sedimentary microfacies changes can come down to three modes (Figure 5-12): (1) The evolution of lake basin clay—sheet sand—lake basin clay: this evolution mode is the most common. The top and bottom of a cycle are characterized by thick and stable mudstone deposits, where thin-layer sheet sands are distributed locally. Main sheet sands and non-main sheet sands are developed in the middle of the cycle, and are distributed widely. (2) The evolution of lake basin clay—sheet

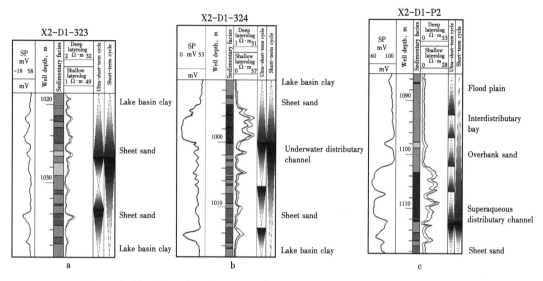

Figure 5-12 Vertical evolution of sedimentary microfacies in short-term cycles

sand—underwater distributary channel—sheet sand—lake basin clay: This evolution mode is characterized by the development of underwater distributary channels. The scale of the river channel gradually decreases upwards, and the underlying sheet sands have a prograding superposition pattern, reflecting the evolution process of A/S ratio from large to small and then to large rapidly. (3) The evolution of lake basin clay—sheet sand—underwater distributary channel—overbank sand—interdistributary bay—flood plain: This evolution mode is dominated by the delta plain subfacies deposits in the ascending hemicycle. Distributary channels are gradually evolved from continuous development mode into stripped and isolated development mode, and are eventually replaced by widespread flood plains. Sheet sand deposits are developed in the base level descending hemicycle, where underwater distributary channel sand bodies are not developed.

Chapter 6 Establishment of Reservoir Geological Knowledge Database

Reservoir geological knowledge database refers to the reservoir geological parameters that are highly summarized after a large number of studies, and can qualitatively or quantitatively characterize the geological characteristics of different types of reservoirs, and have universal significance. It can be used to guide the research, prediction and geological modeling of unknown reservoirs. Fine reservoir description refers to the quantitative and fine research on reservoirs that is carried out with a view to economically and effectively developing an oilfield, enhancing its oil (gas) recovery ratio, and ascertaining its remaining oil (gas) distribution characteristics, laws and control factors after it enters the high water cut period or the ultra-high water cut period. In these studies, the key is to finely depict reservoirs, carry out inter-well prediction, and establish fine geological models. To complete this work, on the one hand, there is need for the refinement of actual oilfield data and technologies; on the other hand, it is necessary to establish a more detailed reference or template than the predicted reservoir, that is, there shall be a geological knowledge database of various reservoirs to guide the description of interwell reservoirs.

Section 1 Main Content of Reservoir Geological Knowledge Database

The reservoir geological knowledge database is an important foundation for fine reservoir description and reservoir modeling. For complex geological bodies, drilling information is always insufficient. The information provided by well data is only a very small part of the real features of geological bodies, just like conjuring up the whole thing through seeing a part of it. Even the 3D seismic data that can provide continuous information cannot reproduce the real features of underground reservoirs due to its low resolution and obvious multiplicity. In this case, in order to reveal the real features of reservoirs, it is necessary to predict complex geological bodies. Therefore, while making full use of the existing drilling data and seismic information, it is necessary to provide a more detailed reference or template than the predicted reservoir. For this reason, it is very meaningful to establish a corresponding reservoir geological knowledge database.

A broad reservoir geological knowledge base includes all qualitative and quantitative knowledge summaries that can characterize the 3D spatial characteristics and geneses of different genesis types of reservoirs and their controlling effects, involving classification and naming of various reservoir sedimentary facies, sedimentary background (basin type, palaeoclimate, palaeontology, palaeogeography, hydrodynamic conditions, etc.), sedimentary characteristics, diagenetic characteristics, reservoir characteristics, hydrocarbon-bearing characteristics, etc. A narrow reservoir geological knowledge base only refers to the parameters that can quantitatively characterize the spatial characteristics, boundary conditions and physical characteristics of various sand body genetic units, as well as various sedimentary models that can achieve qualitative characterization.

According to the current research results, the reservoir geological knowledge database mainly includes lithology and lithofacies library, sedimentary environment and sedimentary microfacies library, geometry library, physical parameter library, diagenetic library, geostatistical parameter library, barrier and interlayer parameter library, etc. And with the deepening of research, the content is continuously added to improve the degree of quantitative characterization (Table 6-1). The types of sedimentary bodies that have been studied in the geological parameter library include meandering rivers, braided rivers, and braided river deltas.

Table 6-1 Main Content of Reservoir Geological Knowledge Database

Parameter library type	Main content	Quantitative degree
Lithology and lithofacies library	Lithofacies type, structural characteristics, bedding type, genetic explanation	Mainly qualitative
Sedimentary environment and sedimentary microfacies library	Microfacies name, microfacies morphology, microfacies scale, microfacies combination law, microfacies internal structure, sedimentary structure, curve shape	Mainly qualitative
Facies model library	Logging facies model, seismic facies model, sedimentary model	Mainly qualitative
Scale library	Length, width, thickness, area, length – width ratio, width – thickness ratio	Mainly quantitative
Physical parameter library	Parameter characteristics, rhythmicity, coefficient of variation, range	Mainly quantitative
Geostatistical parameter library	Variogram function	Mainly quantitative
Barrier and interlayer parameter library	Type, genesis, occurrence, scale	Qualitative and quantitative

Section 2 Establishment Methods of Reservoir Geological Knowledge Database

One method for the establishment of the reservoir geological knowledge database comes from the systematic description of geological bodies with representative significance and universal laws, namely the establishment and anatomy of prototype models. The geological knowledge database established using this method has universal application value for a certain type of sedimentary bodies. The other method is to establish the reservoir geological knowledge database based on the specific analysis of the area with a relatively high degree of research, which is used to guide the description and prediction of areas with the same characteristics and a low degree of understanding. The application of the geological knowledge database established using this method is regional.

At present, there are three main methods for establishing a geological knowledge database at home and abroad, namely, description of field outcrops, modern deposition and physical simulation method, and anatomy of dense well pattern areas. The reliability of fine research on geological outcrops and modern sedimentary bodies is higher. The anatomy of dense well pattern areas has developed rapidly in recent years and has been widely used in oil and gas field development.

I. Establishment of a reservoir geological knowledge database using the method of description of field outcrops

The outcrop analysis technology is the most basic and most important means of geological research, and has gradually become one of the main methods for fine reservoir description and modeling. Outcrop research can obtain more precise geological knowledge and corresponding reservoir prediction methods than dense well patterns. However, limited by geographical conditions, most underground reservoirs have no outcrops and only comparable outcrop observations can be relied on. Therefore, the application of this method is limited to a certain extent.

The description of field outcrops is intuitive, complete, accurate, and convenient for large-scale research. The geological knowledge database system established using this method is precise and highly quantitative. Reservoir parameters are determined through fine geological description and measurement of reservoirs and then indoor analysis and statistics. This method has great theoretical research significance. However, due to being limited by the uncertainty of depositional environments and conditions, this method has certain limitations.

In recent decades, European and American countries have changed their traditional working methods for the purpose of regional geological exploration, and have spent huge sums of money in the fine research on reservoir outcrops for oilfield development. For example, the Department of Energy, USA, surveyed the shelf sand ridge outcrops in the Powder River Basin, Wyoming (Tomutsa et al., 1986); An expert group (Heresim group) organized Britain, France, the Netherlands, Norway, etc. studied delta outcrops in Yorkshire, England, and provided a knowledge database for the establishment of the geological model by the North Sea Brent group (Archer and Hancock, 1980; Rudkiewiz, 1990); Mayer et al. (1993) compared the surface outcrops of Muddy J sandstones with the geological characteristics of underground reservoirs in Colorado, USA; Dreyer et al. (1993) analyzed the lithofacies and fluid flow units through the correlation of the strata of Tilje Formation in the river-dominated fan delta of the central shelf of Norway using the outcrop data of the Ridge Basin in California, USA; Lowry et al. (1993) performed reservoir simulation for the research on the sequence outcrops of the river-dominated delta front in Ferron Sandstone Member of Mancos Shale Formation in the central-east part of Utah. One of the most influential studies is the outcrop investigation of Gypsy sandstones near Tulsa, Oklahoma, which was invested by BP. The research fund was millions of dollars. The entire research project included: outcrop survey, dozens of shallow wells in the coverage zone, seismic exploration and radar exploration, and drilling of 5 deep experimental wells. These outcrop studies were designed to establish geological models of oil-bearing series scale and also geological models of sand body scale.

Domestic research on outcrops and modern deposition for the purposes of development and application of oilfields and the establishment of reservoir geological models began in the 1980s. For example, the outcrop survey of the braided river delta and distributary channel sand bodies in the Yousha Mountain of Qinghai (Lin Kexiang et al., 1995; Lei Bianjun et al., 1998), the outcrop survey in the braided river delta in Fuxin Basin (Wang Jianguo, Wang Defa, 1995), the outcrop survey of the modern braided river deposition in the Yongding River, Daihai modern braided river deposition, and fan delta, and the research on the point bar of the modern meandering river deposition in the Juma River, etc., have all contributed to the macroscopic description and microscopic description of heterogeneity of different types of reservoirs. At present, the latest and most detailed outcrop research in China is the 9th Five-Year Plan key scientific and technological research project of CNPC— "Fine Description of and Application Research on Reservoir Outcrops" undertaken by RIPED. It is the first comprehensive anatomical research project on fan delta and braided river reservoir outcrops in China. Mu Longxin and Jia Ailin et al. (2000) established a detailed reservoir geological knowledge database of braided river and fan delta depositional systems by studying the braided river outcrops in Datong, Shanxi and the fan delta deposit outcrops in Sangyuanyingzi, Luanping. The reservoir geological knowledge database is of very good theoretical reference significance to the study of distribution prediction of reservoirs in similar depositional environments.

II. Establishment of reservoir geological knowledge database using modern deposition and physical simulation methods

Modern deposition investigation is a traditional and effective method of geological work. It is convenient for carrying out the research on onshore depositional systems (such as alluvial fans and fluvial deposits), but it is inconvenient for observing some underwater sedimentary bodies. The physical simulation of a depositional system is to reproduce the deposition process in the laboratory and to study the various properties of the reservoirs in the simulated sedimentary bodies referring to various geological parameters during the formation of sedimentary bodies. For example, a large number of reservoir sand body models of different sedimentary types can be obtained through lake basin water tank simulation experiments. Such models are similar to outcrops, and their greatest advantage includes convenient measurement (slice cutting and sampling at will), detailed recording of deposition processes, and clear genesis mechanism. Such models are of great significance to the study of sedimentology and the determination of the parameters of macroscopic distribution law of reservoirs, but their application to reservoir evaluation parameters (physical properties, fluid characteristics) and specific oilfield modeling is very limited.

Physical simulation realizes the forward modeling of sedimentary bodies, and allows people to clearly understand the various changes in the sedimentary body formation process. However, the application of this method has certain limitations due to the long period of physical simulation of sedimentary bodies, large expenditures and limited simulation conditions, as well as generally poor comparability between simulated geological bodies and real geological bodies.

III. Establishment of geological knowledge database using the method of dense well pattern area anatomy

The dense well pattern here can be a dense well pattern in the research area, or a comparable dense well pattern in mature oilfields. The geological knowledge database established by dense well pattern anatomy has lower accuracy than that established based on outcrop data or modern deposition data, but it can be used to guide reservoir prediction research in relatively sparse well pattern areas. Especially through the use of horizontal well data and production logging data, the accuracy and credibility of dense well pattern anatomy have been greatly improved.

Carrying out fine reservoir geology research (including stratigraphy research, tectonics research, sedimentology research, reservoir evaluation research, etc.) and making full use of the static and dynamic data of mature development blocks are the economical and effective method for establishing a reservoir geological knowledge database for oilfield coverage areas. A lot of practice has proved that the extensive applications of methods and technologies such as high-resolution sequence stratigraphy division and correlation, fine research on sedimentary microfacies, detailed development performance analysis, large scale industrial mapping for small intervals, etc. are enough to obtain reliable reservoir prototype geological models.

The advantage of the geological knowledge database established by dense well pattern anatomy is: detailed research on underground conditions can be carried out according to a lot of dynamic and static data of dense well patterns. However, due to the limitation of well spacing, there is a need for other information for inter-well reservoir prediction.

Section 3 Establishment of Reservoir Geological Knowledge Database Using Geological Outcrop Data

With the deepening of petroleum exploration and development, exploration and development work becomes more and more difficult. How to drill as few wells as possible in exploration; even under the condition of a discovery well, reservoirs are described basically correctly and a conceptual reservoir model is established. How to establish a 3D fine geological model of an old oilfield during the development stage and to ascertain the distribution of underground remaining oil in the high water-cut period has become a world-class problem. The key to all of these is how to predict reservoirs more precisely, so the foundation and methods for fine reservoir prediction must be established. Traditional geological research methods and existing oilfield data are not enough to solve such problems. Field geological outcrop data and modern deposition data have the advantages such as intuitiveness, completeness, accuracy, etc., so they are regarded as some of the most important research data for understanding the underground geological bodies. Taking the field observation of Luanping fan delta outcrops as an example, this section describes the establishment of a reservoir geological knowledge database using outcrop data.

I. Field outcrop selection standard

The selection of outcrops is the key to outcrop observation. In general, the selection of outcrops must meet the following basic conditions.

(1) Reservoir types are representative. Braided river and fan delta are two representative sedimentary facies types in the fluvial and lacustrine delta reservoirs in China, and both occupy important positions in the reservoirs of different series of strata in different types of basins in China. It is of great significance to the study of their prototype geological law.

(2) The characteristics of sedimentary types are obvious. Only outcrop reservoirs with typical sedimentary characteristics have the basic conditions for carrying out the research on prototype geological law, and the established geological knowledge database can be applied more extensively. This is one of the key factors for outcrop selection.

(3) Facies belts are distributed completely on outcrop sections and there are plentiful geological phenomena. Outcrops must have relatively complete facies belts distributed and plentiful geological phenomena, so that the heterogeneous characteristics and change laws of reservoirs at different levels can be understood and mastered in detail and accurately.

(4) The selected outcrop deposition type has strong comparability with the oilfield. According to current studies, there are many factors that control the distribution of remaining oil, including the heterogeneity of reservoirs, fluid characteristics, pressure systems, etc., but the most important is still the heterogeneity of reservoirs. Therefore, the established geological knowledge database can be applied most widely only by selecting outcrops with good comparability with oilfields.

(5) Facilitate further research and construction. The establishment of the geological knowledge database must be based on enough and detailed field outcrop research results, and large-scale field surveys are required. If necessary, drilling, logging and other operations shall also be carried out. Therefore, it is very necessary to choose outcrops favorable for construction and survey conditions.

(6) Traffic conditions are convenient. The outcrop research used to establish the geological knowledge database is different from the general outcrop survey and description, and quite a lot of field equipment is needed for construction on outcrops every day. Therefore, convenient traffic conditions are one of the prerequisites for the research work.

Luanping Basin is located in the northern part of Hebei Province, and belongs to Luanping County in terms of administrative divisions. The geographical coordinates are $117°15'-117°30'$ east longitude and $45°50'-41°00'$ north latitude. The research area is 165km away from Beijing, and has convenient traffic conditions.

The selected section belong to the Upper Jurassic-Lower Cretaceous, including Sangyuanyingzi fan delta plain-front section, Yangshugoumen fan delta front section, railway bridge shore lake facies section and railway station shore-shallow lake section (Table 6-2). The total east-west length of the section is about 10km. Sangyuanyingzi main section extends 1.3km from east to west. The entire section almost includes various microfacies and sand bodies of fan delta, has been subjected to repeatedly multi-stage superimposition, and is an ideal place for research on field fan delta outcrops.

Table 6-2 Overview of Luanping fan delta section

Name	Sedimentary facies belt	Section direction(Length)	Outcrop cause	Outcrop conditions
Sangyuan section	Fan delta plain front	EW(1300m)	Highway cutting	Good
Yangshugou section	Fan delta front	EW(500m)	Natural outcrop	Good
Railway bridge section	Shore lake	NS(200m)	Railway cutting	Common
Railway station section	Shore-shallow lake	NS(1000m)	Railway cutting	Good

II. Division of hierarchical interfaces of outcrop reservoirs and distribution characteristics of barriers and interlayers

(I) Division of hierarchical interfaces of outcrop reservoirs

The division of hierarchical interfaces is a stratigraphic division method developed in recent years, and its core content is the division of hierarchical interfaces according to the scale and level of sedimentary units. For a high-level stratigraphic unit, its hierarchical interface is consistent with its time interface. In this research field, there are currently different division schemes, and different factions hold their own opinions. They include the division method represented by Miall method where the level of the stratigraphic interface is consistent with the numerical order (that is, Level 1 interface is the smallest) and the division method represented by Normark method where the level of the stratigraphic interface is opposite to the numerical order (that is, Level 1 interface is the largest). This book not only refers to the advantages and disadvantages of different division methods, but also considers structural division schemes and habits in the division of hierarchical interfaces. Finally, the division scheme where the level of the stratigraphic interface is opposite to the numerical order has been determined in this book. According to high-level hierarchical interfaces, stratigraphic division and correlation can be carried out to guide the establishment of the reservoir skeleton model. According to low-level hierarchical interfaces, reservoir architectural structure units can be divided in sand bodies or composite sand bodies, and the structure and heterogeneity of sand bodies can be anatomized.

According to the research on fan delta outcrops, eight levels of hierarchical interface have been divided (Table 6-3), and the correlation scope and application principle of the interfaces at all levels have been pointed out. This is the basis for guiding the establishment of a fine reservoir skeleton model.

Level 1 interface: the bottom interface is the beginning of basin depostion and is in direct contact with bedrocks; it is a typical lithologic mutation surface, which is easy to identify not only from core logging, but also from seismic data.

Level 2 interface: it defines the different superimposition patterns of depositional systems (such as fan deltas or braided rivers), and is generally a flooding surface or a lake flooding surface.

Level 3 interface: the interface between different sedimentary bodies. The thickness between

Level 3 interfaces on Luanping fan delta outcrops is generally about 20m.

Level 4 interface: it limits the superimposed bodies of multi-stage channels; in addition, the author deems that Level 4 interface is the relatively reliable minimum correlation unit in downhole stratigraphic correlation, and the cross-well correlation of the lower-level interface is less reliable.

Level 5 interface: it is the interface that limits single channels, natural levees and crevasse splays.

Level 6 interface and Level 7 interface: they are interfaces among layer systems, layers, lamina groups and laminas, and are one of the main geological factors for studying in-layer heterogeneity. Level 6 interface and Level 7 interface can cause local environment of remaining oil.

Level 8 interface: undetermined.

Table 6-3 Division of hierarchical interfaces of depositional systems

Level	Sedimentary body	Lateral extension	Comparability
1	Basin filling complex	10-100km	Can be tracked regionally on the section, can be compared regionally on electrical logging curves, can be identified seismically
2	Depositional system complex	3-5km	Can be tracked regionally on the section, can be compared regionally on electrical logging curves, can be identified seismically
3	A single depositional system	3-10km	Can be tracked regionally on the section, can be compared on electrical logging curves within the scope of an oilfield, can be identified 3D-seismically
4	Channel complex	50-200m	Can be compared on the section; can be compared under the condition of small well spacing
5	Single-channel natural levee, crevasse splay	20-100m	Can be identified and compared on the section; difficult comparison downhole
6	Series of strata, layer	5-10m	Can be identified on the section; cannot be compared downhole
7	Lamina group, lamina	2-5m	Can be identified on the section; cannot be compared downhole
8	Undetermined		

The division of hierarchical interfaces is an open system. According to research needs, a large-scale interface of any Level can be defined as a Level 1 interface, and then it is subdivided level by level. In addition, the micro-scale of interfaces is infinite, and there are micro-scales such as grain space, mineral orientation, crystal orientation, etc. after lamina. As long as it is needed in research, it can be subdivided level by level.

There are basically two type division schemes for barriers and interlayers: (1) the division scheme characterized by lithology. (2) the division scheme characterized by sedimentary environment. For example, the former is for the division of mudstone interlayers and argillaceous siltstone interlayers, and the latter is for the division of shore-shallow lake interlayers and drape interlayers. It can be said that the two schemes have their own advantages. The first division scheme is closer to production actuality, and is currently widely used. The second division scheme is closer to theoretical research. In addition, there is a division scheme characterized by thickness and continuity for the division of barriers and interlayers.

(Ⅱ) Research on prediction methods of barriers and interlayers

The prediction of barriers and interlayers is actually one of the content of reservoir prediction. The prediction methods of barriers and interlayers are basically consistent with those of reservoirs. In summary, they can be roughly divided into two types: depositional mechanism prediction and geostatistical prediction. The depositional mechanism prediction method is mainly based on depositional mechanism, and interlayers are classified according to their genetic type; then their relationship with sand bodies and their distribution scope are estimated in terms of genesis. The statistical prediction method is to make a statistical analysis of the frequency and density of interlayers at all levels in terms of wells or outcrops, and then in terms of research area or inter-well.

1. Depositional mechanism prediction

Depositional mechanism prediction is generally the prediction according to different deposition types starting from the deposition type of interlayers in a single well or several wells in the case of few wells or few data. In addition, foreign scholars have done a lot of work in the application of sedimentary theory for interlayer prediction. For example, Geehan *et al.* (1985) summarized the discontinuous distribution of interlayers after studying the Cast Legate sandstone outcrops in Utah.

2. Geostatistical prediction

Geostatistical prediction is the prediction performed based on this law after summarizing a certain statistical law according to the frequency, density and change characteristics of interlayers. The data background required by geostatistical prediction is relatively detailed, and the research work shall be generally carried out in dense well pattern areas or on outcrops.

3. Characteristics of barriers and interlayers of Luanping fan delta outcrops

1) Characteristics of barriers

The barriers are composite fan delta sedimentary bodies with multiple cycles and multiple stages, and the reservoirs are basically distributed in thin layer shape. The distribution of the barriers is very extensive no matter whether areally or on the section. There are widely distributed shore-shallow lake mudstone barriers between different subsequences; in addition, there are also barriers in a subsequence that are formed by the transgression and regression of lake water, the migration of fan deltas, and the swing of river channels. Therefore, the barriers in this area have the characteristics such as diverse genetic types, and the barriers can be divided into the following types in terms of lithology.

(1) Grayish green mudstone barriers. Such barriers are mostly lacustrine mudstones in fan delta TST or HST, and shore-shallow lake mudstones, silty mudstones and some argillaceous siltstones between underwater channels. The thickness of the barriers varies greatly. There are barriers of more than 10m in thickness and also thin barriers of only 1-2m in thickness, but their distribution scope is relatively wide, and they are basically distributed in the outcrop scope. It is speculated that they are also developed stably within a larger scope. Such interlayers are Type I interlayers in this area.

(2) Red and varicolored mudstone barriers. Compared with fan delta and shore-shallow lake deposits, most above-water fan delta plains are in an oxidizing environment. The barriers deposited in this environment mostly show an oxidized color. In terms of genesis, they are mainly inter-fan deposits, floodplain deposits and inter-channel deposits. The first two are large in scale, most of them are 3-5m thick, and some of them are above 5m thick. The deposits of the last type are thin, and their thickness is generally between 1 and 2m.

(3) Gray marl barriers. The supply of debris outside the basin is relatively small, and under arid to semi-arid climate conditions, the carbonates in the lake environment undergo chemical precipitation and form such barriers with the argillaceous debris supplied offshore. A layer of gray marlstone interlayer, about 10cm thick, has been found in the B reservoir group of the CV subsequence group in this area. The rock layer is distributed on the whole outcrop and its thickness is very stable. In fact, such rock layer is often sandwiched in grayish green mudstone layers, forming a stable barrier with gray-green mudstones. This type of lithologic barrier rarely occurs in lacustrine depositional systems, so it is of little significance as a barrier. However, due to its large area distribution, stable thickness and easy identification on logging curves, it can be generally used as the main marker horizon in stratigraphic correlation.

2) Interlayer features

Interlayers refer to impermeable barriers or low-permeability layers within the sand bodies of reservoirs, and are divided into two types such as physical interlayers and lithologic interlayers. This area is a fan delta depositional system, and the main microfacies forming sand bodies are near-shore water channel, far-shore channel and braided channel. In terms of depositional mechanism, the rivers are all braided rivers. Theoretically, the internal interlayers of the sand bodies in the braided river depositional system are not developed. Actual research has also confirmed this point. The interlayers divided by lithology mainly include the following types.

(1) Argillaceous rock interlayers. Such interlayers include three major types such as mudstone interlayers, argillaceous siltstone interlayers and silty mudstone interlayers, and have the most lithology types. Their extension scope is narrow, but they have a significant impact on vertical connectivity and permeability.

(2) Carbonaceous mudstone interlayers. Thin carbonaceous mudstone interlayers are common in the upper part of channel sands or overbank sand bodies. The layer thickness is generally less than 10cm, and the carbon chips are horizontally oriented, which has a certain influence on the heterogeneity of sand bodies.

(3) Discontinuous boulder clay layers. Discontinuous boulder clay layers are often seen in channel sand bodies, especially in the middle and lower parts, and there are some tearing structures. Most of such interlayers appear intermittently and are the product of embankment collapse.

(4) Tight conglomerate layers. There are often tight conglomerate layers with a thickness of 5-10cm inside the main sand bodies of delta deposits. Such interlayers have very low permeability (<0.01mD), and a high frequency of occurrence. They are of great significance to the study of interlayers in fan delta reservoir sand bodies.

III. Establishment of the geological knowledge database of Luanping fan delta

With the research goal of fine reservoir description, six types of geological knowledge databases have been established, including sedimentary model library (description of typical sedimentary models of depositional systems), reservoir lithology and lithofacies library (description of reservoir rock types), sedimentary microfacies library (description of the genetic units of reservoir sand bodies), sand body scale library (description of the geometry, scale and size of sand bodies), reservoir physical parameter library (focusing on the description of the distribution heterogeneity of reservoir physical parameters), and geostatistical parameter library (mainly the structural parameters of the variogram of each facies type, used for geological modeling).

(I) Sedimentary model library

Fan delta is divided into 3 subfacies and 16 microfacies (Table 6-4, Figure 6-1). Fan delta plain is the main part of the sandy and gravelly deposits in the fan delta system, including six microfacies such as mudflow deposit, large braided channel, interchannel, flood plain, flood fan, and overbank deposit. The development of mudflow and braided channel is the most characteristic. The area from the near-shore channel to the outer edge of sheet sand is the fan delta front. Sand body deposits are mainly divided into near-shore channel, far-shore channel, mouth bar, distal bar and sheet sand. Among them, the two types of channels are the main parts of sand body deposits and also the most favorable reservoir development zones. Before the sheet sands in the fan delta is a lacustrine depositional system, which is developed in semi-consolidated sand bodies near the shoreline. Slump occurs due to its own gravity, and secondary transport occurs to the deeper part of the lake water to form slump turbidite bodies.

Table 6-4 Division of fan delta sedimentary facies

Depositional system	Subfacies	Microfacies
Fan delta depositional system	Fan delta plain	Mudflow, braided channel, interchannel, flood plain, flood fan, and overbank deposit
	Fan delta front	Near-shore channel, far-shore channel, flood fan, natural levee, mouth bar, overbank deposit, interchannel, sheet sand
	Front fan delta	Shore-shallow lake deposit, slump deposit

(II) Reservoir lithology and lithofacies library

The geological phenomena of the Luanping fan delta section are very rich. The division of lithofacies is based on the principle from fine to coarse. Firstly, the lithofacies units that can be identified on 41 outcrops have been divided. In addition, considering that the too fine division of lithofacies is not conducive to reservoir modeling and lithofacies cannot be identified on underground cores, 41 lithofacies units have been merged and 20 representative lithofacies types have been summarized.

During the establishment of the fan delta lithology and lithofacies library, the following aspects are mainly summarized:

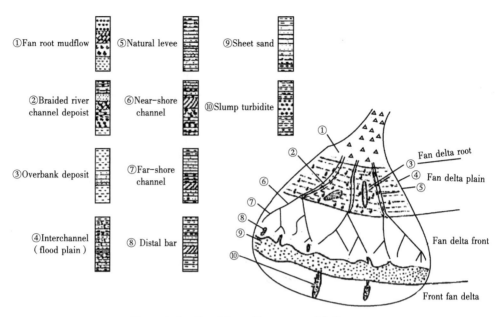

Figure 6-1　Fan delta sedimentary model diagram

(1) Lithofacies type, which is the key to establishing a lithologic and lithofacies library;

(2) Texture characteristics, mainly including grain size characteristics and grain size distribution characteristics;

(3) Bedding types, main bedding types and scale;

(4) Genesis interpretation, giving the possible types of microfacies in different lithofacies;

(5) Other characteristics, mainly including the composition, roundness and color of sorted minerals;

(6) Graphical features, that is, representing the first few items in graphical form.

The fan delta lithology and lithofacies library established accordingly is shown in Table 6-5.

Table 6-5　Fan delta lithology and lithofacies library

Lithofacies type	Texture	Main sedimentary bedding	Other features	Genesis interpretation	Graphical features
Massive cobble conglomerate facies (Gm)	Coarse grained, hybrid texture	Massive bedding	Angular, complex composition	Fan root, braided channel bottom	△△△△△△
Massive fine conglomerate facies (Gmf)	Coarse grained, inequigranular texture	Massive bedding	Sub-angular, complex composition	Mid-lower part of braided channel	△ ○ △ ○
Fine conglomerate graded layer (GVm)	Coarse grained, inequigranular texture	Reverse rhythm, positive rhythm	Sub-angular, complex composition	Appearance of various channels	○○ △△ ○○○ △ ○○ ○○

Continued

Lithofacies type	Texture	Main sedimentary bedding	Other features	Genesis interpretation	Graphical features
Sandy conglomerate facies with trough bedding (SGc)	Coarse grained, inequigranular texture	Large trough bedding	Sub-angular, mainly feldspar and debris	Mid-lower part of channel	
Sandy conglomerate facies with tabular bedding (SGb)	Coarse grained, vinequigranular texture	Large tabular bedding	Sub-angular, mainly feldspar and debris	Middle part of channel	
Sandy conglomerate facies with parallel bedding (SGp)	Coarse grained, inequigranular texture	Parallel bedding	Sub-angular, mainly feldspar and debris	Middle part of channel	
Sandy conglomerate facies with massive bedding (SGm)	Coarse grained, uniform	Massive bedding	Sub-angular, mainly feldspar and debris	Main body of channel	
Sandy conglomerate facies with diluvial bedding (SGh)	Coarse grained, inequigranular texture	Positive and reverse rhythms with parallel bedding	Sub-angular, complex composition	In various channels	
Sandstone facies with massive bedding (Sm)	Medium grained, uniform	Massive bedding	Mainly sub-angular—sub-round, gray, grayish-white	Middle part of near-shore and far-shore channels	
Sandstone facies with trough bedding (Sc)	Mainly medium coarse sands, containing fine gravels and fine sands	Medium trough bedding	Mainly sub-angular—sub-round, mainly gray, grayish-white	Mid-lower part of near-shore and far-shore channels	
Sandstone facies with tabular bedding (Sb)	Mainly medium coarse sands, containing fine gravels and fine sands	Medium tabular bedding	Sub-angular—sub-round, mainly gray, grayish-white	Middle part of near-shore and far-shore channels	
Sandstone facies with parallel bedding (Sp)	Mainly medium coarse sands, containing fine gravels and fine sands	Parallel bedding	Sub-angular—sub-round, mainly gray, grayish-white	Main body of near-shore and far-shore channels	
Fine sandstone facies with waxy bedding (Sfw)	Fine grained	Wavy bedding	Sub-round, gray, grayish-green	Upper part of channel, overbank deposit	
Sandy conglomerate graded layer (SVm)	Large grain size change, inequigranular texture	Positive and reverse rhythms	Sub-angular, complex composition	Channel inside	

Continued

Lithofacies type	Texture	Main sedimentary bedding	Other features	Genesis interpretation	Graphical features
Siltstone facies with small cross bedding (Ssc)	Fine grained and silty sands, containing medium sands	Small cross bedding	Sub-round, grayish-green	Upper part of near-shore and far-shore channels	
Siltstone facies with waxy bedding (Ssw)	Fine grained and silty sands, containing medium sands	Small wavy bedding	Sub-round, grayish-green	Top of near-shore and infralittoral channels	
Siltstone facies with herringbone cross bedding (Ssy)	Fine grained and silty sands, cross containing medium sands	Herringbone bedding	Sub-round, grayish-green	Shore lake	
Heterogeneous lithofacies (M)	Hybrid grain, size from fine gravels to silty sands	Parallel bedding	Sub-angular—sub-round, gray, grayish-green	Channel inside	
Mudstone facies with horizontal bedding (Mh)	Muddy texture	Horizontal bedding	Grayish-green, red, varicolored	Interchannel, flood plain, prodelta clay	
Calcareous mudstone facies (Mca)	Muddy texture	Massive bedding	Gray, grayish-white	Shore-shallow lake	

(Ⅲ) Sedimentary microfacies library

Sedimentary microfacies library refers to the geological knowledge database composed of sedimentary microfacies and their internal texture, structure, morphology and other parameters, and includes the main content below.

(1) The name of genetic units or sedimentary microfacies, which is the basis for the establishment of the library, and the sedimentary microfacies library can be established only through accurate division of genetic units;

(2) The form of genetic units, including the planar form, section form and 3D form of genetic units;

(3) The size of genetic units, including length and width;

(4) The lithology and occurrence of the barrier layer that may appear inside a genetic unit;

(5) The bedding structure inside a genetic unit, especially the main bedding structure type;

(6) Changes in internal grain size rhythm characteristics, changes in grain size and grain sequence;

(7) The change of sorting property on a single profile, mainly describing the change of minerals and grain size;

(8) Logging curve shape, which is one of the more important items in the sedimentary microfacies library. Outcrop logging knowledge can be combined with underground information only through logging interpretation and logging curve shape identification.

Based on the above items, a geological knowledge database of fan delta sedimentary microfacies has been finally established, as shown in Table 6-6.

Table 6-6 Geological knowledge database of Luanping fan delta sedimentary microfacies

Microfacies type	Shape	Large Width, m	Small Thickness, m	Barrier layer	Sedimentary structure	Grain size	Sorting property	Logging curve
Plain debris flow		1-3	10-30	Horizontal argillaceous silty rock	Large trough, massive bedding			
Braided channel		300-500	1.5-8	Horizontal clay, argillaceous silty rock	Large trough, parallel bedding, tabular bedding			
Near-shore channel		200-1000	3-16	Horizontal, waxy clay, clay powder	Parallel bedding and cross-bedding (trough, tabular)			
Far-shore channel		100-500	0.5-5	Horizontal mudstone, wavy clay powder	Parallel bedding, wavy bedding			
Natural levee		50-200	0.5-1.5	Horizontal mudstone	Parallel bedding, wavy bedding			
Mouth bar		30-100	2-4	Wavy clay powder	Cross bedding			
Distal bar		20-70	1-3	Wavy clay powder	Small cross bedding			
Sheet sand		300-2000	0.5-2	Horizontal shale	Massive bedding, parallel bedding, wavy bedding			
Slump turbidite		50-200	2-6	Horizontal shale, inclined mudstone	Massive bedding			
Overbank deposits		100-300	1-1.5	Horizontal mudstone	Wavy bedding			

(Ⅳ) Sandy body scale library

The sand body scale library is the embodiment of the quantification of the geological knowledge database. The type, scale and distribution of sand bodies in the full section of Luanping fan delta are all very important research contents. A total of 132 sand bodies have been identified in the whole section, and they have been numbered and named. According to their distribution, connectivity, and microfacies types, a large number of actual measurements and statistical analyses have been carried out so as to obtain the scale parameters of various sand bodies, including the proportions of various sand bodies (reservoir architecture units) (Table 6-7), the width-to-thickness ratio of the sand bodies and other data information (Table 6-8).

Table 6-7　Statistics of the proportion of sand body types of Luanping fan delta outcrop

Sand body type	Number	Percent, %	Area, m^2	Percent, %
Braided channel	11	8.2	11587.2	14.27
Near-shore channel	16	11.9	19994.8	24.62
Far-shore channel	32	23.9	8426.4	10.38
Overbank deposit	18	13.4	3841	4.7
Sheet sand	20	14.9	2768.6	3.41
Distal bar	15	11.2	1138.2	1.4
Natural levee	8	6	2770.6	3.41
Mudflow	3	2.2	18185	22.39
Small slump	1	0.7	78.4	0.1

Table 6-8　Statistics of the scale of genetic units of Luanping fan delta outcrop

Microfacies type	Average thickness m	Maximum thickness m	Width m	Width after recovery m	Width-to-thickness ratio
Braided channel	3.18	6.38	242.13	286.68	85.70
Near-shore channel	3.81	5.8	233.07	299.63	108.61
Near-shore—far-shore channel	2.88	4.6	321.96	442.48	182.47
Far-shore channel	1.85	2.85	111.09	133.73	85.12
Overbank deposit	1.22	1.94	128.23	152.88	122.36
Sheet sand	1.15	1.75	85.12	111.62	109.27
Natural levee	2.08	3.175	145.37	171.25	109.97
Mudflow	8.2	13.07	425.1	587.42	63.29

(V) Reservoir physical parameter library

Through a large number of sampling and analysis, the distribution characteristics of the physical parameters of different rock types, different types of sand bodies and facies belts in the fan delta have been statistically analyzed (Table 6-9, Table 6-10, and Table 6-11). Due to different diageneses, the absolute value of physical properties has no practical significance for popularization and application, but statistical parameters such as coefficient of variation and range etc. have certain universal significance.

(VI) Geostatistical parameter library

It is mainly the statistics of the structure parameters of the variogram of each microfacies sand body (Table 6-12). When the variogram is established, the facies types are appropriately combined to form totally 6 microfacies including prodelta clay, front sand bar and sheet sand, front near-shore—far-shore channel, overbank deposit, plain braided channel, and prodelta slump turbidite.

Table 6-9 Statistics of physical properties of various reservoir sand bodies in fan delta system

Lithology type	Porosity, %					Permeability, mD				
	Number of samples	Min.	Max.	Avg.	Mean square deviation	Number of samples	Min.	Max.	Avg.	Mean square deviation
Siltstone	12	1.1	10.7	5.31	2.74	9	0.01	0.1	0.05	0.03
Fine sandstone	66	1.5	15.7	7.7	3.02	56	0.01	6.55	0.22	0.36
Medium sandstone	40	2	16.7	9	3.92	39	0.01	8.34	1.03	1.53
Coarse sandstone	40	2.2	20	9.11	4.3	36	0.01	10.9	0.93	1.36
Pebbled sandstone	108	0.3	23.6	9.66	3.55	103	0.01	98.1	2.2	3.7
Conglomerate	142	2.6	24.1	11.04	2.82	125	0.01	134	6.09	10.02
Avg. in the whole area	408	0.3	24.1	9.59	3.44	368	0.01	134	4.97	7.86

Table 6-10 Permeability distribution of reservoirs of main genetic facies in fan delta (unit: mD)

Genetic facies	<0.01	0.01-1	1-5	5-10	10-15	15-20	20-30	30-50	>50
Braided channel	5 /35.7	6 /42.9	2 /14.3	0 /0	0 /0	1 /7.1	0 /0	0 /0	0 /0
Near-shore/Far-shore channel	82 /42.3	73 /37.2	16 /8.2	5 /2.6	2 /1.0	5 /2.6	3 /1.5	2 /1.0	6/3.1
Distributary mouth bar	4 /26.7	11 /73.3	0 /0	0 /0	0 /0	0 /0	0 /0	0 /0	0 /0
Near-end front	50 /73.5	16 /23.5	2 /2.9	0 /0	0 /0	0 /0	0 /0	0 /0	0 /0
Far-end front	27 /81.8	1 /18.2	0 /0	0 /0	0 /0	0 /0	0 /0	0 /0	0 /0

Note: 5/35.7 = frequency/frequency

Table 6-11 Permeability distribution of sandstone layers of different sedimentary subfacies in fan delta (unit: mD)

Subfacies type	Avg.	Min.	Max.	Coefficient of variation	Range	Onrush coefficient	Coefficient of homogeneity
Fan delta plain	98.7	3.5	831.4	1.24	239.6	8.4	0.12
Fan delta front	51	2.7	201.8	0.96	74.8	4	0.25
Front fan delta	79.2	39.3	166.7	0.48	4.2	2.1	0.48

Table 6-12 Structural parameters of variogram of each sedimentary facies of Luanping fan delta outcrop

Facies type	Nugget value	Vertical		Transverse	
		Sill value	Range, m	Sill value	Range, m
Prodelta clay	0	0.3777	39.1	0.246	469.2
Front sand bar and sheet sand	0	0.467	27.1	0.0219	321.8
Front near-shore—Far-shore channel	0	0.434	36.1	0.027	430.1
Overbank deposit	0	0.454	30.1	0.0383	571.4
Plain braided channel	0	0.429	30.9	0.0534	480.9
Prodelta slump deposit	0	0.440	36.8	0.0431	338.2

IV. Geological law of fan delta

(1) Fan delta is characterized by multi-layer deposits, forming typical multi-layer sand body distribution features. Interlayer heterogeneity is obvious, and intralayer heterogeneity in braided river deposits is their main feature.

(2) Interlayers in sand bodies are relatively rare, but the distribution of barriers is relatively stable, and generally they can completely separate the upper and lower sand bodies.

(3) A total of 10 types of genetic unit sand bodies of fan delta reservoirs are divided. Various channel sand bodies constitute the main body of the reservoir framework, accounting for about 65% of the total area. The type of sand bodies determines physical characteristics and the macroscopic appearance of architectural structure. The sand bodies of various channels in the fan delta account for 50% of the total sand bodies and their area percent is 65%. They have good connectivity and are the most important structural elements.

(4) The width-to-thickness ratio of the sand bodies of the genetic units in the fan delta is generally 80-120, which is obviously larger than that of the braided river genetic units.

(5) The physical properties of various sand bodies in the fan delta are quite different, and braided channels and underwater distributary channels have good physical properties. In terms of physical property distribution, front mouth bars and sheet sands are a relatively small range of physical properties and have good homogeneity, while braided channels and underwater distributary channels have a wide range of physical properties and strong heterogeneity. Generally speaking, the larger the grain size, the better the physical properties, and the more serious the heterogeneity.

(6) Because fan delta deposits are layered, there are obvious differences in the distribution of physical properties between sand layers of different microfacies combinations. The fan delta plain is dominated by braided channel deposits, with high permeability but strong heterogeneity. The fan delta front and prodelta are dominated by underwater distributary channels, mouth bars, and sheet sands, with low permeability but weak heterogeneity.

(7) Summarize the statistical probability model of sand body distribution at different scales. The interlayer scale reflects the distribution law of composite sand bodies, and the intralayer scale reflects the distribution law of single sand bodies. Combining them can comprehensively reflect the distribution of sand bodies of different scales. The range of the lateral distribution of a single sand body is about half of its width-to-thickness ratio.

(8) Through the study of outcrop sequence stratigraphy, the principle of merging pseudo-parasequences, the principle of lateral variation of parasequence types, and the principle of gradual change of mudstone thickness in stratigraphic correlation are summarized, which are used to effectively guide downhole stratigraphic correlation and the establishment of the layer model in geological models. The principle of merging pseudo-parasequences means that within a small distance, the thickness of the divided parasequences shall not differ too much. If their thickness difference is a multiple, they shall be combined according to the principle of thickness balance. The principle of lateral change of parasequence type means that with the lateral change of facies,

the type of parasequences also changes. The principle of gradual change of mudstone thickness is: if the thickness change of stratigraphic units is abnormal in single-layer correlation, the existence of deep cut valleys and shallow cut valleys is taken into account. In addition, the width-to-thickness ratio of sand bodies will change regularly with the rise and fall of the base level. Using this law to restrict the random interpolation in modeling can improve the effect of reservoir prediction.

Section 4　Establishment of Reservoir Geological Knowledge Database Using the Method of Dense Well Pattern Area Anatomy

Through the anatomy of a dense well pattern area, the establishment of the reservoir geological knowledge database of an entire block or an oilfield is a research hotspot at present. The establishment of the reservoir geological knowledge database is an important manifestation of the quantification of fine reservoir description. The quantification of the characteristic parameters of sand bodies can improve the accuracy of interwell reservoir prediction and reservoir modeling. Taking Xingerzhong area of Daqing Oilfield as an example, this section describes the establishment of the reservoir geological knowledge database through dense well pattern anatomy.

The reservoir geological knowledge database is a complex knowledge system. According to the data characteristics and reservoir deposition types of the research area, probe into the establishment of the reservoir geological knowledge database through dense well pattern area anatomy. Strictly speaking, the degree of refinement of the reservoir geological knowledge database established using the method of dense well pattern area anatomy is very difficult to compare with that using the field outcrop anatomy method, but it is integrated with the research area, so it has strong comparability. Therefore, it has a good guiding role for relatively sparse well pattern areas and blocks with uneven well patterns.

I. Ideas and methods of establishment of a reservoir geological knowledge database

Generally, most oilfields lack observable outcrop profiles, it is also difficult to select suitable and comparable modern sedimentary bodies, and the most common data available are the core data and dense well pattern data of the oilfields. Therefore, among the above methods for establishing a reservoir geological knowledge database, the method of dense well pattern anatomy is applied mostly widely. Daqing Oilfield is currently a deeply developed mature oilfield. It is feasible to establish a geological knowledge database through fine dense well pattern area anatomy and the extraction of relevant parameters based on a large amount of dynamic and static data.

The data characteristics of the selected Xingerzhong anatomy area are as follows.

(1) Through years of development geology research, the sedimentary model, scale, and genetic characteristics of this set of reservoirs, and the combination of different microfacies have been understood deeply;

(2) There are abundant data on dense well patterns and short-well spacing logging, thus

adding a lot of deterministic information. After multiple times of infill adjustments, the well spacing in Daqing Oilfield is generally 150-200m; in addition, there are some data on wells at the same well site, wells with small well spacing (well spacing less than 50m), etc.;

(3) There are a lot of production performance data, testing data and coring inspection well data, thus providing a basis for fine characterization and verification of the scale and heterogeneity of various sand bodies;

(4) The structure of the research area is simple and faults are not developed, which is conducive to fine stratigraphic correlation and sand body correlation.

According to the specific conditions of the anatomy area, the geological knowledge database of the meandering river delta reservoirs in Xingbei region of Daqing Oilfield is established mainly using the method of dense well pattern anatomy. Moreover, the reliability and predictability of the reservoir geological knowledge database are improved by making full use of outcrop analysis and modern sedimentary description results.

II. Geological knowledge database of reservoirs in Xingerzhong area

The established geological knowledge database of reservoirs in Xingerzhong area includes reservoir lithology and lithofacies library, sedimentary microfacies library, sedimentary model library, sand body scale library, reservoir physical parameter library, and interlayer information library.

(I) Reservoir lithology and lithofacies library

During the establishment of the lithology and lithofacies library, the following aspects shall be mainly summarized: lithofacies type, which is the key to establishing a lithologic and lithofacies library; texture characteristics, mainly including grain size characteristics and grain size distribution characteristics; main bedding structures, bedding types and scale; genesis interpretation, giving the possible types of microfacies in different lithofacies; other characteristics, mainly including sorting property, mineral composition, roundness and color; graphical features, that is, representing the first few items in graphical form. For dense well pattern areas, the information mainly comes from core description and analysis and test results. Lithofacies type, color, bedding structure, and genesis interpretation are the main features of this description. See Table 6-13 for the specific lithology and lithofacies library.

Table 6-13 Lithology and lithofacies library of delta facies in Xingbei region

Lithofacies type	Color	Main sedimentary bedding	Other features	Genesis interpretation	Example photo
Calcareous sandstone facies containing boulder clay	Gray	Massive bedding	Rare, boulder clay, around 2cm	Distributary channel bottom lag deposit	
Fine sandstone facies with oblique bedding	Gray, often brown due to high oil content	Tabular, trough-like cross-bedding	Large bedding scale and difficult core identification	Channel bottom deposit	

Continued

Lithofacies type	Color	Main sedimentary bedding	Other features	Genesis interpretation	Example photo
Massive argillaceous fine siltstone	Gray	Massive bedding	Bad oil bearing property	Channel top deposit	
Sandy mudstone facies	Light gray	Lenticular, wavy bedding	Burrow pores visible	Channel top deposit, interchannel deposit	
Fine siltstone facies with waxy bedding	Gray; light brown in case of containing oil	Wavy bedding	Poor oil bearing property	Non-main sheet sand deposit	
Fine sandstone with flaser bedding	Gray; brown in case of containing oil	Flaser bedding	Saturated oil	Main sheet sand, underwater distributary channel deposit	
Argillaceous siltstone facies with composite bedding	Gray, deep gray	Wavy, lenticular bedding	Bioturbation structure visible	Thin bedded sheet sand, interdistributary bay	
Silty mudstone facies with metamorphotic lamina	Light gray	Metamorphic lamina	Mixed sand and clay	Interdistributary bay	
Ostracoda-bearing massive mudstone facies	Deep gray, black	Massive, horizontal lamina	Charcoal debris visible	Deep lake, semi-deep lake deposit	

(Ⅱ) Sedimentary microfacies library

Sedimentary microfacies library refers to the geological knowledge database composed of sedimentary microfacies and their internal texture, structure, morphology and other parameters, and includes the main content below: the name of genetic units or sedimentary microfacies, which is the basis for the establishment of the library; a reasonable sedimentary microfacies library can be established only through accurate division of genetic units; the form of genetic units, including the planar form, section form and 3D form of genetic units; the lithology and occurrence of the shielding layer that may appear inside a genetic unit; the bedding structure inside a genetic unit, especially the main bedding structure type; changes in internal grain size rhythm characteristics, changes in grain size and grain sequence; the change of sorting property on a single profile, mainly describing the change of minerals and grain size; logging curve shape, which is the main source of information for the study of the identification and spatial distribution of sedimentary microfacies. The geological knowledge database of reservoirs in two subfacies environments such as delta plain and delta front has been summarized on the basis of the previous detailed research according to the research content of the geological knowledge database of these sedimentary microfacies and the characteristics of the research area.

The delta plain subfacies can be further divided into several types of microfacies, such as superaqueous distributary channel, sub-channel, abandoned channel, overbank deposit, interdistributary bay, flood plain, etc. Their detailed description is shown in Table 6-14.

Table 6-14 Geological knowledge database of delta plain sedimentary microfacies in Xingbei area

Sedimentary facies name	Shape		Main lithofacies	Sedimentary structure	Rhythmicity	Curve shape schematic
	Plane	Profile				
Underwater distributary channel			Calcareous sandstone facies containing boulder clay, fine sandstone facies with oblique bedding	Tabular, trough-like oblique bedding		
Sub-channel			Fine sandstone facies with small oblique bedding	Small tabular, trough-like oblique bedding		
Abandoned channel			Sandy mudstone facies	Composite bedding, horizontal bedding		
Overbank deposit			Massive argillaceous fine siltstone, sandy mudstone facies	Massive composite bedding		
Interdistributary bay			Silty mudstone facies with metamorphotic lamina, argillaceous siltstone facies	Composite bedding, metamorphotic lamina		
Flood plain			Mudstone, silty mudstone	Horizontal lamina		

Superaqueous distributary channel   Overbank   Abandoned channel   Inter-channel   Flood plain

The delta front subfacies can be further divided into sedimentary microfacies types, such as underwater distributary channel, mouth bar, front sheet sand, interdistributary bay, lake basin clay, etc. Their detailed description is shown in Table 6-15.

Table 6-15 Geological knowledge database of delta front sedimentary microfacies in Xingbei area

Sedimentary facies name	Shape		Main lithofacies	Sedimentary structure	Rhythmicity	Curve shape schematic
	Plane	Profile				
Underwater distributary channel			Fine sandstone facies with flaser bedding, massive argillaceous fine siltstone facies	Argillaceous lamina, oblique bedding		
Mouth bar			Fine sandstone facies with flaser bedding, fine siltstone facies with wavy bedding	Sand ripple bedding, argillaceous lamina		
Front sheet sand			Fine sandstone facies with wavy bedding, fine sandstone with flaser bedding, argillaceous silty sand with composite bedding	Composite bedding, wavy bedding, rhythmic bedding		
Interdistributary bay and lake basin mud			Silty mudstone facies with metamorphotic lamina, massive mudstone facies with ostracode	Composite bedding, metamorphotic lamina, horizontal lamina		

Legend: Underwater distributary channel | Mouth bar | Main sheet sand | Non-main sheet sand | Thin bedded sheet sand | Interdistributary bay and lake mud

(Ⅲ) Sedimentary model library

The generalized physicochemical model of the depositional environment and its deposits on the basis of comprehensive research on the physical, chemical and biological characteristics of modern deposits in a certain environment is the sedimentary model, which includes the spatial form, lithologic combination, sedimentary structure, biological characteristics, dynamic conditions, structural setting, etc. of sedimentary bodies. The core elements here are the spatial form, contact relationship and combination law of sedimentary bodies. Based on the fine anatomy of the research area, this project has drawn on previous research results and outcrop analysis results so as to establish the sedimentary model library of reservoirs in the area.

1. Delta plain distributary channel overbank sand model

The delta plain is dominated by the development of superaqueous distributary channels. The channel sand bodies are mutually cut, continuously superimposed, and distributed in a lenticular shape with flat top and convex bottom in mudstones. Overbank deposits are not developed (Figure 6-2). The continuously distributed distributary channel sand body is a combination of multiple single-channel sand bodies, and can be divided through the identification of abandoned channels, sand body thickness change analysis and logging curve comparison, so as to depict the shape of a single channel sand body. A single meandering channel is composed of point bar sand bodies, and multiple lateral accretion bodies are usually developed inside it, thus forming a semi-connected body due to being blocked by lateral accretion layers.

2. Delta front distributary channel sheet sand model

The delta front is dominated by sheet sand deposits. Underwater distributary channel sand

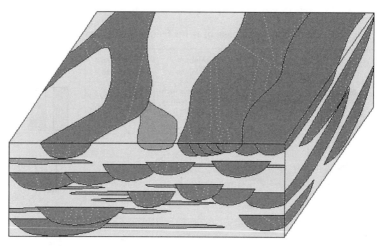

Figure 6-2   Sedimentary model of delta plain distributary channel

bodies are not developed, are distributed in strip shape and pod shape in sheet sands, and have scouring and erosion effects on sheet sands (Figure 6-3). The underwater distributary channel sand bodies have obvious vertical superimposition characteristics, reflecting the relatively stable development position of the channels and the periodic change of the water flow energy. The sedimentary thickness of sheet sands is not large, but they have good continuity. Especially thin-bedded sheet sands are mainly suspended deposits of fine-grained materials, are distributed more widely, and have better continuity.

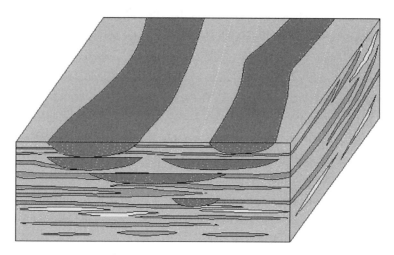

Figure 6-3   Sedimentary model of delta front distributary channel sheet sands

(Ⅳ) Sandy body scale library

The sand body scale library is a parameter library for quantitative description of sand bodies and an important basis for sand body prediction. The sand body scale library includes data information such as the length, width, thickness, length-width ratio, width-thickness ratio, drilled rate, etc. of sand bodies. According to the specific characteristics of sand body development, the parameters will also be different to some extent. For example, the parameters used in the extension range of sand bodies distributed in strip shape, fan shape and round shape are different.

1. Development scale of sand bodies in a single sand layer

Through the identification and comparative analysis of genetic unit sand bodies, the parameters of the genetic unit sand bodies in the anatomy area have been statistically analyzed, reflecting the development status of sand bodies in each single layer (Table 6-16).

Table 6-16 Statistics of genetic unit sand body parameters (part of the data shown)

Strata	Drilled rate	Sand body thickness Min, m	Sand body thickness Max, m	Sand body thickness Avg., m
S11	41.79	0.2	1.6	0.44
S11-1	35.45	0.2	2	0.83
S21	86.19	0.2	2.6	0.94
S21-1	94.78	0.2	2.69	0.78
S21-2	44.4	0.2	2.64	0.92
S22	79.48	0.16	2.03	0.54
S22-1	89.93	0.19	2.5	0.65
S23	97.76	0.2	3.4	0.87
S23-1	97.39	0.17	2.4	0.83
...	...	...	...	...
Avg.	81.14	0.19	2.74	0.87

2. Scale of sand bodies in different microfacies

According to the distribution of sand body thickness in the whole area, the overall thickness of the sand bodies is relatively small, and mainly 0.5-3m, and the sand bodies of about 1m thick are the most developed (Figure 6-4). This is because sheet sand deposits are the most developed in most layers in the anatomy area, and such sand bodies have a small thickness and a wide

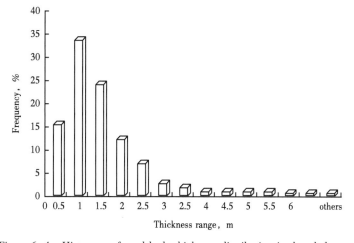

Figure 6-4 Histogram of sand body thickness distribution in the whole area

distribution, while thick channel sand bodies are developed only in the middle of Sa-2 sand group of Saertu reservoir group and Pu-1 sand group of Putaohua reservoir group.

The statistics of the scale of sand bodies in different microfacies show (Table 6-17): the superaqueous distributary channel sand bodies have a thickness of 4-6m, a width of about 500m, and a width-thickness ratio of about 100; the underwater distributary channel sand bodies have a thickness of 2-3m, a width of about 400m, and a width-thickness ratio of about 200; the thickness of the main sheet sands is 1-2m; the thickness of the non-main sheet sands is 0.5-1.5m; the thickness of the thin-bedded sheet sands is about 0.5m. The average drilled rate of sand bodies in different microfacies reflects the planar distribution of sand bodies in each microfacies. The statistics show that the planar distribution range of the sheet sands is the widest, and the thin-bedded sheet sands have a significantly higher drilled rate than other microfacies as well as the widest distribution range and the best continuity. The drilled rate of the main sheet sands is similar to that of the non-main sheet sands, indicating that their spatial distribution continuity is similar. The distributary channel sand bodies have a significantly lower drilled rate than other microfacies sand bodies, and are less distributed on the whole.

Table 6-17 Scale parameters of different types of sand bodies

Sand body type	Thickness m	Width m	Width-to-thickness ratio	Drilled rate %
Superaqueous distributary channel	Superaqueous distributary channel	±500m	±100	7
Underwater distributary channel	2-3	±400m	±200	19
Main sheet sand	1-2	—	—	21
Non-main sheet sand	0.5-1.5	—	—	36
Thin bedded sheet sand	0.5	—	—	24

(V) Reservoir physical parameter library

The analysis of reservoir physical parameters involves rhythmicity, sorting coefficient, variation coefficient, range, etc. Due to the lack of core analysis data at present, the physical parameter analysis mainly relies on logging interpretation results, so the available parameters are limited.

According to the core analysis data of the three inspection wells in the area and neighboring areas (Figure 6-5, Figure 6-6), the porosity of reservoirs in the whole area is mainly 20%-32% with an average of 23.6%; the permeability is mainly 1-400mD with an average of 469.8mD. According to the analysis of different lithologies, medium-coarse sandstones have the best physical properties, with an average porosity of 28.7% and an average permeability of 4243.7mD; fine sandstones have the second best physical properties, with an average porosity of 25.7% and an average permeability of 674.7mD; siltstones have the worst physical properties, with an average porosity of 21.6% and an average permeability of only 35mD.

According to the logging interpretation results, the physical properties of the main microfacies types have been analyzed (Table 6-18). According to the analysis results, the superaqueous distributary channel sand bodies have similar physical properties to the underwater distributary

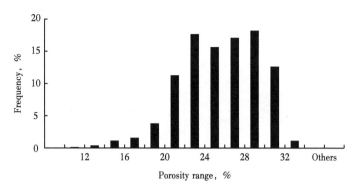

Figure 6-5 Distribution histogram of the porosity from core analysis

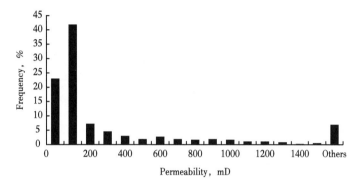

Figure 6-6 Distribution histogram of the permeability from core analysis

channel sand bodies, with high porosity and high permeability, porosity of generally more than 20% and averagely 27.7%, and permeability of more than 400mD; the main sheet sands have a similar porosity (averagely 27%) to the distributary channel sands, but an obviously lower permeability (averagely 172mD). Compared with the main sheet sands, the non-main sheet sands have worse physical properties, with an average porosity of 25% and a permeability of 100mD. The thin-bedded sheet sands have no effective thickness, so they have not been basically interpreted.

Table 6-18 Statistics of physical properties of reservoirs of different microfacies

Main sedimentary facies type	Porosity, %			Permeability, mD		
	Min.	Max.	Avg.	Min.	Max.	Avg.
Distributary channel	21.4	32.2	27.8	16	1930	483.5
Underwater distributary channel	21.6	32.0	27.6	15	1896	431.4
Main sheet sand	19.2	32.3	27	1	1764	172.8
Non-main sheet sands	16.4	32.8	25	1	1652	102.0

(Ⅵ) **Interlayer information library**

On the basis of the previous research, the distribution of interlayers in the area has been further summarized, including their genesis, type, occurrence, scale, etc. In addition, the geological knowledge database of interlayer distribution has been established (Table 6-19).

Table 6-19 Geological knowledge database of the distribution of interlayers in channel sand bodies in Xingbei region

Main interlayer type	Shape	Lithology	Scale	Genesis interpretation
Argillaceous lateral accretion layer		Argillaceous siltstone, mudstone	Thickness 20-30cm, width 100-300m, length 600-800m	Fine-grained materials developed in point bars and deposited during the intermittent period of meandering rivers
Argillaceous thin layer		Silty mudstone	Thickness about 1-50cm, planar extension up to 400m	Deep lake and semi-deep lake argillaceous deposits developed in sheet sands
Argillaceous lamina		Dark gray and black mudstone	Decimeter level in most cases, small scale, large quantity	Argillaceous lamina formed in bedding structures under low-energy hydrodynamic conditions

The thickness of the sand bodies in the research area is not large, which are characterized by thin alternate layers of sandstones with mudstones. However, various interlayers are still developed inside the sand bodies, including three types such as argillaceous lateral accretion layers, argillaceous thin layers and argillaceous laminas.

Argillaceous lateral accretion layers are developed in the thick-bedded channel sand bodies of Putaohua reservoir group, and are the lateral accretions of mudstones inside point bars. According to the previous research, the argillaceous lateral accretion layers in the point bar sand bodies are the argillaceous deposits deposited during the intermittent period of meandering rivers due to river energy reduction and covering lateral accretion bodies; the lithology of the argillaceous lateral accretion layers is mudstones and argillaceous siltstones, which are distributed diagonally. The argillaceous lateral accretion layers have a dip angle of about $1.0°-2.5°$, a thickness of about 20-30cm, a width of up to 100-300m, and a length of up to 600-800m. Usually 1-2 interlayers are developed in a single channel sand body, the interlayer frequency is 0.4 layer/m, and the interlayer density is about 0.3. The argillaceous thin layers are mainly developed in the sheet sands, and are located in their contact part. The argillaceous thin layers are developed mostly horizontally and their lithology is lacustrine silty mudstone and mudstone deposits. The thickness of the argillaceous thin layer varies from a few centimeters to several tens of centimeters, and can reach a maximum of 50cm. Their planar extension range varies greatly and their shape is irregular. The argillaceous laminas are developed in both channel sand bodies and sheet sands; the lithology of the argillaceous laminas is dark gray and black mudstones; their scale is small and mostly decimeter level; there are a lot of argillaceous laminas.

Chapter 7 Prediction of Interwell Reservoirs with Seismic Data

Seismic data have a good application effect in the prediction of interwell reservoirs in the exploration stage. After entering the development stage, the resolution of seismic data is difficult to meet the requirements of high-precision reservoir prediction, so the application of seismic data is limited. However, due to the high lateral resolution of seismic data, geologists are always reluctant to give up the utillization of seismic information. Therefore, in the past ten years, geophysicists and geologists have been continuously probing into the application of seismic data in fine reservoir prediction. It can be said that with the improvement of the quality of 3D seismic data and the development of processing and interpretation methods, seismic data have been widely applied in reservoir prediction and modeling in the development stage.

In terms of working ideas, the use of seismic information in the development stage shall emphasize the constraints of geological knowledge; that is, long-term accumulated geological knowledge is used to constrain and guide the interpretation of seismic data. In terms of methods, in addition to reprocessing and re-interpretation (including seismic attribute analysis, logging constraint inversion, seismic frequency division technology, etc.) for the purpose of improving the resolution of original seismic data, the application of seismic data in the development stage also includes high-precision 3D seismic, cross-well seismic, 4D seismic, etc. for the development stage. This chapter focuses on several typical seismic prediction methods.

Section 1 Reservoir Prediction with Seismic Attributes

The application of seismic attributes first originated in the 1960s. After a rapid development stage in the 1980s, it was gradually matured in the 1990s, forming a variety of methods and technologies such as multi-dimensional attribute (dip, azimuth, and coherence) analysis, automatic seismic facies analysis, etc. They play an important role in reservoir prediction.

I. Classification of seismic attributes

Many scholars have carried out researchs on the classification of seismic attributes, and classified seismic attributes from different perspectives. Nevertheless, it is very difficult to establish a complete list of seismic attributes. Brown (1996) summarized the seismic attributes into 66 types in four categories such as time, amplitude, frequency, and attenuation. Quincy Chen (1997) summarized seismic attributes into 91 types in eight categories such as amplitude, waveform, frequency, attenuation, phase, correlation, energy, and ratio. In terms of the basic definition of seismic attributes, they are physical quantities that characterize the geometric shape, kinematic characteristics, dynamic characteristics and statistical characteristics of seismic waves, and have a clear physical meaning. Therefore, it is natural and reasonable to classify seismic attributes by the time, amplitude, frequency, phase or statistical characteristics in seismograms. Generally, seismic attributes can be divided into several categories such as time, amplitude, frequency, coherence, attenuation, etc., and each category is divided into instantaneous time window, single-trace time window, and multi-trace time window attributes according to calculation methods. According to the data volume extracted by seismic attributes, they can involve pre-stack seismic attribute analysis and post-stack seismic attribute analysis. Generally, most seismic attribute analyses are made on post-stack seismic data. In terms of the practical application of seismic attributes and according to different research objectives, series of strata, and lithologic changes combined with the geological significance of seismic attributes, it is relatively accurate to divide seismic attributes into the types such as amplitude statistics type, frequency (energy) spectrum statistics type, phase statistics type, complex seismic trace statistics type, sequence statistics type and correlation statistics type (Table 7-1).

Table 7-1 Classification of seismic attributes

Classification	Identification of faults and fractures (vugs)	Stratigraphic sequence identification	Lithologic identification			
			Fluvial and lacustrine sand body	Alluvial fan and delta sand body	Carbonate rock	Igneous rock
Amplitude statistics type	Instantaneous amplitude, total amplitude, ratio of adjacent peak amplitudes, root-mean-square amplitude, average amplitude, average absolute amplitude, maximum peak amplitude, average peak amplitude, maximum valley amplitude, maximum peak amplitude	Absolute amplitude, maximum peak amplitude, average peak amplitude, maximum valley amplitude	Instantaneous amplitude, absolute value amplitude, ratio of adjacent peak amplitudes, main amplitude, maximum peak amplitude	Absolute value amplitude, sum of absolute value amplitudes, main amplitude	Root-mean-square amplitude, average absolute amplitude, maximum peak amplitude, average peak amplitude, maximum valley amplitude, maximum peak amplitude, total amplitude, average amplitude	Root-mean-square amplitude, average absolute amplitude, maximum peak amplitude, average peak amplitude, maximum valley amplitude, maximum peak amplitude, total amplitude, average amplitude
Frequency (energy) spectrum statistics type	Instantaneous frequency, dominant frequency, peak frequency, average energy, total energy	Instantaneous frequency, dominant frequency peak	Specific frequency band energy, effective bandwidth, varc length, dominant frequency peak, center frequency rating	Instantaneous frequency, effective bandwidth, specific frequency band energy, dominant frequency peak, dominant frequency sequence, average zero cross point frequency	Effective bandwidth, arc length, average zero cross point frequency, dominant frequency sequence	Instantaneous frequency, dominant frequency, peak, average zero cross point frequency, dominant frequency sequence

Continued

Classification	Identification of faults and fractures (vugs)	Stratigraphic sequence identification	Lithologic identification			
			Fluvial and lacustrine sand body	Alluvial fan and delta sand body	Carbonate rock	Igneous rock
Phase statistics type	Instantaneous phase cosine	Instantaneous phase, apparent polarity, response phase, instantaneous phase cosine	Apparent polarity, instantaneous phase, response phase	Instantaneous phase, apparent polarity, response phase	Instantaneous phase, instantaneous phase cosine	Instantaneous phase, instantaneous phase cosine
Complex seismic trace statistics type	Average reflection intensity, average instantaneous phase, average instantaneous frequency, reflection intensity slope	Average reflection intensity, average instantaneous phase, average instantaneous frequency	Waveform classification, reflection intensity slope, average reflection intensity, average instantaneous frequency, instantaneous frequency slope	Waveform classification, average reflection intensity, average instantaneous frequency, instantaneous frequency slope, average instantaneous phase	Waveform classification, average instantaneous frequency, average reflection intensity, average instantaneous phase	Waveform classification, average instantaneous phase, average reflection intensity, average instantaneous frequency
Sequence statistics type	Energy half time, ratio of positive and negative samples, number of crests, number of troughs	Energy half time, ratio of positive and negative samples, number of crests, number of troughs	Energy half time, ratio of positive and negative samples, number of crests, number of troughs	Energy half time, ratio of positive and negative samples, number of crests, number of troughs	Energy half time, ratio of positive and negative samples, number of crests, number of troughs	Energy half time, ratio of positive and negative samples, number of crests, number of troughs
Correlation statistics type	Coherence, similarity coefficient, correlation kurtosis, average SNR, correlation length, correlation component	Coherence, concentrated correlation, average correlation, similarity coefficient, correlation length, correlation component	Concentrated correlation, average correlation, similarity coefficient, correlation kurtosis, correlation length, correlation component	Correlation maximum, correlation minimum, similarity coefficient, correlation kurtosis, average SNR	Coherence, average SNR, correlation length, correlation component, correlation maximum, correlation minimum	Coherence, average SNR, correlation length, correlation component

From the perspective of attribute extraction methods, it is reasonable to divide seismic attributes into interface attributes (attributes extracted along structural layers) and volume attributes (attributes extracted with two structural layers or time intervals as the top and bottom of the time window). And this is understood more easily for geological research. Interface attributes mainly reflect the planar change law of geologic bodies and can be used to predict the planar distribution of structures, sedimentary facies and parameters. Volume attributes are the distribution of parameters in 3D space, reflect the changes of geologic bodies in 3D space, and can be used to establish 3D structural models, sedimentary facies models and parameter models of geologic bodies, and planar mean values can also be extracted from volume attributes for cross-well prediction. Du Shitong (2004) elaborated interface attributes and body attributes.

(Ⅰ) **Interface attributes**

Interface attributes are seismic attributes related to the interface, which are obtained along the 3D horizon surface in a 3D data volume, and provide information about changes along the interface or between two interfaces. The picking methods include instantaneous attribute picking, single-trace time window attribute picking and multi-trace time window attribute picking.

Instantaneous attributes are the attributes picked at the location of the arrival of seismic waves based on the analysis of complex seismic traces.

The single-trace time window attribute is picked along a variable time window. During the picking process, the position and length of the time window are variable during sliding between traces. The upper and lower boundaries of the variable time window are determined by interpreted seismic horizons. The seismic attribute picking time window can also be slid along an interpreted horizon, and a fixed time window length is taken above or below the horizon. As a special case, a time slice can be regarded as a seismically interpreted flat interface with no ordinary meaning to determine the location of the attribute picking time window. The attribute picking result is generally assigned to the midpoint of the time window. When the attribute picking time window slides between the traces so as to change the length, it is necessary to use the average normalization algorithm for the seismic attribute, and to reduce the calculation result to a fixed time window length so as to ensure meaningful comparison between the traces.

The fixed-scale or variable-length time window for single-trace time window attribute picking can also be used for multi-trace time window attribute picking. To pick some multi-trace time window attribute, it is required to define a trace number and trace mode boundary in addition to defining upper and lower boundaries so as to form a time window. The seismic attribute picking result is assigned to the middle trace position and time window midpoint in the specified trace mode. A new attribute plane can be obtained by repeating the above picking process for each middle trace position.

(Ⅱ) **Volume attribute**

Compared with 2D seismic data, the advantage of 3D data is that it can cause multi-trace 3D seismic wave attribute picking at the arrival position along a sliding time window defined by the model framework of different spatial traces. This is the picking of a volume attribute. The seismic volume attribute is a complete attribute cube derived from a 3D seismic data volume, and is another type of image of seismic data. This kind of image can be used to reveal seismic features that are difficult to identify using other profile images, such as channel sand bodies, reef blocks, sedimentary features of various stratigraphic sedimentary units, etc., and has important use value. Instantaneous attributes that generate attribute cubes are the attributes picked at the location of the arrival of seismic waves based on the analysis of complex seismic traces. For a data volume, the single-trace time window attribute is an attribute plane generated between two time slices, but the position and length of the time window in this case are fixed. Repeatedly use the fixed time window for attribute picking, and overlap in time according to a certain step length, to generate a new attribute volume. A fixed time window is also used for multi-trace time window attribute picking. The multi-trace time window attribute defined and picked using different spatial trace modes can be used to study reservoir anisotropy characteristics and identify reservoir fractures or fault distribution patterns.

A coherent data volume is a volume attribute data volume picked by multi-trace time window attributes. Coherence is a measure of the similarity or dissimilarity of waveforms in a time window presented in the directions of two longitudinal and two transverse lines. Under this framework definition, there are 8 points. The standard value of correlation is 1, and a smaller value indicates the degree of discontinuity or irrelevance. The typical algorithm is KL transformation, and an 8-point multi-trace time window attribute picking mode is used. The first principal element component of multiple traces characterizes the main feature of the seismic record, while the second principal element component gives the second feature of the remaining quantity in the data. The two have similar images but different numerical ranges. The third principal element represents the third remaining feature index in the data. Coherence calculation is to generate volume attributes for a certain time window, and the center of the time window is on a specific horizon; or an overlapping sliding time window is used to generate a new data cube, called a coherent data volume. Coherent volume attributes or coherent data volumes are very sensitive for detecting seismic wave discontinuities such as faults and unconformities, provide a very effective display method for identifying stratigraphic sedimentary features and faults, and thus are widely applied. Algorithms and software related to coherent data volumes have also developed rapidly, and have become important tools for automated interpretation of 3D seismic data volumes.

II. Geological significance of seismic attributes

Seismic attributes refer to geometrical, kinematic, dynamic or statistical features of seismic waves reflected from mathematical transformation of prestack or poststack seismic data, and includes multiple aspects such as amplitude, velocity, frequency, time, phase, attenuation by absorption, etc. Multiple attribute data can be extracted from each aspect using different algorithms. It is estimated that there are currently as many as 300 types of seismic attributes, and new attributes are constantly emerging. Faced with such a plethora of seismic attributes, it is necessary to fully understand the practical significance of each seismic attribute, and grasp the geological phenomena it can reflect and the geological problems it can be used to solve in order to obtain good application effects.

The application of seismic attributes is based on the spatial changes in the physical properties of reservoirs and the properties of the fluids filled in reservoirs, and the changes in seismic attributes such as seismic reflection velocity, amplitude, and frequency etc. based on geometry, kinematics, dynamics and statistics. A lot of important characteristic information about the changes of formations, faults, fractures, lithology and hydrocarbon-bearing properties can be obtained through the extraction and interpretation of seismic attributes. Some of these seismic attributes have corresponding geological significance, while some do not have clear corresponding geological significance. There are about 30 – 50 types of typical seismic attributes, and their geological significance is summarized in Table 7-2.

Table 7-2 Geological meaning of seismic attributes (according to Hou Bogang, 2004)

No.	Attribute type	Description mode or content	Geological meaning
1	Frequency attribute	Related features analyzed using Fourier spectrum and power spectrum	It reflects changes in formation thickness, lithology, and fluid-containing composition. It is often used to detect frequency-selective absorption due to overlying strata anomalies such as gas saturation or existence of fractures. It can also be used to identify frequency changes due to changes in stratigraphic characteristics and lithofacies etc.
2	Amplitude or energy attribute	Related dynamic characteristics recording energy, maximum amplitude, etc.	It reflects changes in wave impedance, formation thickness, rock composition, formation pressure, porosity and fluid-containing composition in the target layer. It can be used to identify amplitude anomalies or analyze sequence features. It can also be used to track stratigraphic features, such as deltas, river channels, various fans or special lithological bodies, and to identify lithological changes, unconformities, and fluid accumulation, etc.
3	Waveform attribute	Number of crests or waveform area etc. in the statistical time window	It mainly reflects the change law of wave impedance in the target layer, sedimentary sequence, stratigraphic bedding characteristics, ancient denudation surface, ancient structural characteristics, sedimentary process and its continuity, the size of sedimentary basins, etc.
4	Correlation attribute	Using the eigenvalue of correlation function	It reflects the stability of depositional conditions and the smoothness of strata boundaries
5	Resolution attribute	Characteristic quantity related to autocorrelation parameters and bandwidth	Usually there are interlayers with strong reflection performance in a large set of uniform rock formations, then a small amount of clear reflection waves can be seen in the seismic record on a quiet record background, and the resolution parameter of the record is large in this case. When reflective layers are thin alternate layers, and the reflected waves from adjacent reflective layers interfere with each other to form a complex wave group, the resolution parameter of the seismic record is small, and the bandwidth is narrow
6	SNR attribute	SNR is often taken as the ratio $\eta = A_q / \sqrt{E_n}$ of the mean amplitude A_q of effective waves to the square root amplitude E_n of interference.	It reflects the strength of the interference background on the seismic record, and also the change of geological conditions, for example, the regularity of the seismic record of the reservoir part often deteriorates and the SNR decreases

Continued

No.	Attribute type	Description mode or content	Geological meaning
7	Absorption parameter	Characterized by absorption coefficient, attenuation factor, quality factor, etc.	According to the close relationships of strata absorption properties with lithofacies, porosity, hydrocarbon-bearing components, etc., it can be used to predict lithology and the distribution of sandstones and mudstones, and under favorable conditions, to predict the presence of oil and natural gas
8	Autoregression analysis	AR model is generally used	It can be used to predict the existence, type and boundary of oil and gas reservoirs. Autoregression models (limited to 3-5 orders) are used in general

* In addition to the attributes listed in the table, there are a number of attributes that are defined by mathematical methods, but they have no obvious physical meaning, such as the second moment of spectrum and L2 mode etc.

For sedimentary facies interpretation and sand body distribution prediction, Ling Yun (2003) pointed out the five most basic seismic attributes, namely instantaneous frequency, instantaneous phase, instantaneous amplitude, coherent data volume and waveform clustering. The first three are single-trace calculated seismic attributes calculated, and the latter two are multi-trace calculated seismic attributes.

The three single-trace calculated seismic attributes such as instantaneous frequency, instantaneous phase, and instantaneous amplitude are the most classic and well-known. They can be used to reconstruct seismic data and are obviously unique. Although the three instantaneous attributes are used in many aspects, it is difficult to find a good application example to illustrate the direct relationships between them and depositional environment. However, under the condition of relatively maintaining the high resolution processing of amplitude, the relationships between the above three seismic attributes and depositional environment are as follows. (1) The instantaneous frequency is related to the natural frequency of the formation where the information is extracted, and the natural frequency of the formation is related to the thickness (density) of sediment particles. From the perspective of resonance, the resonance frequency is low when sediment particles are coarse, and the resonance frequency is high when sediment particles are fine. In addition, the instantaneous frequency is also related to the tuning effect of the thin layer thickness. (2) The instantaneous phase (phase of the main frequency of seismic waves) is related to the viscosity of the formation where the information is extracted to the main frequency of seismic waves, that is, when seismic waves pass through the strata of different lithologies, the phase of seismic waves will change, so the instantaneous phase is used to detect lithological boundaries sensitively. (3) The instantaneous amplitude is related to the reflection coefficient of the formation where the information is extracted, that is, it is related to the velocity and density of the formation and the properties of the fluids in pores, so it has a direct relationship with AVO and direct oil and gas prediction.

In addition to the above three seismic attributes, there are many other single-trace calculated seismic attributes. They come into being based on the physical characteristics of different geological bodies and the development of visualization, and have different application effects. But fundamentally speaking, many seismic attributes are finally decomposed through the information of the same seismic trace, and simply the relationship between them and geological information is understood and interpreted from different angles. Therefore, they are directly or indirectly related to the above three basic seismic attributes.

The multi-trace calculation theory is to describe the spatial change of geological information by extracting the difference and similarity of seismic information in space. Coherent data volume (difference) and waveform clustering (similarity) are the most basic multi-trace calculation methods. Coherent data volume is used to detect changes in geological information by means of the difference between data, so it is very sensitive to detecting faults, while waveform clustering is used to obtain the spatial distribution of the same geological information by means of the similarity of waveforms. It can be seen that the two multi-trace calculation theories cover the elements of detecting the spatial change of geological information, so they are called multi-trace calculated basic seismic attributes.

Research shows that when the above five basic attributes do not reflect any information on geological anomalies, the results obtained through other seismic attributes are usually unreliable. Conversely, when the above five seismic attributes have information reflections, other seismic attributes may more clearly reflect changes in geological information than the above five basic seismic attributes. Therefore, when seismic attributes are interpreted, the identification relationships between the five seismic attributes and geological information shall be studied first, and then other seismic attributes are studied. This not only saves a lot of calculation time, but also avoids identification difficulties caused by the mixed use of too much attribute information.

III. Key links of seismic attribute extraction

The process of forming seismic attributes from seismic data is called attribute extraction. In applications, fine horizon calibration and reasonable time window selection are the key to seismic attribute extraction, and directly determine the credibility of the extracted seismic attributes.

(I) Target horizon calibration

There are three main methods for horizon calibration with seismic data, namely, the average velocity method, the synthetic seismogram method and the vertical seismic profile (VSP) method. Due to its large error, the average velocity method for horizon calibration is only suitable for initial structural interpretation, while the VSP method for horizon calibration is not widely used due to being limited by the source of data. The synthetic seismogram method for horizon calibration is the most commonly used calibration method, and its effect lies between the former two.

The synthetic seismogram is formed through wavelet convolution using acoustic velocity logging and density logging, and its corresponding relationship with the reflection events on the seismic profile is often not ideal. There are many reasons: the data acquisition methods of acoustic velocity logging and density logging are completely different from those of seismic profile; the travel path of and interference with ultrasonic waves in acoustic logging are completely different from the propagation path and interference background of seismic waves and the development state of multiples; the characteristics of seismic wavelets are not exactly the same as those wavelets used to make synthetic seismograms; seismic wavelets change with depth, and a synthetic seismogram is constant for a specific well section. Therefore, it is necessary to achieve the best matching relationship between the synthetic seismogram and near-well traces through time lapse with the help of the marker horizon in the calibration process. Generally speaking, for areas where the wave impedance interface is obvious, this mode can achieve good results, but for areas where the wave impedance interface is not obvious, the effect of this mode is often not ideal.

(II) **Time window selection**

A reasonable time window shall be firstly selected to calculate seismic attributes. A too large time window contains unnecessary information; a too small time window will cause truncation and loss of effective components. There are two time window selection methods such as fixed time window method and along-horizon sliding time window method. The selection of time window in reservoir prediction shall be based on the following criteria.

(1) If the top and bottom boundaries of the reservoir can be accurately tracked, the top and bottom boundaries are used to define the time window.

(2) If only the top boundary event of the reservoir can be accurately tracked, the time value corresponding to the top boundary event is taken as the starting point of the time window, the length of the time window is fixed, and each trace contains reservoir information and as little non-reservoir information as possible.

(3) When the top and bottom boundaries of the reservoir cannot be accurately tracked, inter-well interpolation and extrapolation can be performed to open a time window taking the trend of a certain marker horizon as the constraint according to the time thickness of reservoir corresponding to a well. The fixed time window method can also be used, and both the starting point and length of time window are fixed values. The length of time window is selected based on the experience of interpretation personnel and it shall contain as little non-reservoir information as possible.

(4) The seismic information above the top boundary of reservoir does not contain the geological information of reservoir, so when the target layer is too thin, the corresponding seismic wave group is too short, and the time window needs to be extended, it can only be extended down appropriately.

(5) Generally, when the lateral continuity of reservoir reflection is good, it is not suitable to use the method of calculating the average reservoir time thickness as the time window length based on near-well traces.

In the process of selecting time window, the well-seismic calibration result is firstly analyzed to ascertain the seismic reflection characteristics of the target horizon. Then the reasonable time window range is determined through multi-well analyses. Finally, the extracted seismic attributes are calibrated and analyzed to establish the relationship between seismic attributes and geological features.

(III) **Seismic attribute optimization**

There are many types of seismic attributes, and selecting the attribute volume most sensitive to the prediction object is one of the key points for geological body depiction. The criteria and purpose of optimizing attributes is to select the combination of seismic attributes that are closely related to a geological body and have poor correlation with each other from a large number of seismic attributes. If the prediction is made with a set of seismic attributes with good correlation, the result is equivalent to the prediction with a single attribute among them, and the result reliability is poor. If a set of seismic attributes with poor correlation are used for prediction, the different attributes have certain mutual verification characteristics, and the prediction result is more reliable. In practice, the attribute optimization is a complicated task. In addition to analysis with multiple methods such as clustering and crossplots etc., the specific issues shall be specifically analyzed according to the geological characteristics of the research area and the research object so as to achieve the optimal combination of seismic attributes.

IV. Seismic attribute analysis methods

With the improvement of mathematics and computer computing capabilities, the ability to extract seismic attributes continues to increase, and the role of seismic attributes in predicting reservoir characteristics continues to increase. Therefore, seismic attributes have been widely applied in the study the characteristics of reservoirs. The general process of seismic attribute analysis includes horizon calibration, horizon tracking, attribute optimization and statistical prediction. Horizon calibration and horizon tracking are the foundation of seismic data application and the key to establishing the corresponding relationship between seismic data and geological bodies, and are of universal significance to seismic data application. Attribute optimization and statistical prediction are the core of seismic attribute analysis.

The selection of seismic attributes refers to selecting a reasonable subset of seismic attributes from the set of seismic attributes. Different seismic attributes have different geological meanings. Therefore, it is necessary to select the seismic attributes that are sensitive to the problem to be solved. The principles below shall generally be followed in the selection of sensitive seismic attributes. (1) Try to choose attributes with physical meaning as much as possible, which is conducive to establishing the relationship between reservoir physical properties and seismic attributes, while the sensitivity of abstract attributes may only be an accidental phenomenon. (2) The summation of periodically changing data to determine seismic attributes shall be avoided. For example, simple amplitude summation is not as good as the summation or the sum of squares of absolute amplitude values. (3) Attributes that reflect the same physical meaning in statistical analysis (such as the summation of absolute amplitude values and root mean square amplitudes) cannot simultaneously appear in a set of sensitive attributes. There are many methods for selecting seismic attributes, which can be divided into the following types.

(I) Selection method with expert's experience

The selection method with expert's experience is a qualitative method with large human factors. The attributes that have the largest influence on reservoir prediction can be selected directly based on the knowledge of experts, or attributes can be optimized according to the optimization principle specified by experts. Generally speaking, oilfield experts have a good understanding of the geological conditions and reservoir characteristics of a certain area, and can analyze seismic attributes comprehensively with geological data, logging dat, and drilling data, and propose favorable attribute volumes. The attributes selected using this method are generally of clear geological significance and high reliability, and are often used as the basis for the application of other methods.

(Ⅱ) **Mathematical theory method**

There are many seismic attributes, and there are both correlations and differences between different attributes. Sometimes it is difficult to select attributes simply based on empirical judgment. Therefore, some mathematical methods are needed to select a reasonable attribute combination. Typical mathematical methods include cluster analysis method, neural network method, correlation method, attribute contribution method, search method, simulated annealing optimization algorithm, maximum entropy principle method, seismic attribute optimization method of combining rough set theory with self-organizing neural network, etc. Such methods do not need the analysis of all attributes, while mathematical methods are directly used to select some ideal attributes, and they are simply analyzed; then, some attributes that are considered unreasonable are deleted; and finally the few remaining attributes are used to carry out prediction. The advantage of such methods is that the workload of researchers is reduced, there is no need for a deep understanding of the survey area and the meaning of seismic attributes and the result is objective. The disadvantage of such methods is that the reliability is not high, and the selected seismic attributes sometimes have no clear geological significance.

(Ⅲ) **Forward modeling method**

Forward modeling is a basic method of seismic research. Many domestic scholars have studied the reflection characteristics of wedge models and thin alternate layer models, but there are few examples of really applying the research results of these models in actual seismic attribute interpretation, and in more cases, geological models are established based on actual geological conditions. Many seismic attributes are the reflection of comprehensive factors such as structure, strata, lithology, and oil and gas etc. In theory, it is impossible to solve multiple unknowns from one equation at the same time, and only after making assumptions about some of the parameters, can other parameters be obtained. But for a specific area, with the deepening of research, people will learn more and more about the underground geology. Thus, the information can be used to establish models, some of the unknown factors (such as structural models, stratigraphic models, etc.) are determined, and then other unknown factors (such as the lithology and hydrocarbon-bearing properties of reservoirs) are studied. In a relatively stable sedimentary environment, the lateral changes of factors such as reservoir thickness and lithology etc. are relatively small, so their contribution to seismic attributes can be approximated as a constant. In this case, the main factors leading to changes in seismic attributes of reservoirs may be oil and gas or reservoir physical properties. In a complex and changeable depositional environment, there are many factors that affect seismic attributes. Changes in stratigraphic structure and lithology may contribute more to seismic attributes than oil and gas. The higher the degree of research on an area, the more the known data, the stronger the constraints on attributes, and the higher the credibility of applying attributes to predict lithology or hydrocarbon - bearing properties. In other words, specific circumstances need to be analyzed specifically according to actual data, a suitable forward model shall be established, and seismic attributes sensitive to reservoirs are selected.

Section 2　Logging-Seismic Joint Inversion

With the continuous maturity of seismic inversion technology and the transformation of oil and gas exploration ideas (from structural traps to lithologic and stratigraphic traps), seismic inversion has become a core technology for seismic data application. Seismic wave impedance inversion is a special seismic processing and interpretation technology of using seismic data for formation wave impedance inversion, in order to change seismic waveform records into wave impedance profiles that reflect the lithological information of formations. Therefore, after seismic inversion, interface-type original seismic data are converted into strata-type wave impedance information, so as to obtain the true reflection of actual strata and make reservoir prediction more intuitive and operable. In addition, the application of multiple inversion methods can improve the resolution of seismic data to a certain extent, directly compare seismic data with drilling data and logging data, and obtain formation change information through calibration and interpretation.

I. Types of seismic inversion methods

At present, seismic inversion technologies are emerging in endlessly and developing rapidly, and their classification standards are also diverse. Generally speaking, they can be divided into travel time inversion and amplitude inversion in terms of seismic information used in inversion; they are divided into structure inversion, wave impedance inversion and reservoir parameter inversion in terms of geological results of inversion; in terms of seismic data used in inversion, they are divided into pre-stack inversion and post-stack inversion, and this classification method is the most commonly used.

Pre-stack inversion has not been widely promoted, but in recent years it has become a hot spot for scholars at home and abroad, including tomography technology based on travel time, AVO analysis technology based on amplitude, and elastic wave impedance inversion technology.

Great progresses have been made in post-stack inversion in the past 20 years, a variety of mature methods and technologies of post-stack inversion have been formed, and multiple sets of relatively complete commercial software such as Jason, Strata, etc. have emerged. Post-stack inversion mainly includes structural analysis based on travel time and wave impedance inversion based on amplitude information. Due to the different estimation methods of various parameters and the different calculation processes of inversion results in the actual implementation processes, many inversion methods are derived from post-stack seismic impedance inversion. According to the role of logging data, post-stack inversion can be divided into four types: direct seismic inversion without well constraints (such as trace integral inversion), seismic inversion under logging control (such as recursive inversion), logging-seismic joint inversion (such as logging constrained inversion) and logging interpolation and extrapolation under seismic control, which are used in different stages of oil and gas exploration and development. At present, the widely applied logging constrained inversion is a logging-seismic joint inversion method, which expands the impedance frequency range of inversion to low-frequency and high-frequency ends on the basis of seismic frequency band through the combination with logging information, geologic models, etc.

II. Applicability of several seismic inversion methods

(I) Trace integral inversion

Trace integration is a direct inversion method for calculating the relative wave impedance of the formation using post-stack seismic data. It is derived under the condition that formation wave impedance is continuously differentiable with depth, so it is also called continuous inversion. Simply put, trace integration is to integrate the seismic record processed with high resolution from top to bottom, and eliminate direct-current components, and finally obtain an integrated seismic trace. As we all know, the reflection coefficient is $R = \frac{\rho_2 v_2 - \rho_1 v_1}{\rho_2 v_2 + \rho_1 v_1}$; when the wave impedance contrast is not great, $\rho_2 v_2 - \rho_1 v_1 = \Delta \rho v$ and ρv is set to the average value of $\rho_2 v_2$ and $\rho_1 v_1$; then $R \approx \frac{\Delta \rho v}{2 \rho v}$. Therefore, the reflection coefficient is integrated to approximately get: $\int R \mathrm{d}t \approx \frac{1}{2} \int \frac{\Delta \rho v}{\rho v} = \frac{1}{2} \ln \rho v$. Where, R is the reflection coefficient, and ρ and v are the rock density and rock wave velocity, respectively. Therefore, the integral of reflection coefficient is proportional to the natural logarithm of wave impedance ρv, which is a simple concept of relative wave impedance. Thus, the reflection coefficient reflecting the velocity difference between rock formations can be converted into the wave impedance reflecting the characteristic changes of formations themselves, and the lithology can be directly interpreted.

From the above integration principle, it can be seen that the lithology interpretation and the transverse continuous prediction of sand layers do not require any drilling control. Moreover, the calculation method is simple and the recursive cumulative error is small. However, this method has two fatal flaws. (1) The absolute wave impedance and absolute velocity of formations cannot be obtained, and the method cannot be used to quantitatively calculate reservoir parameters. (2) Limited by the inherent frequency band, the vertical resolution of seismic data is low, and the method cannot be used to predict the thickness and lateral distribution of thin sand bodies in development series of strata. Therefore, the trace integration method is only suitable for the establishment of conceptual reservoir models in the exploration stage.

(II) Recursive inversion

The seismic inversion method of recursive calculation of formation wave impedance (velocity) based on reflection coefficient is called recursive inversion, which is also a direct inversion method. The key to recursive inversion is to estimate the reflection coefficient of formations from seismic records and obtain the wave impedance information that best matches the known drilling data. Logging data in recursive inversion mainly play the role of calibration and quality control, and doe not directly participate in the inversion calculation, so recursive inversion is also called seismic inversion under logging control.

The basic principle is to assume that the underground medium is a layered medium, and there are a series of $j+1$ reflective interfaces parallel to each other. The theoretical model of noise-free migration seismic record is:

$$s(i) = r(i) * w(i)$$

Where, $s(i)$ is the seismic record; $r(i)$ is the formation reflection coefficient; $w(i)$ is the seismic wavelet.

Through wavelet deconvolution processing, the reflection coefficient $r(i)$ can be obtained from the seismic record, and then the formation wave impedance or interval velocity can be calculated recursively:

$$Z_{j+1} = Z_0 \prod_{i=1}^{j} \frac{1+r_i}{1-r_i}$$

Where, Z_0 is the initial wave impedance; Z_{j+1} is the wave impedance of layer $j+1$.

Therefore, the wave impedance of the marker horizon can be selected as the reference wave impedance from the interval transit time curve and density curve (the Gardnar formula $\rho = 0.31V^{0.25}$ can be used for conversion when there is no density logging), so as to convert the reflection coefficient obtained from deconvolution into wave impedance.

Recursive inversion is a conversion process of seismic data, and the resolution, SNR and reliability of the results completely depend on the quality of seismic data. Therefore, the seismic data used for inversion shall have features such as wide frequency band, low noise level, relative amplitude preservation and accurate imaging. Logging data, especially acoustic logging data and density logging data, are the comparison standard and interpretation basis for transverse seismic prediction. Before inversion processing, they shall be carefully edited and corrected so that they can correctly reflect the physical characteristics of rock formations.

Recursive inversion relatively completely retains the basic characteristics of seismic reflection, has no multiplicity of model-based seismic inversion, and can clearly reflect the spatial changes of lithofacies and lithology. Moreover, under the condition of relatively stable lithology, recursive inversion can better reflect the physical property changes of reservoirs. Therefore, in the early exploration period when there are few drilling data, and under the condition of thick reservoirs, lithofacies analysis, thickness prediction, reservoir physical property prediction, etc. can be performed by means of recursive inversion data. In the reservoir monitoring stage, recursive inversion data can provide time-lapse seismic inversion velocity difference analysis, and can be used to determine the spatial changes of the pressure and physical properties of reservoirs and judge the migration of oil and gas. The biggest disadvantage of this inversion method is that due to the limitation of seismic frequency bandwidth, the resolution is relatively low, and cannot meet the needs of research on thin reservoirs, so the method is limited in the application of fine reservoir prediction in the middle and late stages of development.

(Ⅲ) **Logging constrained inversion**

Logging constrained inversion is currently the most widely used inversion method, and is a model-based wave impedance inversion technology. The basic principle is: logging data have a high vertical resolution, but are only a narrow view of discrete distribution; the vertical resolution of seismic data is not high, but seismic data can provide information on the spatial changes of strata; the combination of the two can complement each other. A reliable high-resolution initial geological model can be established based on the data with high vertical resolution and under the control of laterally traceable seismic data. Perform forward modeling of the initial model and calculate the synthetic seismogram; then compare the calculation error between the synthetic seismogram and the actual seismogram to obtain the model modification parameters (perturbation); modify and update the initial model repeatedly, so that the synthetic seismogram is closest to the actual seismogram in the sense of least squares. The final high-resolution wave impedance model data is the inversion result. This inversion method is essentially logging-seismic joint inversion. The specific process is shown in Figure 7-1.

The formula used in logging constrained inversion can be summarized as:

$$M = M_0 + [G^T \cdot G + G_n \cdot C_m^{-1}]^{-1} \cdot G^T \cdot (S - D) \qquad (7-1)$$

Where, M is the updated model; M_0 is the initial model; G is the sensitivity matrix, or Jacobi operator, which is the matrix composed of a series of partial derivatives; C_n is the noise covariance matrix; C_m is the model covariance matrix; S is seismic data; D is the calculated seismic data; $S-D$ is called residual deviation or residual.

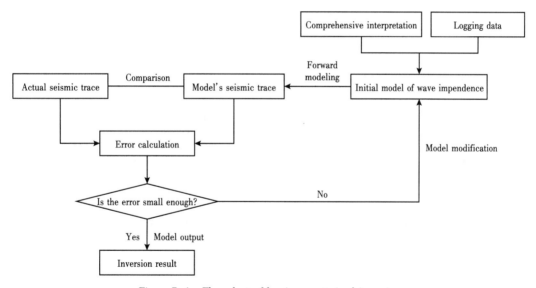

Figure 7-1 Flow chart of logging constrained inversion

$M-M_0$ is the amount of model modification or perturbation, which is calculated by Equation (7-1) based on the residual error. After model modification each time, repeat the above calculations until the residual is small to a certain extent; then terminate the iteration to obtain the corresponding wave impedance model.

Logging constrained inversion organically combines seismic data with logging data, breaking through the traditional seismic resolution. In theory, the same resolution as that of logging data can be obtained through logging constrained inversion. It is a key technology for fine reservoir description in the oilfield development stage. Its accuracy depends not only on the geological characteristics of the research target, the number of drilled wells, their distribution and the resolution and SNR of seismic data, but also on the accuracy of the processing process. Logging constrained inversion involves the following three key links: analysis of geophysical characteristics of reservoirs, seismic wavelet extraction and establishment of the initial wave impedance model. The biggest disadvantage of this method is that it has multiple solutions. The multiplicity mainly depends on the degree of conformity between the initial model and the actual geological conditions. Under the same geological conditions, the more the drilled wells, the higher the accuracy. Under the current condition that it is difficult to improve the resolution of seismic data, it is the key to reduce multiplicity to establish an initial model that can better reflect the true underground conditions.

III. Logging constrained inversion guided by geological understanding

Logging constrained inversion technology organically combines seismic data with logging data and the inversion result contains both logging information and seismic information, breaking through the limitation of traditional seismic resolution. In theory, the same resolution as that of logging data can be obtained through logging constrained inversion. It is a key technology for the prediction of thin reservoirs and fine reservoir description in the oilfield development stage. The basic process is: carry out extrapolation and interpolation for well points under the control of the horizon from seismic interpretation and form the initial wave impedance model. Then the initial wave impedance model is continuously updated using the conjugate gradient method, so that the synthetic seismogram of the model has the best similarity with the actual seismogram. Here the wave impedance model is the inversion result.

The initial wave impedance model is also called a geological conceptual model. The initial wave impedance model is obtained by determining the weight distribution relationships among wells trace by trace according to the distance between the interpolation point and each sample well in the stratigraphic framework model. This model is relatively rough, mainly reflects regional geological background, lacks detailed change characteristics, reflects the trend control of sample wells on the attribute model, and implements the idea of conforming to the laws of geology. The process of building the initial wave impedance model is actually the process of correctly combining seismic interface information with logging wave impedance information. For seismic data, the key is to correctly interpret wave impedance interfaces that play a controlling role. For logging data, the key point is to give appropriate wave impedance information to the formation between wave impedance interfaces.

Establishing an initial wave impedance model as close as possible to the actual geological conditions is an effective way to reduce the multiplicity of the logging constrained inversion result. Starting from this point, the core of strengthening the guiding role of geological understanding lies in the establishment of the initial model. After years of exploration and development, various oilfields have entered the middle and late stages of exploration and development, various data are extremely abundant, and the understanding of the characteristics of reservoirs is also deep. Therefore, it is necessary to integrate as much existing geological knowledge as possible in the inversion process, and especially the knowledge of reservoir sedimentary facies, so that the inversion result can be closer to the actual geological conditions. Therefore, first collect and sort out previous research results, understand the sedimentary background and development characteristics of the sedimentary system in the research area macroscopically, and achieve a preliminary understanding of the sedimentary facies types and distribution in the research area; then establish the sedimentary microfacies model of the area through the use of core data, logging data, and seismic data of the area. On the one hand, establish a logging facies model and divide sedimentary microfacies of single wells; on the other hand, the planar distribution law of sedimentary microfacies is obtained through the attribute analysis of seismic data. Finally, on the basis of a certain understanding of the macroscopic distribution of sedimentary microfacies, fine interpretation is performed, and a horizon framework that can reflect the macroscopic distribution of sedimentary microfacies is established. The horizon framework is used to control the initial model, and thus the initial wave impedance model with facies meaning is established. The advantage of this initial model is that it is closer to the actual geological conditions and can reflect the distribution of sedimentary facies in the target interval from a macroscopic perspective. The model derived solely from well interpolation lacks control over facies boundaries, and the distribution of inter-well reservoirs is entirely the result of interpolation. Without the intervention of geological knowledge, the obtained inversion results often lack regularity. Through the establishment of facies control model, the interference of non-reservoir areas can be eliminated, the distribution pattern of reservoirs can be better revealed, and the distribution range of reservoirs can be predicted.

Section 3 Seismic Frequency Division Technology

Seismic waves are composite waves of signals of different frequencies. Stratum, lithology and noise have different impacts on the components of each frequency band in seismic signals, that is, there is frequency selectivity. Seismic waves have obvious frequency band characteristics, and any improvement in frequency components contributes to the improvement of resolution. The basic idea of the frequency division technology is that the seismic reflections generated by geological bodies with different thicknesses correspond to different discrete frequency components in the frequency domain, thick layers correspond to low frequencies, and thin layers correspond to high frequencies. The key to frequency division processing is to extract geological bodies with different frequency components separately. The complete geological characteristics of geological bodies can be obtained through the combination of different frequencies. The seismic frequency division technology is suitable for the processing and interpretation of 3D seismic data. Its resolution is higher than the resolution capability which can be achieved by the conventional seismic main frequency technology, and it has some advantages in determining reservoir boundaries and predicting reservoirs.

Ⅰ. Frequency division interpretation technology principle

The frequency division interpretation technology uses short-time discrete Fourier transform to transform seismic data to frequency domain. An actual seismic wave is often a composite response of multiple thin layers underground. However, the complex harmonic reflection generated by the layer group composed of these thin layers is unique in the frequency domain. The seismic reflection generated by each thin layer has a specific frequency component corresponding to it in the frequency domain, so the frequency component can indicate the time thickness of the thin layer. The frequency division interpretation technology mainly generates two types of data volumes: tuning volume and discrete frequency energy volume. The length of the time window is very important to the frequency response of amplitude spectrum. In general, the seismic wave $s(t)$ is regarded as the sum of the convolution of the wavelet $w(t)$ with the reflection coefficient sequence $R(t)$ and the noise sequence $n(t)$ (Figure 7-2):

$$s(t) = w(t) * R(t) + n(t)$$

Short-time Fourier transform (STDFT) is a time-frequency analysis method with a fixed time window:

$$H(f,\tau) = \int_{-\infty}^{+\infty} h(t)g(t-\tau)\exp(-2\pi ift)\,dt$$

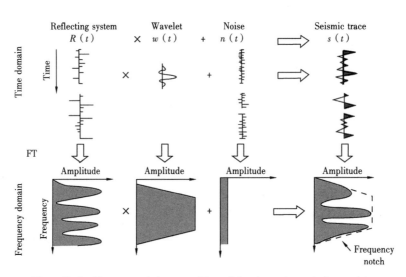

Figure 7-2 The spectral decomposition of the short-time window and its relationship with the convolution model

$g(t)$ is the window function. The allowable condition is $\int_{-\infty}^{+\infty} |g(t)|^2 dt = 1$ The inverse transformation formula is:

$$h(t) = \frac{1}{2\pi}\int_{-\infty}^{+\infty}\int_{-\infty}^{+\infty} H(f,\tau)g(t-\tau)\exp-(2-\pi ift)\,dfdt$$

If the window function is a Gaussian function, the time-frequency resolution is the highest. The Gaussian window function is expressed as $g(t) = \frac{1}{\sqrt{2\pi}} \exp(-\frac{t^2}{2})$, and the standard deviation of the window function $g(t) = \frac{1}{\sqrt{2\pi}\sigma} \exp(-\frac{t^2}{2\sigma^2})$ is taken as 1. The time window of STDFT is fixed and cannot change with the change of frequency, which is not favorable for detecting high and low frequency signals.

The result of frequency division processing is that the amplitude corresponding to each single frequency is tuning amplitude. The time thickness of the formation is derived according to the Rayleigh criterion: the tuning thickness of the formation $\Delta z = \frac{\lambda}{4}$, ($\lambda$ is the wavelength, Δz is the thickness). On the seismic time profile, $\Delta z = vt/2$, where t is the two-way travel time, v is the velocity, and $v = \lambda f$, where f is the frequency. Then the time thickness of the formation $T = 1/(2f)$.

It should be noted that in the interpretation of frequency division, the selection of the time window range of the target layer is very important. The difference in frequency response between large time window amplitude spectrum and small time window amplitude spectrum is huge. The frequency response of large time window amplitude spectrum is similar to that of a wavelet, and can often cause white noise or flattening. The function of the geological body subjected to time-frequency conversion in a small time window is like a filter on the reflected wavelet, and the amplitude spectrum is no longer of white noise. Small time window spectrum decomposition can reduce the geological randomness of the sampled formation. A small time window seismic data is spectrally decomposed to generate a tuning 3D volume in small time window frequency domain. The tuning 3D volume is usually composed of thin-layer interference, wavelet overlapping and noise. Assuming that resonance changes along the flattening layer, by equalizing the wavelet for each frequency slice, the wavelet can be subjected to the smallest noise whitening, thereby eliminating the influence of the wavelet. In the main frequency range, a relatively high SNR generates a clear thin-layer tuning image, and noise interference can be ignored. The tuning of amplitude and frequency can be vividly reflected through the full frequency range. Interpretation personnel observe planar tuning characteristics by analyzing the frequency slices of interest, and recognize the structure and mode of a geological body during its deposition process, thereby predict the lateral variation of the geological body and the longitudinal distribution of heterogeneous anomalies. After determining the tuning 3D volume based on the horizon, multiple discrete frequency energy volumes in the longitudinal direction can be obtained by calculation. Spectral components are decomposed into multiple common frequency component data volumes or common sample point data volumes, and the change and extension of different frequencies in the target volume in different time and space are observed and analyzed through the animation display of characteristic slices of amplitude and phase frequency. The longitudinal and transverse distribution of heterogeneous bodies is objectively predicted based on the comprehensive analysis and calibration of data including actual seismic data, logging data, drilling data, coring data, etc.

II. Frequency division processing and interpretation method

(I) Frequency division denoising

Different noises are distributed in different frequency bands. In view of the characteristics of noise distribution on different frequency bands, different denoising methods are adopted, which can thus achieve a good denoising effect and also effectively protect the effective signals of other frequency bands. The specific denoising method is: the adaptive surface wave attenuation technology is used to suppress surface wave interferences; the frequency division abnormal noise attenuation technology is used to suppress various abnormal amplitude interferences; the monochromatic noise suppression technology is used to suppress single-frequency interferences such as 50Hz power frequency interferences etc.; multiples are suppressed using the Radon transform multiple attenuation technology.

(II) Frequency division energy compensation

Seismic waves propagate in the earth medium, which are a function of time, and frequency and decay with time and frequency. Conventional energy compensation methods include conventional spherical divergence and absorption compensation method, conventional inverse Q filtering method, conventional spectral whitening method, single-trace deconvolution method, and multi-trace deconvolution method. The advantage of frequency division compensation is to solve the problem of compensating frequency attenuation. The low-frequency energy absorption curve is obtained by polynomial E exponential curve fitting, more accurate compensation can be made for spherical divergence and earth absorption, the relative relationship of amplitude can be well maintained, and the waveform consistency of seismic data can be improved.

(III) Frequency division static correction

The static correction error is actually a low-pass filter, which has a great inhibitory effect on high frequencies. If the static correction error exceeds 1/4 cycle, the frequency component after participating in superposition will not be strengthened, but will be weakened, so that the main frequency of the signal is reduced, the bandwidth is reduced, and the resolution of the signal is reduced. The 1/4 cycle of the frequency 10Hz is 25ms, for which the static correction error of 1ms is not a big problem. However, the 1/4 cycle of 80Hz frequency is only 3.125ms, and the error of 4ms destroys the in-phase principle of superposition. Therefore, different frequency components have different accuracy requirements for static correction. The low-frequency correction amount is large, and the high-frequency correction amount is small. The application of the same frequency range cannot take into account both high and low frequencies. Therefore, it is necessary to perform frequency division static correction. Frequency division static correction can solve the problem of different static correction amounts for high, medium and low frequencies, improve the accuracy of velocity analysis, obtain more accurate dynamic correction amount, ensure the homogeneity of superposition, protect high-frequency information, and improve the SNR and resolution of data.

(IV) Frequency division inversion

Frequency division inversion is an inversion method proposed only in recent years. Su Shuchun and Wang Xiaohua were the first to directly use the term "frequency inversion". In addition to the use of "frequency division inversion" in previous studies, similar terms such as "multi-scale inversion" and "multi-resolution inversion" were often used. The purpose of frequency division inversion is to perform a targeted inversion of each frequency band, better highlight the rock formation thickness and lithology information corresponding to each frequency band, avoid the influence of noise in other frequency bands, and further improve inversion accuracy. According to the selectivity of formation, lithology and noise to seismic signal frequency, the different formation thickness and lithology information carried by different frequency components is fully utilized, the influence of noise in other frequency bands is avoided, and targeted inversion research is conducted, which is the starting point of frequency division inversion. Through frequency band division inversion, the reflection coefficient information contained in each frequency band is inverted, and the information is added so as to get the reflection coefficient information of the whole seismic frequency band (the same is true for wave impedance inversion). In fact, each frequency band of seismic signals does contain reflection coefficient information. Frequency band division inversion to obtain the reflection coefficient information contained in each frequency band has practical physical meaning.

Section 4 Development Seismic Technology

Development seismics is the general term for geophysical work carried out in the development stage that has been gradually recognized in recent years. It was also called production geophysics, development and production geophysics, oil and gas reservoir geophysics, or reservoir geophysics previously. The core of development seismics is to solve problems in the development stage of oil and gas fields using seismic data. In addition to typical fine reservoir prediction and structural description, the monitoring of development process is also included. The main development seismic technologies include high-precision 3D seismics, vertical seismic profile (VSP), crosswell seismics, microseismic detection, seismics while drilling, time-lapse seismics, multi-wave multi-component seismics, etc. At present, high-precision 3D seismics is applied widely, and VSP and cross-well seismics are applied in some oilfields, while microseismic detection, seismics while drilling, time-lapse seismics and multi-wave multi-component seismics have not been widely applied.

The development seismic technology is another rapidly developing new technology in the fine description of reservoirs, in addition to geological understanding and modeling methods. In the mid-1970s, with the rapid development of digitization and computer technology and the emergence of interactive interpretation workstations, new geophysical technologies became practical methods for oil and gas field development. In the 1980s, development seismics gradually became an independent subject, monographs on development seismology were published, and seminars were held. Since the 1990s, development seismics has received extensive attention from scholars at home and abroad. China has also begun to vigorously study development seismics and multiple related monographs have been published.

At present, most of the world's major oilfields have entered the middle and late stages of development, and the requirements for the precision of reservoir research are getting higher and higher. Therefore, it can be said that the development seismic technology is a discipline of reservoir geophysics that emerged in response to the needs of production practice. It is a seismic observation, processing, and interpretation technology for specific small–scale reservoirs that require fine structure description based on seismic exploration. That is, on the basis of multi-disciplinary data involving drilling, logging, petrophysics, oilfield geology, reservoir engineering, etc., the target bodies—reservoirs in the process of reservoir development and production are subjected to the fine depiction of geometric structures and physical structures, the geological conditions of reservoirs are predicted transversely, and the changes of reservoirs in the process of oil and gas development are monitored dynamically. The development seismic technology is playing an increasingly important role in solving of fine structural research problems, sequence correlation, reservoir identification, reservoir distribution and parameter calculation, remaining oil distribution, reservoir performance monitoring, etc.

The main research content of development seismics includes time-lapse seismics, crosswell seismics, VSP, etc. at present. (1) The VSP technology is mainly used to interpret near-wellbore fine structures, describe lithology and obtain fracture parameters. This technology can cooperate with the interpretation of seismic data in the well area so as to provide parameters for the description and geological modeling of near-wellbore reservoirs. (2) The crosswell seismic technology can be used to predict the lateral changes of crosswell geological conditions through crosswell seismic imaging (including crosswell tomography and crosswell reflection wave imaging), and provide services for crosswell geological modeling and fine reservoir description. At present, the technology plays a significant role in the thermal recovery of heavy oil (including steam flooding and fire flooding, etc.). It is mainly used to monitor the front of steam flooding (fire flooding) and determine the connectivity, fractures, and faults of reservoirs, and plays an important role in finding remaining oil in old oilfields. (3) The time-lapse seismic technology is also called 4D seismic technology and time-delay seismic technology. The basic principle is: carry out a 3D observation every certain time interval, and perform cross-equalization processing of 3D data volumes observed at different times, so that those reflection waves that are not related to reservoirs have repeatability, while retaining the difference between reflection waves related to reservoirs. By subtracting from the initial basic observation data volume, the changes of reservoirs over time are determined. Time-lapse seismics plays a special role in finding remaining oil in old oilfields, reservoir performance monitoring, fine reservoir description, etc., and has currently received widespread attention from oil companies in the world.

At present, the development seismic technology has been greatly improved from the perspective of exploration, but from the perspective of development, it is still immature and there are many problems that cannot be solved with this technology. For example, the research on reservoir prediction with the development seismic technology basically only serves the stage of reservoir development evaluation, there has been very little research on targeted reservoir prediction with the development seismic technology for remaining oil potential tapping in the late development stage of old areas, and the prediction of thin alternate reservoirs has not yet been achieved very well. Moreover, the current seismic prediction accuracy of reservoirs often cannot meet the needs of deployment of development adjustment wells. Therefore, the development seismic technology is still in the initial stage of its development at present. From the perspective of input-output benefit ratio, except that 3D seismic fine survey method and pre-stack depth migration imaging method are successful, others are still being tried and studied.

Chapter 8 Description of the Internal Architecture of Genetic Units

The description of the internal architecture of genetic units is a key point in fine reservoir description. With the deepening of oilfield development, the key concern of fine reservoir description is the internal heterogeneity of reservoirs. Various interlayers and barriers and seepage differences result in the formation of main enrichment areas for remaining oil distribution in areas which cannot be swept by injected water; therefore, the key to research is to identify genetic unit sand bodies, depict their internal structure features, and describe the features of the smallest level of heterogeneity as much as possible.

At present, the main method used to depict the internal structure of sand bodies is reservoir architecture analysis. The architecture refers to the geometry, size, direction and mutual configuration relationship of the internal architecture units of reservoirs, which is also called structural unit, architectural structure, architecture unit, etc. in geological terms. Therefore, reservoir architecture analysis is also called reservoir architectural structure analysis and hierarchical structure analysis. The reservoir architecture analysis method originated from the study of fluvial sedimentary bodies. Miall (1985) put forward the reservoir architecture analysis method of fluvial facies for the first time at the 3rd International Conference on Fluvial Sedimentology. In 1985, he published the article "Analysis of Architecture Elements—A New Method of Fluvial Facies Analysis", which comprehensively introduces the concepts of architecture elements and architecture interfaces etc. in the method. This represents the birth of the theory of reservoir architecture analysis method. In the early stage when Miall proposed the concept of river sand body architecture unit and the division of internal hierarchical interfaces, the importance of the reservoir architecture analysis method was not fully understood, or it was considered too cumbersome. With the continuous deepening of oilfield development and the requirements of producers for EOR, researchers have gradually found that the heterogeneity of sand bodies has caused a large amount of movable oil to be retained in sand bodies without being recovered. A full understanding of the complexity of these internal architectures will help increase initial production and also the success rate of enhanced production projects (Miall, 1990). This economically significant issue has strongly promoted the development of this field. And it is required to finely divide the different levels of architecture units, interfaces and thin interlayer types in sand bodies so as to study the distribution and change of porosity and permeability in sand bodies. At present, in addition to the leading research on river sand bodies, such studies have also been carried out on tidal sand bodies, submarine fan channel sand bodies, fan deltas and braided rivers etc. (Miall, 1987; Mu Longxin, Jia Ailin, 1997), and the division scheme of research level for clastic sedimentary architecture units has been proposed. This chapter focuses on the theoretical methods and applications of reservoir architecture analysis.

Section 1   Theoretical Methods of Reservoir Architecture Analysis

Ⅰ. Architecture interface and architecture unit

The architecture method emphasizes the grading system of a research object, and depicts sedimentary bodies hierarchically. Therefore, a hierarchical interface and the sedimentary unit defined by the interface must be established, that is, architecture interface and architecture unit. The architecture interface is a series of sedimentary interfaces that divide sedimentary bodies, which can be the changing surface of lithology, grain size, and sedimentary structure etc. Usually there are different sedimentary characteristics or argillaceous barriers and interlayers are developed above and below the interface, reflecting the beginning of a new phase of sedimentary unit. The architecture interface is hierarchical, has different development scales and time scales, and can be developed regionally throughout a basin, representing the beginning and end of a geological time, e.g. the boundary surface of a channel group or an ancient valley group. The architecture interface can also be developed locally, representing the beginning and end of a sedimentary event, such as the sedimentary interface of a point bar or a diara, or even the interface of as small as a laminaset.

The concept of architecture unit is actually about what is the most basic element in the sedimentary environment. The definition of architecture unit cannot be too small. If a very fine sedimentary unit (such as a single cross bedding) is used as a architecture unit, it will increase the difficulty of architecture description, it will also be very difficult to find regularity in terms of deposition and reservoir properties, and architecture analysis cannot be favorably carried out in many cases due to data restrictions. On the contrary, the architecture unit cannot be too rough. Otherwise, it will obscure the understanding of reservoir heterogeneity characteristics that affect the distribution of remaining oil. At present, scholars from various countries have basically reached the same view on this issue; that is, microfacies units are taken as the most basic sedimentary architecture units, and then based on this, they are divided and studied level by level. They mainly correspond to the sedimentary units defined by levels 3-7 architecture interfaces. In fact, there is a relatively uniform standard for dividing the architecture units of all sedimentary environments at present. Taking Miall's research as an example, the division of the architecture units of a fluvial system is introduced below.

Miall divided a fluvial sedimentary system into 8 basic architecture units. Individual lithofacies can be identified in the field and in cores (Table 8-1) using Miall's (1977, 1978) lithofacies diagrams. The 8 basic architecture units are composed of these lithofacies aggregates (Table 8-2). The scale of these architecture units varies. The depth of river channels ranges from a few decimeters to tens of meters, and the length of a large river core sandbar complex can reach several kilometers.

Table 8-1 Lithofacies division

Lithofacies code	Lithofacies	Sedimentary structure	Interpretation
Gms	Massive, matrix-supported gravel	Graded bedding	Mudflow deposit
Gm	Massive or original bedding gravel	Horizontal bedding, imbricate structure	Longitudinal sand bar, lag deposit, sieve deposit
Gt	Layered gravel	Trough cross bedding	Small channel filling
Gp	Layered gravel	Tabular cross-bedding	Longitudinal sand bar delta
St	Medium to coarse sand containing medium gravel	Single or groups of trough cross-beddings	Sand dune (low flow regime)
Sp	Medium to coarse sand containing medium gravel	Single or groups of tabular cross-beddings	Tongue-shaped transverse sand bar, sand wave (low flow regime)
Sr	Fine to coarse gravel	Wave mark	Waviness
Sh	Fine to coarse sand containing medium gravel	Horizontal lamellation or parting lineation	Planar laminar flow (upper flow regime)
Sl	Fine to coarse sand containing medium gravel	Low angle ($<10°$) cross bedding	Scouring-filling, scoured sand dune, retrograde sand dune (sand wave)
Se	Erosion and scouring with internal debris	Primary cross bedding	Scouring-filling
Ss	Fine to coarse sand containing medium gravel	Wide and shallow scouring	Scouring-filling
Fl	Sand, silty sand, clay	Fine lamina	Floodplain or concave slope flood deposit
FSc	Silty sand, clay	Lamina-shaped to massive	Floodplain swamp deposit
Fcf	Clay	Massive, sandwiched with freshwater mollusks	Floodplain swamp deposit
Fm	Clay, silty sand	Massive, mud crack	Floodplain or drape deposit
C	Coal, calcareous clay	Plants, clay film	Swamp deposit
P	Carbonate rock	Edaphogenic	Ancient soil

Table 8-2 Architecture units in fluvial deposits

Architecture unit	Symbol	Main lithofacies composition	Geometry and interrelationship
Channel	CH	Random combination	Finger-shaped, lenticular or mat-shaped; upper concave eroded basement; large changes in scale and shape; general existence of the second internal erosion surface
Gravel bar and bed form	GB	Gm, Gp, Gt	Lens-like, blanket-like; usually plate-like body; sandwiched with SB

Continued

Architecture unit	Symbol	Main lithofacies composition	Geometry and interrelationship
Sediment gravity flow	SG	Gm, Gms	Tongue-shaped, mat-shaped, usually sandwiched with GB
Sand bed form	SB	St, Sp, Sh, Si	Lens-shaped, mat-shaped, blanket-shaped, wedge-shaped; existing in river channels, crevasse splay, sand bar top, small sand bar
		Sr, Se, Ss	
Downstream aggradational large bed form	DA	St, Sp, Sh, Si	Lenticle located on a flat basement or a channel basement; sandwiched with a convex third-level interface inside and in the top
		Sr, Se, Ss	
Laterally aggradational deposit	LA	St, Sp, Sh, SI, Sr、	Wedge-shaped, mat-shaped, tongue-shaped, with the characteristics of internal lateral accretion surfaces
		Se, Ss, G and rare F	
Lamina sand sheet	LS	Sh, SI, a small quantity of St, Sp, Sr	Mat-shaped, blanket-shaped
Overbank fine-grained deposit	OF	Fm, Fl	Thin to thick blanket-shaped; generally sandwiched with SB; possibly filled with abandoned channel deposit

(1) Channel (CH). Separated by a flat or concave erosion surface, there are many such channels in a river system. Large channels often contain complex fillings, which are composed of one or more other architecture unit types.

(2) Gravel bar and bed form (GB). Tabular or cross bedding gravels form a simple longitudinal sand bar or transverse sand bar.

(3) Sediment gravity flow deposit (SG). It mainly refers to the gravel deposit formed by mudflow and lithofacies Gms (Table 8-1) is the main lithofacies.

(4) Sand bed form (SB). The lithofacies formed by low flow regime bed form include St, Sp, Sh, Si, Sr, Se and Ss (Table 8-1), and their mutual combinations form a series of architecture units with different geometric shapes. The architecture unit that best reflects SB is flat sheet sand, which is located at river bottom, sand bar top or crevasse splay.

(5) Downstream aggradational large bed form (DA). It is similar to SB, but such architecture unit (DA) has the characteristic that the inside and top interface is convex upward. The various components of DA are interrelated under hydrodynamic conditions, indicating that the paleocurrent direction is parallel or sub-parallel to the tilt direction of the interface. It can be seen that such architecture unit represents complex sand bar deposits developed by downstream accretion. From the topographic relief of the top interface, it can be inferred that the minimum water depth is only a few meters.

(6) Laterally aggradational deposit (LA). Such deposits are a collection of many DAs. The included angle between the paleocurrent direction indicated by the bed form and the dip direction of the internal accretion is large, indicating that the architecture unit is developed through lateral accretion, which is the well-known point bar. There is a transition type between the LA and DA

architecture unit types, especially in rivers with multiple channels.

(7) Lamina sand sheet. It is a collection of many SBs. The lithofacies mainly include Sh and Sl. This combination indicates that the upper water flow regime is flat bed form, usually a temporary river.

(8) Overbank fine-grained deposit (OF). It is composed of mudstones, siltstones and a small amount of sandstones formed in flood plains and abandoned channel environments. Palaeosol, coal, pond deposits and evaporites are also very important components.

There is currently no unified standard for the grading system. The habit of geological researchers is to follow the division scheme of erathem, system, series, formation and member in the division of large strata. In order to help the stratigraphic mapping of reservoir units, the strata below member level are usually divided into member, sub-member, belt, flow unit or lithological unit. These units shall be tracked as extensively as possible. This stratigraphic subdivision is helpful for description and simulation of reservoirs.

Since the concepts of architecture interface and architecture unit came into being, there have been many opinions. Allen (1966) pointed out that by using a set of strata contact surfaces with hierarchical sequences, the interior of a sand body can be divided into genetically connected strata. Fluid fields in environments such as rivers and deltas can be divided into levels. The grading system he proposed is used to help explain the changes in the paleocurrent data collected in different areas ranging from a single layer to a large outcrop or group of outcrops. The grading system includes 5 levels, namely, small wave mark, large wave mark, sand dune, channel, and a "comprehensive system" representing the sum of the above 4 isopolar changes. Miall (1974) added the scale of the entire fluvial sedimentary system to this concept and collected some data to illustrate the correctness of this concept. Jackson (1975) pointed out that bed form can be divided into three levels according to its time scale and actual scale: micro-bed form refers to a small waviness structure, with its time scale ranging from a few seconds to a few hours; medium bed form refers to large-scale deposits (decimeter to meter level), many of which are mainly formed during what Jackson calls dynamic events. Dynamic events such as hurricanes, seasonal floods, spring tides or eolian sandstorms transport a large amount of sediments disproportionately within a geological instant at that time, and essentially the sedimentary system still remains unchanged between dynamic events. Examples of medium bed form are: underwater dunes and sand waves formed in various environments, tongue-shaped sand bars in rivers, transverse sand bars, longitudinal gravel bars, etc. Such medium bed form is at least an order of magnitude larger in volume than the micro-bed form, indicating the importance of water flow during the formation process. A giant bed form refers to the long-term accumulation of sediments caused by main structures, landform and climate, and is generally composed of superimposed micro-bed form and medium bed form sediments. Examples include river point bars, aeolian arm-shaped rhythm layers, tidal deltas, continental shelf sand ridges, etc. The height of the giant bed form in a channel is almost the same as the depth of the channel, and its length is about the same as the channel width. It is at least an order of magnitude larger than the medium bed form, and can have a complex 3D geometry. Giant bed form represents the deposition and erosion in hundreds to thousands of years. These concepts, complementary to architecture interface and architecture unit, together constitute the grading system of a sedimentary system.

II. Architecture interface grading

The fluvial system grading system is the most complete for architecture interface grading and has been promoted and applied. The establishment of the fluvial sedimentary interface grading system was firstly from the research by Allen (1983) on the Welsh Devonian brown sandstone. The research is the most definite attempt to establish the concept of a grading system for dividing fluvial sedimentary interfaces. Allen described three-level interfaces: Level 1 interface is the layer system interface defined by McKee and Weir (1953); Level 2 interface is a combination of various sedimentary unit boundaries delineated by Level 1 interface. They can be equivalent to the layer system interface defined by McKee and Weir, but do not include the combination of sedimentary units composed of more than one lithofacies; Level 3 interface is equivalent to the main interface defined by Bridge and Diemer (1983).

There are two sedimentary system grading schemes, namely, the scheme where the number sequence is consistent with the level and the scheme where the number sequence is opposite to the level. The following is an example of Miall's grading system.

(I) Level 1 interface

Level 1 and 2 interfaces represent the interfaces within micro-bed form and medium bed form sediments. The definition of Level 1 interface is the same as that by Allen (1983). They are equivalent to the interfaces of staggered layer systems (Figure 8-1). There is no or little internal

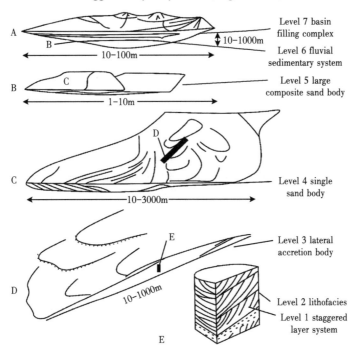

Figure 8-1 Interface grading system of the fluvial sedimentary system (according to Miall, 1988)

erosion on these interfaces, and they represent a series of actual continuous deposits of similar sand bed forms. The subtle changes in the layer system and a small amount of erosion may be caused by activation when the water level changes (Collinson, 1970), or may be the result of changes in the direction of sand bed forms (Haszeldine, 1983). These interfaces may not be very remarkable on cores, but the existence of activated surfaces can be identified by the erosion and pinch-out of the staggered forest beds above the layer system bottom surface (Table 8-3, Table 8-4, Figure 8-1, Figure 8-2).

Table 8-3 Comparison of scale grading system between fluvial sedimentary system and turbidity sediment system (according to Miall, 1990)

Turbidity sediment system (Mutti and Normark, 1987)		Fluvial sedimentary system (Miall, 1987)		Formation level	Time range, year	Bed form level (Jackson, 1975)
Level	Example	Level	Example			
1	Basin filling fan complex	Undetermined	Basin filling scouring (or river) complex	Group, supergroup	$10^6 - 10^7$	
2	Turbidity sediment system, fan	6	Fluvial sedimentary system, fan	Formation and member	$10^5 - 10^6$	
3	Fan, channel—natural levee complex	5	Main channel	Tongue-shaped body	$10^4 - 10^6$	Giant bed form
4	Single channel filling	4	Point bar, side bar	Layer	$10^2 - 10^3$	Giant bed form
5	Lithofacies and bedding pattern erosion and filling (Composition equivalent to the scale of levels 1 to 3 in the fluvial sedimentary system)	3	Giant bed form growth		$10^0 - 10^1$	Medium bed form
		2	Layer system group similar to lithofacies		$10^{-2} - 10^{-1}$	Medium bed form
		1	Lithofacies unit		$10^{-5} - 10^{-3}$	Small bed form

Table 8-4 Scale range of sedimentary units in alluvial sand bodies (taking the Westwater Canyon Member of Morrison Formation near Gallup, New Mexico as an example) (according to Miall, 1988)

Interface level	Lateral extension of sedimentary unit	Sedimentary unit thickness, m	Sedimentary unit area, ha	Genesis	Underground mapping method
7	200km×200km	0-30	$4×10^7$	Member or sub-member, controlled by concealed structures	Comparison of regional electrical logging curves
6	1km×10km	10-20	10^4	Sheet sand body of channel genesis	Comparison of regional electrical logging curves in an oilfield, 3D seismics

Continued

Interface level	Lateral extension of sedimentary unit	Sedimentary unit thickness, m	Sedimentary unit area, ha	Genesis	Underground mapping method
5	0.25km×10km	10-20	2500	Banded channel sand body	Unless the well spacing is very small, mapping is difficult; 3D seismics
4	200m×200m	3-10	40	Giant bed form unit (lateral accretion, downstream accretion)	The dip angle of Level 3 and 4 interfaces can be identified on cores
3	100m×100m	3-10	10	Activation of giant bed form	The dip angle of Level 3 and 4 interfaces can be identified on cores
2	100m×100m	5	10	Layer system group similar to lithofacies of staggered layers	Core lithofacies analysis
1	100m×100m	2	10	Single staggered layer system	Core lithofacies analysis

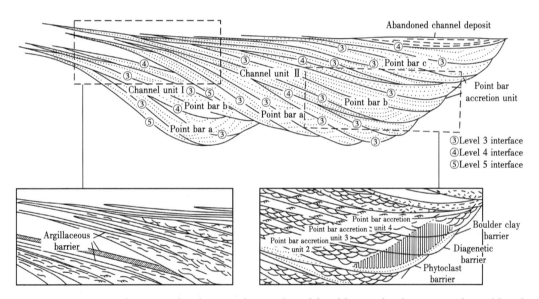

Figure 8-2 Corresponding internal architecture framework model and barrier distribution map obtained based on the study of Mesozoic meandering channel sand bodies in the eastern margin of Ordos Basin
(according to Jiao Yangquan, 1995)

(Ⅱ) Level 2 interface

According to the definition by McKee and Weir (1953), Level 2 interface is a simple layer system group interface. It defines the boundaries of the micro-bed form group or medium bed form group, and symbolizes the change of flow conditions or flow direction, but has no obvious time gap (Figure 8-1). The lithofacies above and below the interface are different, but the interface usually has no obvious signs of bedding surface truncation or other signs of erosion, and has only the small changes that appear in the above-mentioned Level 1 interface. Such interface can be distinguished from Level 1 interface by the change of lithofacies on cores.

(Ⅲ) Level 3 interface

When architecture reconstruction shows that there are two types of giant bed forms including lateral accretion and downstream accretion, Level 3 and 4 interfaces can be determined (Figure 8-1). A single sedimentary unit (layer or architecture unit) is bounded by an interface of Level 4 or higher level.

Level 3 interface is a cross-cut erosion surface in the giant bed form, has a small dip angle (generally less than 15°), and is dominated by underlying cross - beddings with angular truncation. It can cut through more than one interlaced layer system, and is usually covered with internal debris boulder clay. The lithofacies combination above and below the interface is similar.

Level 3 interface can be developed at the top of a small sand bar or sand bed form sequence, and is covered with mudstones or siltstones, indicating the minimum flow rate. Subsequent strata usually have a layer of internal debris boulder clay at the bottom, which is composed of clastics of covered fine-grained sediments. These features are easily identified on cores.

These interfaces represent water level changes or changes in the sand bed form direction in the giant bed form (Figure 8-1), and they symbolize large-scale "activity" or "accretion". On cores, these interfaces can be identified by their gentle inclination. The area of the unit defined by Level 3 interface is generally less than $0.1 km^2$. If there is a core, when the well spacing is very small, it is theoretically possible to compare Levels 3 and 4 underground interfaces.

(Ⅳ) Level 4 interface

Level 4 interface represents the upper interface of the giant bed form, and is generally flat to upward convex. The truncation of the underlying bedding surface and Level 1 to 3 interfaces has a low angle or may be partially parallel to the upper interface, indicating that they are lateral or downstream aggradational interfaces. The shape of such interface is often reflected by Level 3 interface of the underlying giant bed form unit. The unit below the interface is often covered with an argillaceous caprock. It may be appropriate to determine the second type of Level 4 interface, that is, the bottom scouring surface of a small channel. For example, in the case of a large rapid river channel, a high interface is the boundary. The best clue to the existence of Level 3 and 4 interfaces underground is their low sedimentary dip angle, which shall be identifiable on cores, and may also be obvious on the formation dip log. On small outcrops or a single core, it may be very difficult to distinguish between Level 3 and 4 interfaces. The low dip angle and drape breccia or mudstone lenticles are similar. The best way to distinguish these interfaces is that if the lithofacies combination above and below them is different, it indicates the change in the type of units (giant bed form).

Levels 2, 3 and 4 interfaces in this classification are all included in Level 2 classification by Allen (1983). Level 3 and 4 interfaces are equivalent to the "secondary" interface classified by Bridge and Diemer (1983).

(Ⅴ) Level 5 interface

Level 5 interface refers to the interface of large conglomerates such as channel filling complex etc. Generally, it is flat to slightly concave upwards, but it can be characterized by local erosion-filling topography and retained gravel layer at the bottom.

(Ⅵ) Level 6 interface

Level 6 interface is the boundary surface that defines a channel group or paleo-valley group, and the mappable stratigraphic unit such as member or sub-member is bounded by Level 6 interface (Figure 8-1).

The interfaces that may be easiest to identify and compare downhole are Level 5 and 6 interfaces, which is due to their very wide lateral extension and basically simple, flat or slightly curved channel-like geometry. The performance of Level 4, 5, 6 and 3 interfaces can be very similar in cores. They can be distinguished by carefully comparing cores that are very close together. In a fully developed oil and gas field, the well spacing may be only a few hundred meters or less, and this goal is easy to achieve.

The identification and comparison of various interfaces is obviously helpful to explain the complexity of fluvial sedimentary systems, and may be particularly useful in identifying and proving giant bed forms. There are still many aspects to study on this issue.

Even on the very good outcrops, the correct classification of interfaces is not always easy. The following three principles can make this task easier. (1) The interface of any level can be truncated by the interface of the same level or higher level. (2) Since an interface usually records an erosion event, it is often more logical to determine the interface based on the characteristics after the erosion event rather than the previous ones. (3) The level of a low level interface can be changed transversely.

In the division of interfaces, both positive and negative sequences are allowed, but the key is to correctly determine the formation mechanism and significance of the geological bodies between different levels of interfaces.

Section 2 Reservoir Architecture Research Steps and Examples

Sedimentary facies analysis is indispensable in the application of reservoir architecture research methods. Sedimentary facies research and reservoir architecture research are complementary to each other. They have both similarities and some differences. Generally, the research methods of sedimentary facies include core facies marker identification, seismic facies identification, logging facies identification, sedimentary facies combination analysis, sedimentary facies evolution, etc., which are also required for reservoir architecture analysis. Especially large-scale reservoir architecture analysis is consistent with sedimentary facies analysis. In terms of characterizing the internal characteristics of small-scale sedimentary bodies, reservoir architecture analysis has its own unique methods and means, which is equivalent to the depiction of barriers and interlayers of reservoirs and a single reservoir unit partitioned by barriers and interlayers. For reservoir architecture, it is to characterize the internal structure of reservoirs through the identification of reservoir interfaces and the division of reservoir architecture units.

The idea of reservoir architecture is to describe the characteristics of reservoirs hierarchically, and the architecture of a level has a restrictive effect on the next level. Therefore, the idea of reservoir architecture research is also level – by – level refining and description. For different research objects, the research methods for reservoir architecture analysis are different. The research ideas and methods for field outcrops are basically consistent with those for modern sediments, while the research ideas and methods for underground geological bodies are different from those for field outcrops.

I. Reservoir architecture analysis steps

(I) Architecture research on field outcrops and modern sediments

The research on field outcrops and modern sediments is an important basis for architecture research and for obtaining prototype models of different geological body architecture styles. Similar to the general outcrop description, the research on field outcrops involves outcrop survey and selection, fine measurement, sampling analysis, and system anatomy, and can be divided into the following steps.

1. Outcrop selection

There are two purposes for selecting outcrops: one is to select representative outcrops in order to anatomize a certain type of geological bodies; the other is to select outcrops around a basin with equivalent horizons to study the underground conditions. In either case, try to choose complete outcrops that have good traffic conditions, are convenient for observation and have rich deposition phenomena.

2. Interface identification and division

Based on changes in lithology, structure and texture, interface levels are established and independent sedimentary units are divided.

3. Divide lithofacies and identify architecture units

Through the identification of lithofacies, sedimentary sequences are established, architecture units are identified, and their characteristics are described.

4. Measure the scale of architecture units

Measure the length, width, thickness, dip direction, dip angle, distribution frequency, etc. of architecture units.

5. Sample analysis

Systematically sample key architecture units and analyze the change law of their physical properties.

6. Comprehensive establishment of architecture models

Comprehensively consider the sedimentary characteristics, physical property distribution and combination law of architecture units, and establish architecture models.

(II) Architecture research in dense well pattern areas

Generally, reservoir architecture analysis includes five steps: determination of sedimentary system type, establishment of fine isochronous stratigraphic framework, combined microfacies analysis, single microfacies (single sand body) identification, and internal architecture anatomy of genetic units.

1. Determination of sedimentary system type

Different types of sedimentary systems have big differences in their development laws and reservoir characteristics. Determining the sedimentary system type is the key to architecture analysis. Therefore, it is necessary to ascertain the sedimentation type of the research object at first. For example, the development process of meandering rivers has the characteristics such as lateral migration and straightening of bends, thus forming a combination of point bars and abandoned channels. There are multiple lateral accretion bodies developed inside a point bar, which are partitioned by lateral accretion layers so as to form semi-connected bodies. Braided rivers are characterized by the development of diaras. Under the alternate control of flood period and low water period, near-horizontal silt layers are developed in diaras. It can be seen from this that it is vital to determine the sedimentation type.

2. Establishment of fine isochronous stratigraphic framework

The sedimentary units defined by the isochronous stratigraphic framework have genetic links. Therefore, on the basis of isochronous stratigraphic division, the genetic interpretation of single sand bodies can be performed under the guidance of the theory of sedimentology, which also provides a basis for correlation of single sand bodies. For outcrop profiles, isochronous interfaces are identified and laterally tracked, isochronous geological bodies are compared, and architecture units are identified.

3. Combined microfacies analysis

Generally, a completed sedimentary microfacies map reflects the distribution law of combined microfacies, reveals the spatial distribution trend of single sand bodies, and can provide constraints for the further division of single sand bodies.

4. Identification of single microfacies (single sand bodies)

In combined microfacies, a single microfacies is divided and a single sand body is identified through the application of multiple methods.

5. Anatomy of the internal architecture of genetic units

The identification of the interlayers in a single sand body is the core of its anatomy; therefore, determining the genetic type, development mode, and identification mark of the main interlayers is the focus of the internal architecture anatomy of a single sand body.

II. Example of the analysis of reservoir architecture in dense well pattern areas

Taking Xingerzhong area of Daqing oilfield as an example, the anatomy method of fluvial reservoir architecture is described (Figure 8-3). The anatomy object is the main pay formation—Pu-1 Member in Daqing Oilfield, which mainly develops distributary channel sand bodies. Despite years of waterflooding development, it is still the main enrichment layer of remaining oil in Daqing Oilfield. Therefore, understanding such sand body structure is the key to remaining oil prediction.

The main method adopted is: guided by the theory of reservoir architecture analysis, a single sand body and its internal structural characteristics are described on the basis of the study of

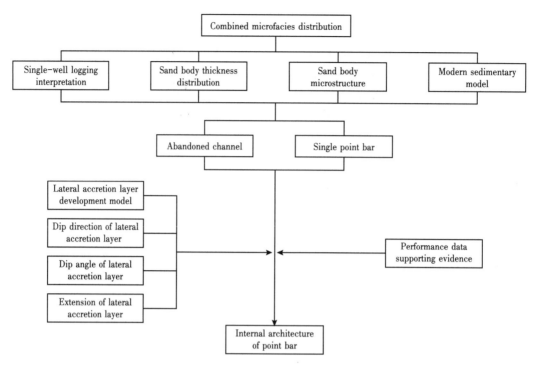

Figure 8-3 Typical channel sand body anatomy method

composite microfacies according to the idea of hierarchical anatomy. First, according to the sedimentary model of a meandering river, the point bar sand bodies that constitute the channel deposits are identified, and a single channel is divided into multiple point bar bodies; then, according to the sedimentary characteristics of the point bar, the internal structure of the point bar sand bodies is anatomized, and the distribution model of the lateral accretion bodies and lateral accretion layers inside the point bar sand bodies is established; finally, the model is verified based on the performance data.

(Ⅰ) **Analysis of reservoir architecture elements**

Delta plain subfacies is the continuation of a fluvial sedimentary system, and its sedimentary microfacies division still uses the terms of fluvial sedimentary microfacies in many aspects. The delta plains in this region mainly develop meandering river deposits. The architecture elements mainly include channel, point bar, abandoned channel, and overbank deposits, which are equivalent to the architecture units defined by Level 4 and 3 interfaces classified by Miall. Architecture elements are closely related to the microfacies units. Chapter 7 has described some microfacies and their combinations. The following focuses on the deeper levels of architecture elements that reflect the internal structural characteristics of microfacies combinations.

1. Channel sand body deposits

Channel sand bodies are the uppermost sedimentary units in a river, have large thickness, and are characterized by a typical positive rhythm sequence where the grain size becomes small upwards vertically and the sedimentary scale becomes small upwards. A classic meandering river is characterized by a single curved channel. The channel has a large depth and a small width-depth ratio, the two sides of the channel are asymmetrical, the concave bank is eroded and the convex bank is subjected to lateral accretion, forming sand bodies distributed widely areally. Generally, the bottom of the channel sand bodies is a scouring surface, boulder clay deposits are occasionally visible above the scouring surface, and upwards, medium and fine sandstones gradually become fine sandstones, siltstones and pure mudstones (overbank deposits), showing an obvious dualistic structure. From bottom to top, there are parallel beddings, trough cross beddings, climbing beddings, and ripple beddings. The lithology of the top is mudstones with horizontal beddings.

Supper aqueous distributary channel sand bodies are extremely developed in Pu-1 Member in Xingerzhong area, which are mainly high-curved meandering river deposits. The grain size is relatively small, and there is no medium sandstone developed. The lithology is mainly fine sandstones and siltstones, and boulder clay is occasionally visible in the bottom. The relative amplitude values of SP and GR are large. The shapes of the logging curves include bell shape, box shape, and a combination of bell shape and box shape, where box shape predominates. The amplitude difference of the microlog curve is large. Based on the classic sedimentary model of point bars, architecture elements such as point bars, lateral accretion surfaces, lateral accretion layers, and lateral accretion bodies etc. can be further identified (Figure 8-4).

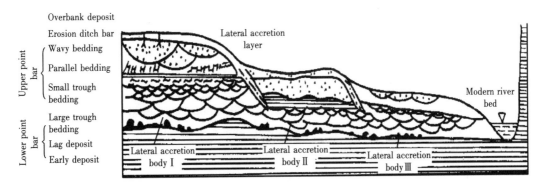

Figure 8-4 The concept of "deposition pattern" of point bars (according to Xue Peihua, 1991)

1) Point bar

Point bars are the main body of meandering river reservoirs. Meandering river point bars are formed by lateral accretion of rivers. Sediments are affected by spiral flow at meandering bends, causing unbalanced sediment transport and scouring and erosion on concave banks; then convex banks accept deposits and point bars are formed (Figure 8-5).

Point bar sand bodies are sand-rich belts of meandering rivers, and are often composed of several lateral accretion bodies, between which argillaceous interlayers are developed. Due to the

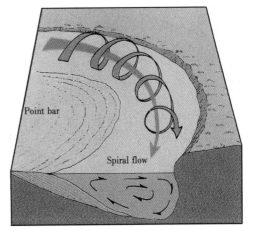

Figure 8-5 Development pattern of spiral flow and point bar (according to W K Hamblin, 1998)

different flood energy, the preservation of interlayers is different. The existence of these interlayers makes point bar sand bodies appear as a slight return on logging curves. The average thickness of point bar sand bodies in the anatomy area is 4-5m, and there are often 1-2 interlayers. The SP and GR curves are generally box-shaped, and the microlog curve has a large difference in amplitude.

2) Lateral accretion surface

The erosion surface existing between lateral accretion bodies is a special scouring and erosion surface formed by lateral erosion (non-tectonic action). Sedimentary compensation is made on this erosion in the late stage, so it can be called lateral accretion surface. The lateral accretion surface is the lateral accretion isochronous surface of point bar sedimentary bodies, and is characterized by undulations and being inclined to the migration direction of channels. The dip angle varies greatly, and is generally 0°-30°. The upper and lower horizons of the lateral accretion surface are different in sedimentary structure, lithology, and occurrence.

3) Lateral accretion layer

The lateral accretion layer is the most important element among the three elements of a point bar sand body, and refers to the argillaceous layer deposited on the lateral accretion surface. The lithology is mainly fine-grained sediments such as clayey, silty or organic silt. The lateral accretion layer shows obliquely inserted mud wedge shape on profile, and arc shape on plane in most cases. The thickness varies greatly, ranging from centimeter level to meter level. The dip angle corresponds to the lateral accretion surface.

4) Lateral accretion body

The lateral accretion body is the main body of a point bar sand body, the product of the periodic flooding of rivers, the isochronous unit in the point bar sand body and the basic sedimentary formation unit of the point bar sand body. From the beginning of flood peaks to their retreat, in the whole process of hydrodynamic conditions of a river from strong to weak, the river is subjected to lateral accretion so as to form a lateral accretion body. The end sign is usually the appearance of a lateral accretion layer, and the inside of the lateral accretion body has positive rhythm characteristics. It is crescent-shaped on plane and shows regular imbricate shape in space. A point bar is formed from superimposition and combination of some lateral accretion bodies, and is controlled by the difference of hydrodynamic actions each time.

2. Abandoned channel deposit

In the course of evolution, when a whole channel or a certain section of channel loses its function as a route for surface water circulation, the original active channel becomes an abandoned channel, forming an oxbow lake on the landform (Figure 8-6). Abandoned channel is an important sign to identify the boundary of a single channel,

Figure 8-6 Image map of the wetland of Swan Oxbow of the Yangtze River

and is also an important basis for judging the lateral accretion direction of point bar lateral accretion bodies. The location of an abandoned channel on plane is generally adjacent to a channel and is in the convex bank margin of a single meandering belt sand body. An abandoned channel can be formed by sudden abandonment and gradual abandonment; therefore, there are two types of filling in the upper half of the abandoned channel on the profile, namely, filling by mud or alternate deposition of sand and mud. The bottom horizon of the abandoned channel in a single well is equivalent to that of channel sands, and the top horizon of the abandoned channel is lower than the top horizon of channel sands. The logging curves of a sudden abandonment type abandoned channel show that the bottom of the SP and GR curves is finger-shaped or bell-shaped, and their upper part is close to the baseline. The bottom logging curves of a gradual abandonment type abandoned channel are basically the same as those of the sudden abandonment type abandoned channel, and the filling part of the upper abandoned channel is toothed, reflecting the alternate filling of sand and mud. The identification marks of abandoned channels will be described in more detail below.

(II) Identification of point bar sand bodies

Point bars are the most important component of meandering river sand bodies. Especially for high-curvature meandering rivers, point bar sand bodies are usually developed continuously. The maximum identification of single point bar sand bodies is the focus of the description of single sand bodies of meandering rivers.

1. Identification method

Meandering river point bars are formed by the action of spiral flow on sediments carried by a river. One side of the channel is gradually scoured and eroded so as to form a concave bank; the other side continuously accepts sediments and accumulates laterally so as to form a convex bank, i. e. point bar deposits are formed. Meanwhile, with the development of river point bars, the curvature of the channel continues to increase until its realignment, so that the originally deposited channel loses surface water circulation capacity and becomes an abandoned channel. Therefore, according to the laws of nature, abandoned channels and point bars always coexist. Areally, an abandoned channel is generally located on the convex edge of a single point bar sand body, which is very obvious in modern sedimentary observations (Figure 8-7). The coexistence relationship between point bars and abandoned channels shows that the identification of a single point bar body is inseparable from that of an abandoned channel. This also determines the inevitable method for single point bar identification: based on well data, single-well interpretation is firstly performed for point bars and abandoned channels; then abandoned channels are reasonably combined comprehensively using production performance data and point bar development models as well as the method of model fitting, so as to divide a single point bar sedimentary body.

2. Identification mark

Point bar identification marks shall take account of both the sedimentary characteristics of a point bar itself, and the development characteristics of an abandoned channel. The specific identification marks can be considered from the following aspects.

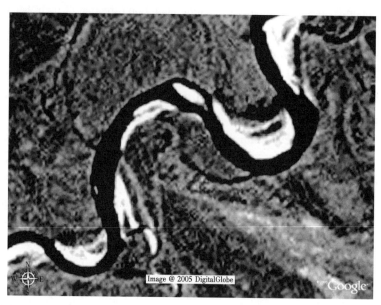

Figure 8-7 Relationship between abandoned channel and point bar development (quoted from Google Earth)

1) Point bar sand bodies

The most important feature of point bar deposits is that they have a positive rhythm longitudinally. The grain size of deposits gradually decreases upwards, and the scale of sedimentary structures also decreases upwards, reflecting the sedimentary process of gradual weakening of water flow energy, finally evolving into natural levee or overbank fine-grained material deposits, forming a typical "dualistic structure". Usually this kind of point-bar vertical sedimentary sequence is incomplete, and the relatively coarse-grained sedimentary materials in the lower part of the point bar are the best preserved. In the case of different river energy, the grain size and bedding type of point bar deposits are different. The lithology of point bar bodies in the area is dominated by fine sandstone and siltstone deposits, boulder clay is occasionally developed in the bottom, trough beddings, wavy beddings, and parallel beddings are visible. Due to the high oil-bearing level, the beddings are usually not clear.

2) Shape of logging curves of point bar sand bodies

The physical properties of point bar sand bodies are good, the relative amplitude difference of their logging curves is large, and their GR and SP curves are smooth and full, and show bell shape, box shape and a combination of box-bell shape (Figure 8-8). Lateral accretion bodies are developed in point bars. A point bar in a single well is composed of several lateral accretion bodies vertically. Argillaceous interlayers (lateral accretion layers) with oblique bedding surfaces are developed between lateral accretion bodies, showing a certain return on logging curves. Especially the microlog curve characteristics are the most obvious. For each single lateral accretion body developed in a point bar, the SP and GR curves are mostly full box-shaped, and have slightly positive rhythm characteristics.

3) Thickness distribution characteristics of point bar sand bodies

The formation of a point bar is an obvious process of "concave erosion and convex accretion",

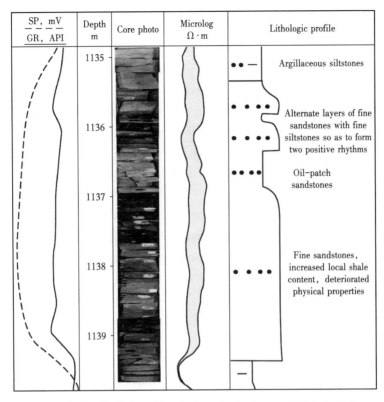

Figure 8-8   Vertical positive rhythm of point bar (well X2-1-J29)

which is formed from accretion of several lateral accretion bodies in sequence until the channel is abandoned. In general, a point bar body is generally lenticular, and the thickness of the point bar sand body is relatively small on the side close to the abandoned channel. However, among the various microfacies sand bodies of meandering rivers, the point bar sand body, as the main body of channel sand bodies, has the largest thickness, while the thickness of the overbank microfacies sand bodies such as natural levee crevasse splay, etc. is much smaller. The statistical results of the area show that the thickness of point bar sand bodies is 4-6m, with an average of about 5m, while the thickness of overflow sand bodies is generally less than 2m. Therefore, point bar sand bodies can be classified according to the difference in their thickness. In addition, compared with overbank sand bodies, channel point bar sand bodies belong to high-energy deposits with good physical properties and high oil content.

4) Distribution characteristics of abandoned channels

From the perspective of sedimentology, the genesis of an abandoned channel is closely related to the formation of a point bar in a meandering river. The appearance of an abandoned channel in a meander belt represents the end of the development of a point bar, and the development of the last abandoned channel represents a diversion of the river. Therefore, point bars are always distributed next to abandoned channels, and abandoned channels are the boundaries of point bars. The abandonment process of a channel is extremely complicated, and is completely accidental in many regions; for example, timber blockage, flooding, human intervention, etc. may lead to channel abandonment, which thus leads to certain difficulties in the identification of abandoned channels.

(1) Logging facies model of abandoned channels.

The identification of abandoned channels in a dense well pattern area of an oilfield is a prediction in itself, because under the current technical conditions, the distribution of abandoned channels cannot be determined directly according to well data. In order to conduct a detailed study of abandoned channels, it is necessary to establish the identification model of abandoned channels. Therefore, on the basis of the interpretation of the genetic environment of abandoned channels and the calibration of the litho-electric relation ship between coring wells, the logging curve identification model of abandoned channels is established. There are two cases such as gradual abandonment type and sudden abandonment type (Figure 8-9). ① gradual abandonment type. After a channel is abandoned, it is in a static water environment, but it is connected to the upper part of the original channel, and often accepts some suspended fine-grained material deposits from the channel during the flood period; therefore, some alternate layer deposits of sandstones with mudstones will appear in the top of the channel. On the logging curves, the lithology of the lower part of the channel is coarse-grained deposits; the microelectrode curve has a large amplitude difference, showing well-developed channel sand body deposits; the upper part is zigzag-shaped, showing alternate layer deposits of sandstones with mudstones. ② Sudden abandonment type. After a channel is abandoned, it is almost in a completely closed still water environment, is isolated from the original channel, and does not accept any deposits from the original channel. On the logging curves, the bottom of the abandoned channel has an equivalent horizon to the adjacent channel, and is thin, corresponding to the bottom deposits of the adjacent channel; in addition, the upper part is mudstone deposits. Due to the different locations of abandoned channels

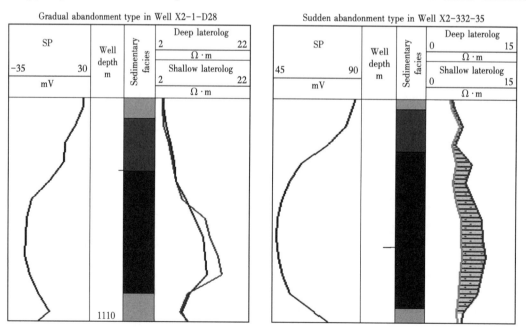

Figure 8-9 Logging curve characteristics of abandoned channels

encountered in drilled wells, the characteristics revealed by the curves are also different. Some channels may be completely abandoned, where sandstones are basically underdeveloped; some channels are partially abandoned, and there are some alternate layer deposits of sandstones with mudstones in the upper part; some channels are rarely abandoned, similar to channel deposits. Therefore, it is necessary to comprehensively distinguish channels in planar combinations.

(2) Planar combination method of abandoned channels.

According to the different hydrodynamic conditions of rivers, the degree of preservation of abandoned channels is also different. For small and medium-sized distributary channel sand bodies with low curvature, the lateral cutting of different single channels is relatively weak, the preservation degree of abandoned channels is relatively high, and abandoned channels are continuously distributed and can be tracked between wells. For large high-curved distributary channels, the hydrodynamic force is relatively strong, and the lateral tangential overlapping of different single channels is relatively strong. Therefore, abandoned channels are preserved incompletely, most of them are distributed intermittently, and it is difficult to track and compare the original appearance of abandoned channels.

The lithology of abandoned channels is dominated by mudstones and silty mudstones. For a wide-band-shaped meandering river sand body composed of multiple meander reaches, an abandoned channel can be distributed inside the channel sand body or in its margin. When thick fine-grained deposits appear in the channel sand body, it generally indicates the existence of an abandoned channel. However, in many single wells, abandoned channel deposits are distributed above point bar sand bodies. It is very difficult to make a distinction between the fine-grained deposits of abandoned channels and those of the floodplains on logging curves; therefore, it is generally very difficult to directly determine abandoned channel deposits in a single well. Yue Dali (2006) predicted the distribution of abandoned channels based on the relative depth of sand tops. The fine-grained deposits of an abandoned channel are thickened from a point bar to the abandoned channel on the cross-section of the abandoned channel. When the top surface of a single layer is flattened, the relative thickness (referred to as the relative depth of the sand stone top) between the top surface of the point bar sand body and the single layer surfaces tends to increase (Figure 8-10). Similarly, the relative depth of the sand top is relatively large in the deposition part of the abandoned channel on plane, while the thickness of the sand body is relatively small.

Generally speaking, a channel may be abandoned many times. Different single abandoned channels will cross and merge with each other on plane. Moreover, some abandoned channels are preserved intact, and some abandoned channels are preserved incompletely, which makes the planar combination and distribution of abandoned channels complicated. Therefore, the planar combination of abandoned channels shall be comprehensively identified according to multiple types of information such as changes in the logging curve shape of all wells, the flow direction of rivers, the superimposition direction of sand bodies, the horizon change of sand bodies, etc. The basic principle is as follows: the overall direction of an abandoned channel is consistent with the strike direction of a river, the overall shape is taken as "S" shape, and the planar prediction of the abandoned channel is performed according to the principle of nearby combination; the affiliation and bending direction of different abandoned channels that are close to each other are determined according to the information on sand body horizons, the development situation of inter-river sand bodies, the evolution trend of sand body thickness, etc.; multiple types of abandoned channel forms are combined according to different abandonment modes such as channel neck cutting, trough cutting, breach and diversion, etc.

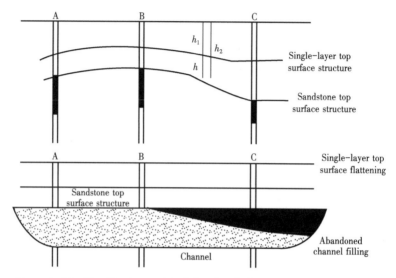

Figure 8-10 Identification model of abandoned channels based on the relative depth of sand top (according to Yue Dali, 2006)

3. Division of point bars

Taking the P13 single layer in the bottom of Pu-1 Member with the most developed channel sand bodies in this area as an example, the single point bars in it are divided. The focus of single point bar division is to determine the boundary of the point bar, and the boundary of the point bar in a composite channel is the junction of the point bar with the abandoned channel. Therefore, the division of point bars inside a channel sand body and the identification of abandoned channels are carried out simultaneously. The identification and division are performed under the guidance of modern river sedimentation models according to the information on single well facies analysis, point bar sand body thickness, relative depth of sand top, etc.

Firstly, pre-interpretation of point bars and abandoned channels for each single well are performed according to the logging response of genetic units. Point bars show bell shape or box shape on logging curves. Due to the presence of lateral accretion layers, the curves will be of return phenomena. Single wells are interpreted based on these characteristics and the interpretation results are shown on the plan view (Figure 8-11). As shown in Figure 8-11, the interpretation of point bars is generally not a big problem, but the interpretation of abandoned channels has a certain degree of multiplicity.

Then, the possible distribution locations of point bars and abandoned channels are analyzed according to the thickness distribution of the sand body and the relative depth of the sand top

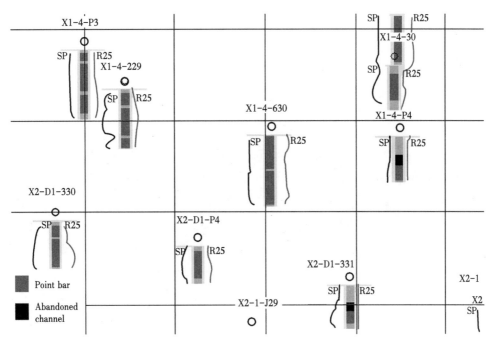

Figure 8−11　Interpretation of P13 single-layer point bar and abandoned channel (part displayed)

(relative depth map of the sand stone top = structure map minus micro-structure map of the sand body top surface). According to the sand body thickness contour map (Figure 8−12), the thickness of the P13 single-layer sand body is not large, but it is distributed continuously. It is generally divided into east and west blocks, and there are several thin sandstone strips inside each block. For example, it is most obvious in wells X1-4-D27—X2-1-P1—X2-2-P1; according to this, the continuous sand body is divided into two high thickness value areas such as left and right ones. The top structure map of P13 (Figure 8−13) and the top microstructure map of the sandstone of this layer (Figure 8−14) are worked out. Subtract the two to obtain the relative depth map of the sand top (Figure 8−15) (the red indicates a small relative depth of the sand top, and a small mudstone thickness; the blue indicates a large relative depth of the sand top, a large mudstone thickness, and high possibility of an abandoned channel). It can be seen from Figure 8−12 that the relative depth of the sand top in wells X1-4-D27—X2-1-P1—X2-2-P1 is large, while the relative depth of the sand top on both sides is small, and there is no obvious continuous sand-top thick mudstone distributed inside the two parts. According to comprehensive analyses, it can be deemed that abandoned channels are distributed along wells X1-4-D27—X2-1-P1—2-2-P1. In the same way, the distribution of abandoned channels inside the sand body can also be identified so as to obtain the distribution map of abandoned channels in the P13 single layer (Figure 8−16). As shown in Figure 8−17, the high-curved abandoned channel is distributed in the composite channel sand body, which is "divided" into multiple point bar sand bodies. Compared with this modern sedimentary phenomenon, the division of the P13 small layer is comparable, so it shows that the division of this layer is reasonable.

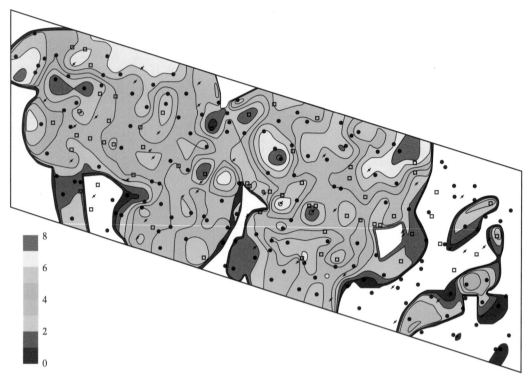

Figure 8-12 Thickness contour map of sand bodies of P13 single layer

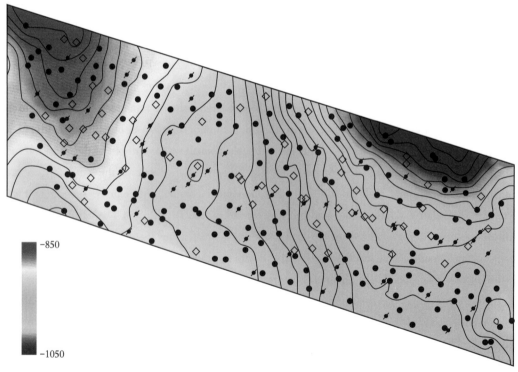

Figure 8-13 Top structure map of P13 single layer

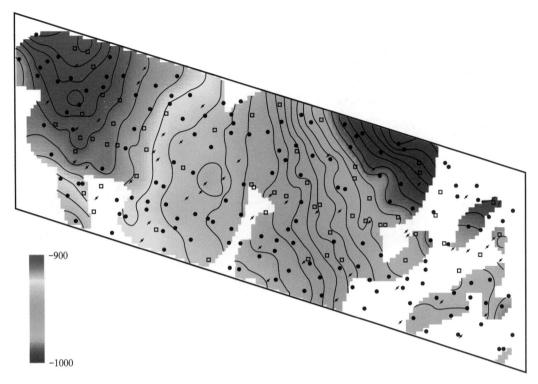

Figure 8-14　Top microstructure map of sand bodies of P13 single layer

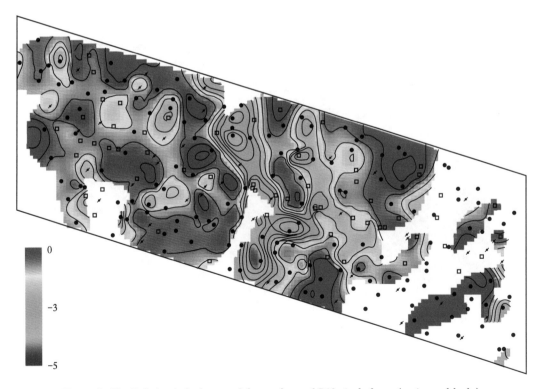

Figure 8-15　Relative isobath map of the sand top of P13 single layer (main sand body)

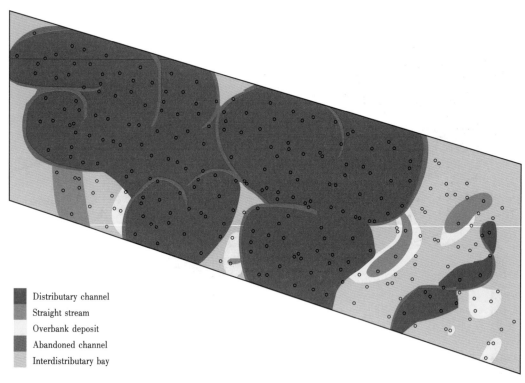

Figure 8-16　Distribution map of point bars of P13 single layer

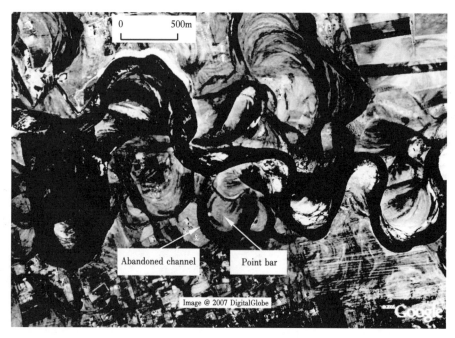

Figure 8-17　Distribution map of abandoned channel deposits in the meander reach of the Hailar River (quoted from Google Earth)

(Ⅲ) Fine anatomy of the interior of point bars

The internal structural features of a point bar can be further anatomized on the basis of the identification of a single point bar according to the idea of gradual anatomy and hierarchical analysis. The spatial distribution model of lateral accretion layers and lateral accretion bodies is established through model fitting taking single-well interpretation data as the control points and the lateral accretion model of meandering river point bars as the guide according to the analysis results of production performance data.

1. Identification of lateral accretion layers in a single well

For a point bar sand body, lateral accretion layers are the main type of interlayers inside the sand body, are developed most extensively, and have the most complex distribution form and occurrence. Therefore, the understanding of the distribution pattern of lateral accretion layers is the key to finely anatomizing the internal structure of a point bar sand body. There are the cores of only one inspection well in this area. According to the description results of the cores of Pu-1 Member, the lithology of the lateral accretion layers in the area mainly includes mudstones, silty mudstones, and argillaceous siltstones, and the thickness of the lateral accretion layers is generally 0.2-0.3m. The microlog curve returns obviously, and its amplitude difference is reduced. The GR and SP curves also return, and the return of the SP curve of some interlayers is not obvious (Figure 8-18). Based on these characteristics, the lateral accretion layers in single wells in the area are interpreted.

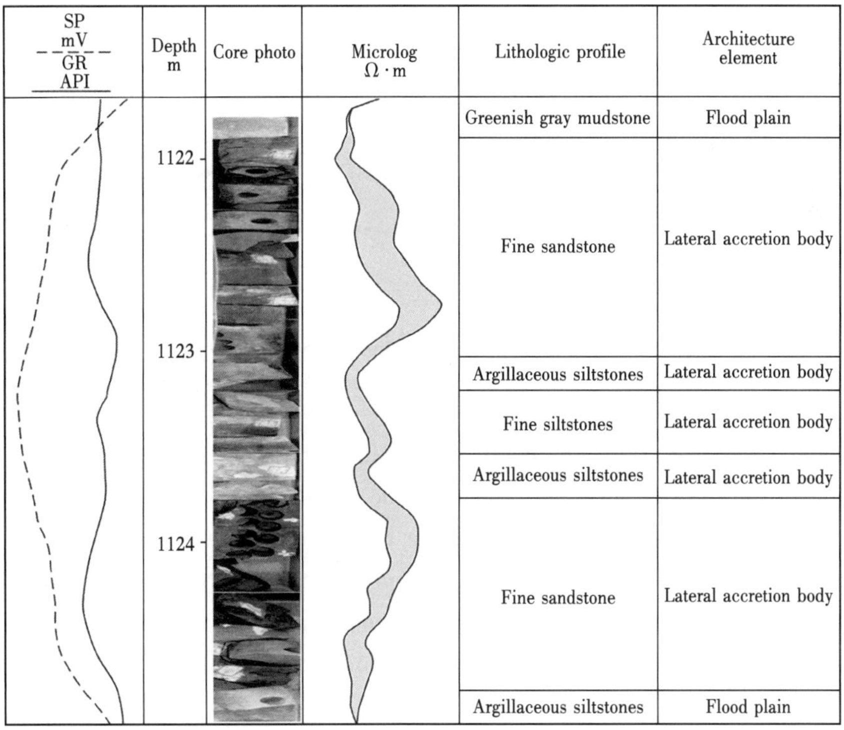

SP mV / GR API	Depth m	Core photo	Microlog Ω·m	Lithologic profile	Architecture element
				Greenish gray mudstone	Flood plain
	1122			Fine sandstone	Lateral accretion body
	1123			Argillaceous siltstones	Lateral accretion body
				Fine siltstones	Lateral accretion body
				Argillaceous siltstones	Lateral accretion body
	1124			Fine sandstone	Lateral accretion body
				Argillaceous siltstones	Flood plain

Figure 8-18 Distribution characteristics of the interlayers in the coring well

2. Correlation of lateral accretion layers

The key to the study of the internal architecture of a point bar is the correlation of lateral accretion layers. The thickness of lateral accumulation layers is small, their distribution continuity is poor, and most of them are developed obliquely, so their spatial correlation is difficult. Reasonable inter-well correlation of lateral accretion layers is the core of understanding the internal structural characteristics of point bars. The current effective method is model fitting based on comprehensive information. Taking the P13 single layer as an example, the correlation principle and method for lateral accretion layers are described below.

1) Correlation principle

(1) Inclined correlation model of lateral accretion layers.

The sedimentary model of point bars was studied by scholars as early as the 1970s, and it has basically been used today, and confirmed by modern sedimentation and simulation experiments. In recent years, there are a lot of literature on the lateral accretion bodies and lateral accretion layers inside point bars. Many scholars at home and abroad have established a variety of distribution patterns of lateral accretion layers based on outcrops and modern deposits (Figure 8-19). These sedimentary models show that lateral accretion layers are developed obliquely, the dip angles of lateral accretion layers are not the same in different regions, and their correlation models are different to some extent; however, in general, the principle of inclined correlation must be followed in the correlation process.

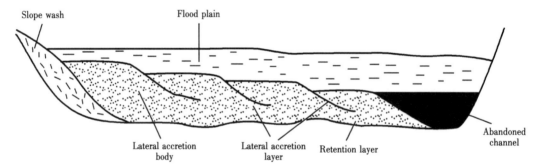

Figure 8-19 Development model of the lateral accretion layers inside the meandering river point bar
(according to Xue Peihua, 1991)

(2) Lateral accretion layers are inclined to the direction of an abandoned channel.

According to the lateral accretion process of the point bar (Figure 8-20), the channel is gradually eroded and migrated toward the concave bank, and the lateral accretion bodies on the convex bank are developed in imbricate shape, forming the main body of the point bar. The lateral accretion layers are the fine-grained sedimentary compensation on the lateral accretion surfaces after floods, so the lateral accretion layers tend to follow the lateral accretion body direction, that is, towards the abandoned channel direction.

(3) Dip angle of lateral accretion layers.

According to previous studies of modern deposits and outcrops, the dip angle of lateral accretion layers inside a point bar is generally 5°-30°. The research conducted by Jiang Xiangyun

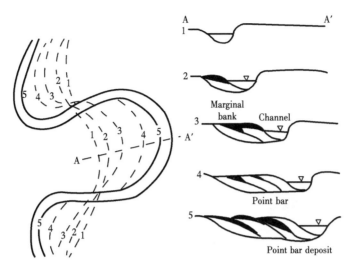

Figure 8-20 Schematic diagram of the lateral accretion process of the meandering river point bar
(according to Yin Yanyi et al., 1998)

(2007) on the single point bar of Ng522 single layer in the west of the seventh block of Gudao Oilfield indicates that the horizontal width of the lateral accretion body is up to 200m, and the thickness of a single complete normal cycle sand body is up to 13.3m, and the corresponding lateral accretion layer has a small dip angle and is only 3°–5°. The thickness of the single positive cycle sand body in this area is 3–7m, with an average of about 5m. It is inferred that the dip angle of the lateral accretion layers may be smaller. According to the comparison of wells with small spacing, the distribution range of the dip angle of the lateral accretion layers in the area can be obtained.

There are many close wells with a well spacing of about 50m in this area, and most of the lateral accretion layers between these close wells are corresponding. Therefore, suitable small spacing wells can be selected for the correlation of lateral accretion layers; moreover, according to the difference in elevation of the lateral accretion layer between two wells and the well spacing, the dip angle of the lateral accretion layer is calculated through a trigonometric function. The following principles shall be followed in the selection of small spacing wells: (1) try to choose a well pair perpendicular to the abandoned channel direction; (2) the lateral accretion layers developed in the wells are obvious, the number of the lateral accretion layers is as small as possible, and their contrast is strong; (3) the selected lateral accretion layer is as close as possible from the marker horizon. As shown in Figure 8-21, Well X1-40-629 and Well X1-40-P4 are two close wells, with a well spacing d of 58m. The curves of the two wells have similar shape, and they have a nearly parallel lateral accretion layer. Because the lateral accretion layer is inclined, the elevation of the same lateral accretion layer is different in the two wells. According to the distance between the lateral accretion layer and the marker horizon, the lateral accretion layer elevation difference Δh between the two wells can be obtained as 1.83m. According to the trigonometric function formula $\tan\theta = \Delta h/d$ (θ is the dip angle of the lateral accretion layer, (°); Δh is the elevation difference of the lateral accretion layer between the two wells, m; d is the well spacing, m), the dip angle of the lateral accretion layer can be calculated to be 1.8°. By analogy, calculations from multiple small spacing data of the area show that the dip angle of the lateral accretion layers in this area is about 2.5°.

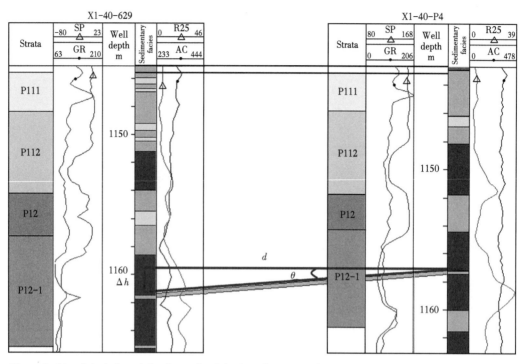

Figure 8-21 Calculation of the dip angle of the lateral accretion layer using the data of small spacing wells

(4) The extension width of lateral accretion layers.

The extension range of a lateral accretion layer can be reversely derived according to its dip angle. Using the meandering river parameter relation summarized by Leeder in 1973, the width of the lateral accretion layer is calculated:

$$W_2 = 2W/3$$

$$\tan\beta = h/W_2$$

Where: W_2—The width of a single lateral accretion body, which is equivalent to the maximum extension width of a lateral accretion layer, m;

W—The width of an active channel, which refers to the width of the active river in the riverbed, m;

h—The depth of the active channel, which is equivalent to the thickness of the positive rhythm sand body, m;

β—The dip angle of the lateral accretion layer, which is calculated according to the above fomrula, (°).

The average thickness of the channel sand bodies in this area is about 5m, and the dip angle of the lateral accretion layers is about 2.5°, so the extension width of the lateral accretion layers can be calculated to be about 120m. According to the comparison of the wells, it is basically consistent with the estimated width. The thickness of the channel sand bodies in this area is small, which reflects that the depth of the channels is not large. Therefore, the lateral accretion bodies developed in the area are flat and the dip angle of the lateral accretion layers is relatively small.

2) Correlation result

According to the previous correlation principles, the internal lateral accretion layers of the point bars in the area have been correlated so as to finely anatomize the internal structural characteristics of the point bars. In addition, the established correlation result has been verified according to the production performance data of the area and the tracer monitoring data of the ASP flooding test well area.

This area develops high-curvature meandering river deposits with frequent migration and swing, sand bodies are developed widely, and there are many point bars. The well-preserved point bar sand body has been selected as an example for anatomy, and the distribution model of the lateral accretion bodies and lateral accretion layer insides point bars in the area has been established (Figure 8-22, Figure 8-23).

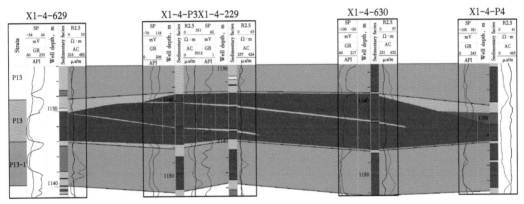

Figure 8-22 Correlation profile of lateral accretion layers

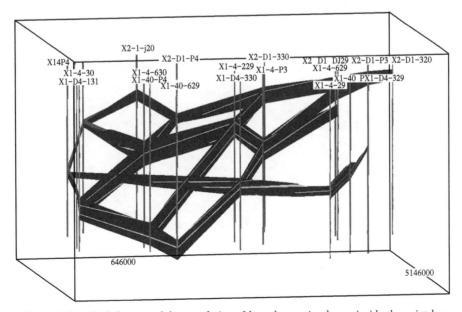

Figure 8-23 Grid diagram of the correlation of lateral accretion layers inside the point bar

(Ⅳ) 3D modeling of reservoir architecture

At present, it is difficult to model a single sand body. Especially for single sand bodies with inclined mudstone interlayers like meandering river point bars, there has always been a lack of effective modeling algorithms. However, with the development of computer technology, the modeling of a single sand body can also be performed using the existing reservoir modeling application software through human-computer interaction.

The purpose of modeling of a single sand body is to describe the distribution of interlayers in the sand body and reveal its internal heterogeneity. Since the interlayers inside the sand body are thin, the modeling grids shall be small enough. The grid size in this modeling is set to 5m×5m×0.1m. Based on the fine anatomy of the internal structure of a single sand body, single-well interpretation is carried out; in addition, different lateral accretion bodies are numbered to ascertain the relationship of development stages. Then according to the scale of occurrence of lateral accretion layers and lateral accretion bodies, the inter-well correlation relationship is established, lateral accretion layers and lateral accretion bodies are correlated, and a 3D model is established.

Figure 8-24 is the hollowed-out display of the lateral accretion body model. (1) In the figure is the hollowed-out display of lateral accretion layers, and the correlation relationship is consistent with the correlation result of the small spacing wells. (2) In the figure is the hollowed-out display of lateral accretion bodies (pink, purple, and light blue represent three-stage lateral accretion bodies), reflecting the lateral superposition pattern of the three-stage lateral accretion bodies from right to left. The dark blue on the far left is abandoned channel deposits. This architecture model of a single sand body can more precisely reflect the 3D spatial structure features of the sand body and reveal the heterogeneous structure in the sand body.

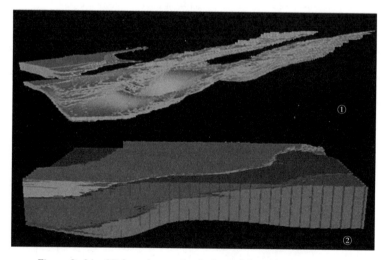

Figure 8-24　3D lateral accretion body model inside the point bar
① Hollowed-out display of lateral accretion layer;　② Hollowed-out display of lateral accretion body

Chapter 9 Description of Reservoir Physical Parameters and Fluids

The description of reservoir physical parameters is mainly a quantitative description of reservoir heterogeneity, focusing on the distribution of permeability. Fluid description focuses on the distribution form and distribution law of remaining oil in the fine reservoir description stage. This chapter is divided into two sections to describe the reservoir heterogeneity description technology and remaining oil description technology.

Section 1 Reservoir Heterogeneity Description Technology

The study of reservoir heterogeneity is the core content of reservoir description and characterization. Oil and gas reservoirs have undergone the combined effects of deposition, diagenesis, and late tectonism in the long geological history, so that there is an obvious difference in the spatial distribution of reservoirs and their internal various geological attributes. This difference is called reservoir heterogeneity.

Reservoir heterogeneity is absolute, unconditional, and infinite, while reservoir homogeneity is relative, conditional, and limited. Reservoirs can be approximately regarded as homogeneous only in a certain heterogeneous layer, under certain conditions, and within a limited range. Therefore, absolute homogeneity does not exist. Of course, the heterogeneity of marine reservoirs is weaker than that of continental reservoirs. The vast majority of oil and gas reservoirs discovered in China come from continental strata, and most of them are developed by water injection. Reservoir heterogeneity will directly affect the movement law of oil, gas, and water in reservoirs, as well as the distribution and development effect of remaining oil and gas.

I. Classification of reservoir heterogeneity

Qiu Yinan (1992) divided the heterogeneity of clastic reservoirs into four types based on the scale of heterogeneity and the practicality of development and production. This classification method has been widely applied.

(1) Interlayer heterogeneity: including the cyclicity of series of strata, the degree of heterogeneity of permeability between sand layers, the distribution of barriers, and the distribution of special types of layers.

(2) Planar heterogeneity: including the degree of connectivity of the genetic units of sand bodies, the changes in planar porosity and permeability, the degree of heterogeneity, and the orientation of permeability.

(3) Intralayer heterogeneity: including granularity rhythmicity, bedding structure sequence,

the degree of permeability difference, the distribution position of high permeability sections, the distribution frequency and size of discontinuous thin argillaceous sandwich layers in a layer, the distribution of other impervious sandwich layers, the ratio of horizontal permeability to vertical permeability in the whole layer, etc.

(4) Pore heterogeneity: it mainly refers to the heterogeneity of microscopic pore structure, including the size and uniformity of pores and throats, and the configuration relationship and connectivity of pores and throats.

II. Reservoir heterogeneity research technology

Based on Qiu Yinan's classification scheme in combination with the reservoir heterogeneity research results obtained by domestic and foreign scholars, this book divides reservoir heterogeneity into two major categories such as macroscopic heterogeneity and microscopic heterogeneity. Macroscopic heterogeneity includes intralayer heterogeneity, planar heterogeneity and interlayer heterogeneity.

(I) Macroscopic heterogeneity

1. Intralayer heterogeneity

Intralayer heterogeneity refers to the vertical change in reservoir properties in a single sand layer. Intralayer heterogeneity is a key geological factor which directly controls and affects the swept volume of injected fluids in the reservoir of a single sand layer vertically. From the perspective of reservoir engineering, the intralayer heterogeneity of reservoirs mainly refers to two aspects: (1) the location of the highest permeability section in the layer and the degree of difference in the permeability between the sections in the layer; (2) the ratio of macroscopic vertical permeability to macroscopic horizontal permeability of a single sand layer. They are important factors for determining fluid channeling. The intralayer heterogeneity exhibited by these two aspects is controlled by many geological features.

1) Granularity rhythmicity

Granularity rhythmicity refers to the vertical change sequence of the size of clastic particles in a single sand layer. Granularity rhythmicity is controlled by depositional environment and depositional mode. Granularity rhythmicity has a very large influence on the vertical change of permeability. The granularity rhythmicity on the profile directly controls the rhythmicity of permeability in reservoirs with small diagenetic changes. Granularity rhythmicity is divided into the following four types.

(1) Positive rhythm: an upward-fining grain sequence with a thick bottom.

(2) Reverse rhythm: an upward-coarsening grain sequence with a thin bottom.

(3) Composite rhythm: a combination of the above two.

(4) Homogeneous rhythm: a sequence with irregular granularity changes vertically.

2) The location of the highest permeability section

It mainly describes that the highest permeability section in a layer is located in the bottom, top or middle. Generally, it is consistent with the above-mentioned granularity types, corresponding to the four types such as positive rhythm, reverse rhythm, composite rhythm, and homogeneous rhythm. Certain microfacies have certain sedimentary sequences. Each sedimentary sequence can always be represented by the vertical combination of several lithofacies. After mastering the sedimentary sequence laws of various microfacies sand bodies, the degree of difference in permeability and the location of the highest permeability section can be identified according to the vertical change sequence of the clastic particle size in a single sand layer, the vertical evolution of sedimentary structure, etc. This requirement can be completely achieved using the technologies currently mastered.

3) Vertical evolution of sedimentary structures and permeability anisotropy

The various beddings in sedimentary structures are composed of differences in the occurrence, arrangement and combination of laminas with different grain sizes. Such differences lead to vertical changes in permeability and also affect the anisotropy of permeability. Through core observation, the following can be described: (1) The lithology, occurrence, combination relationship, distribution law, etc. of various laminas; when possible, the difference in permeability between laminas is measured using a mini-permeameter; (2) The vertical change law of bedding structures; (3) Occurrence of wormholes in argillaceous layers and the nature of their fillers.

In summary, the vertical evolution of sedimentary structures leads to vertical changes in permeability, while the lateral extension and evolution of sedimentary structures lead to the planar orientation of permeability. The anisotropy of permeability is different in different bedding structures.

The permeability anisotropy in parallel beddings is mainly manifested in the difference between horizontal permeability (K_H) and vertical permeability (K_V). Generally, K_H is much larger than K_V, so the K_V/K_H ratio is very small.

The permeability anisotropy in oblique beddings is manifested in the difference of permeability along the dip direction of beddings, in the opposite direction of the dip direction of beddings and in the direction parallel to the strike direction of laminas. The permeability along the dip direction of beddings is the largest, the permeability in the opposite direction of the dip direction of beddings is the smallest, and the permeability in the direction parallel to the strike direction of laminas is between the two.

The permeability anisotropy in cross beddings is the strongest. Weber (1982) proposed a method to calculate the permeability anisotropy in trough cross-beddings (Figure 9-1). The method indicates that the ratio of the permeability $K_{//L}$) in the direction parallel to laminas to that ($K_{\perp L}$) in the direction perpendicular to laminas can reach 3.0 in unconsolidated layers, while the ratio is even larger in consolidated sandstones. According to the research by Emmett *et al.* (1971) on a certain reservoir in Wyoming, $K_{//L}/K_{\perp L}$ can reach 4.0 This difference in permeability has a large impact on fluid seepage.

4) Discontinuous thin interlayers (barriers) in a layer

Discontinuous thin interlayers (barriers) in a layer can play the role of an impermeable barrier or a very low permeability high resistivity layer in fluid flow, and thus have an extremely large impact on the oil displacement process. Discontinuous thin interlayers (barriers) in a layer

Formula

$$\frac{1}{K_\alpha} = \frac{\cos^2\alpha}{K_{/\!/L}} + \frac{\sin^2\alpha}{K_{\perp L}}$$

$$\frac{1}{K_x} = \frac{d}{LK_B} + \frac{1}{K_\alpha}$$

$$\frac{1}{K_y} = \frac{d}{WK_B} + \frac{1}{K_{/\!/L}}$$

$$\frac{K_x}{K_y} = A_H$$

$$K_R = \sqrt{K_x K_y} = \text{Radial permeability to wellbore}$$

$$\frac{H+d}{K_V} = \frac{H}{K(90-\alpha)} + \frac{d}{K_B}$$

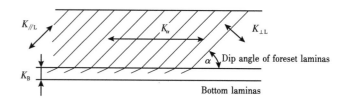
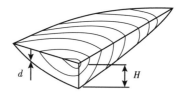

Figure 9-1　The calculation formulas of permeability in different directions in trough cross beddings (according to Weber, 1982)

are also an important factor that directly affects the ratio of the macroscopic vertical permeability to the macroscopic horizontal permeability of a single sand layer from top to bottom, and sometimes may result in direct blockage, form injectant slugs and worsen oil displacement effects.

(1) Interlayer type. Generally classified by lithology, interlayers mainly refer to argillaceous rocks and fine/silty rocks. In addition, interlayers also include various siliceous – cemented, calcareous – cemented, kaolin – cemented strips generated in the diagenesis process, granular sutures caused by strong compaction, and bitumen or heavy oil-filled strips.

(2) The thickness, distribution range and occurrence of various interlayers (as far as possible to establish a relationship with facies belts).

(3) The occurrence frequency and density of interlayers. Interlayer frequency refers to the number of interlayers per unit thickness of the rock formation, expressed in (layers/m). Interlayer density refers to the ratio of the total interlayer thickness in the section to the total thickness of the statistical sandstone section (including interlayers), expressed in percentage (%).

There are often some discontinuous thin argillaceous and silty interlayers and impermeable calcareous sand interlayers in clastic reservoirs. These interlayers (barriers) have an extremely large influence on the vertical or horizontal permeability of a whole sand body. However, the change law of these interlayers (barriers) is often of the order of magnitudes of tens of meters or even several meters. Generally, it is very difficult to control the change law of these interlayers (barriers) using wells in a development well pattern of several hundred meters, and the change law must be predicted based on sedimentary facies analysis. Such interlayers have different scales due to their different geneses inside sand bodies in various depositional environments (Figure 9-2). For example, even if interlayers are extremely thin in the "E" section of each turbidity current event in a turbidity current sand body, they can be distributed very extensively; the width of the abandoned shale-filled interlayers in a channel sand body cannot exceed the width of the ancient channel; there are more and more stable argillaceous interlayers in a delta outer front sand body than in a delta inner front sand body. But it should be said that it is still at a qualitative to semi-quantitative level at present, and needs to be determined according to specific oilfield examples and experience.

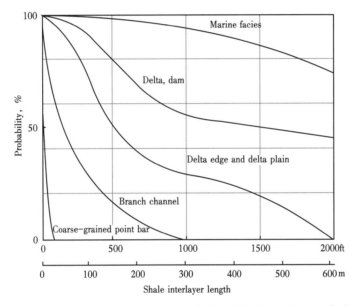

Figure 9-2 Function diagram of the continuity of shale (silty) interlayers with depositional environment changes (according to K. J. Weber)

In view of the current level of prediction of in-layer interlayers, several possibilities can be estimated in the early reservoir evaluation stage: the worst possibility, the best possibility, and the possibility with the highest occurrence probability; moreover, geological models that specifically reflect various possibilities of in-layer interlayers are established, and then a sensitivity analysis is carried out through numerical simulation so as to provide a basis for development strategy decisions.

Under the condition that a development well pattern has been completed, logging interpretation results are used as much as possible to identify in-layer interlayers, and the comparison results of well groups with small well spacing are used as the circumstantial evidence for sedimentary facies analysis and prediction.

5) The ratio of vertical permeability to horizontal permeability

This ratio has a large impact on the water washing effect in the waterflooding development of reservoirs. If the ratio of vertical permeability to horizontal permeability (K_V/K_H) is low, the vertical percolation capacity of fluids is relatively low; on the contrary, it is relatively high.

The K_V/K_H of cores is obtained by laboratory measurement. The K_V/K_H at the bedding scale is obtained by a set of aforementioned equations proposed by Weber (1982) for calculating the permeability anisotropy in cross beddings. According to the paper of S. H. Begg et al. at the SPE 1985 annual symposium, the calculation formula for the influence of in-layer interlayers on the vertical permeability at the sand body scale is as follows:

$$K_V = \frac{1 - F_s}{(1 + fd)(\frac{1}{K_v} + \frac{fd}{K_H})}$$

Where: K_V—Full-layer vertical permeability, mD;

F_s—Interlayer density (that is, the proportion of the thickness of the impermeable interlayers on the section);

f—Frequency of full-layer interlayers (number of interlayers per meter);

K_v—vertical permeability of sandstones, mD;

K_H—horizontal permeability of sandstones, mD;

d—Half of the average interlayer length, m.

The interlayer length can be obtained after sedimentary facies analysis. The horizontal permeability at the whole layer scale is calculated from the statistics of homogeneous sections of the whole layer.

6) Intralayer permeability heterogeneity

The intralayer permeability heterogeneity is usually reflected by some statistical indicators, and is statistically analyzed using core analysis data to the greatest extent possible (calculated using the value of a single sample for a uniform section; calculated by dividing relatively small statistical units of homogeneous sections for a non-uniform section). If core data are not representative, the permeability value ($\geqslant 5$ points/m) from continuous logging interpretation can be used for statistics.

(1) Permeability variation coefficient.

Permeability variation coefficient $K_V = \dfrac{\delta}{\overline{K}}$

$$\delta = \sqrt{\sum_{i=1}^{n=1} (K_i - \overline{K})^2 / n}$$

Where: K_V—permeability variation coefficient;

K_i—The permeability value of a sample in the layer;

$\overline{K}$—Average permeability of all samples in the layer;

n—The number of samples in the layer.

In general, when $K_V < 0.5$, it reflects that the degree of heterogeneity is weak; when $K_V = 0.5-0.7$, it reflects that the degree of heterogeneity is medium; when $K_V > 0.7$, it reflects that the degree of heterogeneity is strong, i.e. serious heterogeneity.

(2) Permeability ratio.

Permeability ratio $J_K = \dfrac{K_{max}}{K_{min}}$

Where: J_K—permeability ratio;

K_{max}—The maximum permeability in the layer, which is generally expressed by the permeability value of the relatively homogeneous section with the highest permeability in the sand layer;

K_{min}—The minimum permeability in the layer, which is generally expressed by the permeability value of the relatively homogeneous section with the lowest permeability in the sand layer.

The closer the J_K value of the reservoir is to 1.0, the more homogeneous the reservoir is.

(3) Permeability rush coefficient.

Permeability rush coefficient $T_K = \dfrac{K_{max}}{\overline{K}}$

Where: T_K—permeability rush coefficient;

K_{max}—maximum permeability in the layer;

$\overline{K}$—average permeability of all samples in the layer.

When $T_K < 2$, the degree of heterogeneity is weak; when T_K is 2-3, the degree of heterogeneity is medium; when $T_K > 3$, the degree of heterogeneity is strong.

2. Planar heterogeneity

Planar heterogeneity refers to the heterogeneity caused by the geometry, size, and continuity of a reservoir sand body, the spatial distribution of parameters such as porosity, permeability, etc. in the sand body, as well as the spatial distribution of porosity and permeability. These factors are directly related to the planar swept degree of injectants.

1) Geometry and all-direction continuity of sand bodies

(1) Geometry of sand bodies.

Sand bodies deposited in various environments generally have their corresponding geometries. The geometry of sand bodies is generally classified using their length-width ratio: the length-width ratio of sheet sand bodies is close to 1:1, and their width-thickness ratio is >1000; the length-width ratio of potato-shaped sand bodies is ≤3:1, and their width-thickness ratio is >100; the length-width ratio of shoestring sand bodies is >20:1, and their width-thickness ratio is >30. Actual classification can be supplemented by geometry naming of genetic significance, such as frond, fan, water channel type, etc.

(2) All-direction continuity of sand bodies.

This is the main content of quantitative description of sand body scale that is directly related to development engineering. General description: all-direction length of sand bodies (m); the degree of control under a certain well pattern (the percentage of the thickness of sand bodies drilled in two wells at the same time); drilled rate, i.e. the percentage of the number of wells where sand bodies are drilled to the total number of wells. In case of highlighting lateral continuity, the width (m) or width-thickness ratio of sand bodies shall be described; the ratio of the actual width of sand bodies to the established well spacing.

2) Sand body connectivity

The composite sand body formed by the mutual contact and connection of sand bodies of various genetic units in the vertical and horizontal directions is called "connected body". Connected bodies further enhance the continuity of reservoirs. This is an important content of studying the planar heterogeneity of reservoirs. According to the connection form of sand bodies, they can be divided into: multilateral sand bodies, which are connected to each other mainly laterally (Figure 9-3a); multilayer sand bodies (or superimposed sand bodies), which are connected to each other mainly vertically (Figure 9-3b); isolated sand bodies, which are not connected with other sand bodies (Figure 9-3c). The description of connected objects usually includes the following aspects.

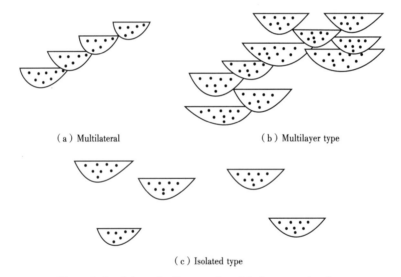

Figure 9-3 Schematic diagram of sand body connection form

(1) The coordination number of a sand body. It refers to the number of sand bodies in contact and connection with a sand body. As shown in Figure 9-4, the coordination number of No. 1 sand body is 1; that of No. 2 sand body is 2; that of No. 3 sand body is 4.

(2) Degree of connectivity. It refers to the percentage of the area of the connected part between a sand body and another sand body in the area of the sand body, or is expressed as the percentage of the number of connected wells to the number of wells controlled by the sand body.

(3) Connected body size. Connected body size refers to how many genetic unit sand bodies are included in a connected body, or the total area or total width of the connected body.

(4) The percolation capacity of sand body contact position. In recent years, with the

deepening of reservoir geology, it has been discovered that sand bodies are in contact and connection with each other, including cutting and scouring type contact connections, which are not necessarily connected channels for fluid flow. This is mainly determined by the percolation capacity of the contact position. Due to the enrichment of boulder clay or calcareous gravels on the overlying scouring surface, or the existence of a mudstone cladding layer, the scouring contact surface between sand bodies may form an impermeable or low-permeability interface. There is no quantitative description method for this yet at present. When the above-mentioned geological phenomena that may damage the connectivity of the sand body contact surface are found in actual work, they shall be verified by interference testing and described using qualitative classification methods. For example, the permeability of Type I contact position is very good; the permeability of Type II contact position is poor; Type III contact position has no percolation capacity, etc.

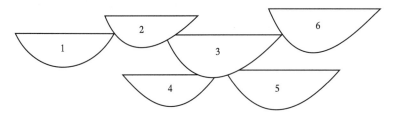

Figure 9-4 Schematic diagram of the coordination number of sand bodies

Many conceptual models have been established for the geometry of sand bodies in various depositional environments. The geometry of two reservoir sand bodies belonging to the same sedimentary facies can be completely similar. However, if their sedimentary scales differ by orders of magnitude, their actual scales can also differ by orders of magnitude. Therefore, in addition to predicting its geometry, it is more important to study the actual sedimentary scale of a reservoir sand body in the research on its continuity.

The transporting agent of various sedimentary systems in continental lake basins is mainly rivers, and lakes have relatively small water body capacity. Therefore, predicting the scale of ancient rivers in each sedimentary system in continental sedimentary basins is a very important link. Moreover, according to the actual conditions of China, the continuity of all sand bodies in delta front belts (including river-delta and Gilbert-type fan delta) is good, and the sand bodies can generally reach the scale of kilometer. When sand bodies reach such a scale, continuity is no longer the main controlling factor for determining the injection-production well pattern for development. On the contrary, the lateral continuity of various river sand bodies and channel-type sand bodies including distributary channel sand bodies in delta plains, underwater distributary channel sand bodies in fan deltas, fan-middle channel sand bodies in sublacustrine fans, etc. tends to be the order of 100 meters or less. In this case, the scale of sand bodies is a key factor for determining the injection-production well pattern. Therefore, predicting the sedimentary scale and continuity of these sand bodies is an important issue to be solved in the analysis of reservoir sedimentary facies.

To solve practical problems encountered in the field of petroleum geology, the scale of sedimentary sand bodies must be studied using drilling data, that is, the 3D distribution of sand bodies must be predicted using the sedimentary phenomena that can be described on the one-dimensional profile. Typical methods are as follows: (1) The empirical formulas established by modern deposition; (2) Measured values or probability values of similar sedimentary outcrops in the area; (3) Empirical formulas and empirical relationships obtained under dense well pattern conditions in similar depositional environments, and it is best to use the empirical values of adjacent development areas in the same basin; (4) For channel-type striped sand bodies, their scale can also be estimated using the method of drilled probability at different well spacing.

After determining the scale and size of the genetic units of various microfacies sand bodies, it is also necessary to analyze the degree and mode of connectivity between genetic unit sand bodies. The sand bodies connected by various means finally constitute units for fluid flow in the oilfield development process. The continuity of connected bodies is a reservoir characteristic that must be considered when making development decisions.

The degree of connectivity between genetic unit sand bodies depends on the deposition rate and transition frequency of sedimentary bodies and the relative magnitude of basin subsidence rate. When the subsidence rate is less than the first two, the degree of connectivity of sand bodies is good; on the contrary, it is poor.

The vertical profile sand body density method proposed by Allen is a practical method for various striped sand bodies. According to the actual data of China's Mesozoic and Cenozoic lake basins, the empirical figures for the critical value of sand body density for predicting the degree of connectivity of channel sand bodies are given, that is, >50% means large-area connection; <30% means isolated sand bodies; for 30%–50%, a specific analysis is required, and there may be locally well-connected sand bodies. When this method is used, the intervals selected as the statistical units must be reasonable, and the intervals with a roughly equivalent degree of sand body development in the same depositional environment shall be selected as the analysis units.

3) Difference in the planar physical properties of sand bodies

The planar changes of reservoir physical properties (mainly permeability) are mainly determined by the distribution of high-energy and low-energy zones during deposition. The planar microfacies distribution map is the basis for reflecting the planar changes of reservoir physical properties. The distribution contour diagram of physical properties under dense well pattern conditions shall be plotted referring to the planar microfacies distribution map in general.

The planar distribution of microfacies is predicted mainly based on vertical sedimentary sequences and microfacies sequences as well as Walter's law and general depositional models in the early evaluation stage, and then the planar heterogeneity of reservoirs is inferred. Of course, the prediction of sand body scale is a prerequisite.

From the perspective of waterflooding development, the purpose of understanding the paleo-flow direction in the analysis of the planar heterogeneity of sand bodies is to understand whether there is directional permeability.

The permeability orientation of river sand bodies and various channel sand bodies shall be consistent with the paleo-flow direction. The permeability orientation of the coastal dam sand bodies strongly transformed by lake waves is perpendicular to the paleo-flow direction. The identification of the regional provenance direction and sediment transport direction can control the paleo-flow direction on a macroscopic scale. The paleo-flow direction of each specific sand body can deviate greatly from the total paleo-flow direction of the sedimentary system. The paleo-flow direction shall be restored within a small scope to the greatest extent possible in the description of reservoirs to be developed.

The basis for judging the paleo-flow direction includes: ① Sand body geometry; ② The dip direction of beddings; when there is no directional coring, the paleo-flow direction shall be inferred based on the dip direction of the strata in cores; ③ Dip logging interpretation.

3. Interlayer heterogeneity

Interlayer heterogeneity is a general description of a set of hydrocarbon-bearing series of strata with alternate sandstones and mudstones, including the regularity of alternate occurrence of sand bodies in various environments on the section, the development and distribution laws of argillaceous barriers, etc. Interlayer heterogeneity is the basis for determining the development series of strata and the technological strategy of separate layer production.

Interlayer heterogeneity is mainly controlled by sedimentary facies. Most sedimentary systems in China's continental lake basins have short processes, narrow facies belts, and fast facies changes, often forming multiple types of sand bodies superimposed into a set of reservoirs, and making interlayer heterogeneity more serious.

1) Sedimentary cyclicity

Sedimentary cyclicity is a manifestation of ordering and superimposition of reservoir sand bodies and non-reservoir interlayers of different genesis and different properties according to a certain rule. In addition, sedimentary cyclicity is the sedimentary origin of reservoir heterogeneity and also the basis for the division and correlation of reservoir groups.

Reservoir description in the development stage is generally aimed at cycles of lower than Level 3. Cycles can be named according to their sedimentary origin, such as transgressive cycles, regressive cycles, etc. In addition, cycles can also be described based on reservoir grain size, changes in physical properties, thickness development, degree, etc.

(1) Positive cycle: It becomes thinner, physical properties become smaller, and thickness becomes smaller from bottom to top.

(2) Reverse cycle: It becomes coarser, physical properties become larger, and thickness becomes larger from bottom to top.

(3) Composite cycle: Different combinations of positive and reverse cycles.

2) Stratification coefficient

The stratification coefficient refers to the number of sand layers in the described series of strata. Due to facies change, the number of sand layers in the same series of strata on plane will be changed. The stratification coefficient is expressed by the average number of sand layers drilled in a single well, that is, the ratio of the total number of sand layers drilled to the number of statistical wells. Generally speaking, the larger the stratification coefficient, the more serious the interlayer heterogeneity.

3) Sandstone density

Sandstone density refers to the ratio of total sandstone thickness on the vertical profile to total formation thickness and is expressed in percent.

4) Interlayer permeability heterogeneity degree

Due to the differences in the depositional environment and diagenetic changes of sand bodies, the permeability of different sand bodies may differ greatly within a set of reservoirs. The degree of interlayer permeability heterogeneity is the key to dividing the development series of strata and determining the production technology.

The degree of interlayer permeability heterogeneity is usually described by the following indicators.

(1) Interlayer permeability distribution form. It mainly describes the distribution of the average permeability of each sand layer on the profile, and shows the degree of difference in the average permeability of sand layers and the distribution position of the highest permeability layer on the profile.

(2) Interlayer permeability variation coefficient.

(3) Permeability ratio.

(4) Single-layer rush coefficient.

The calculation method of these indicators of interlayer permeability is the same as that of in-layer permeability, but the thickness trade-off value shall be considered.

5) The configuration relationship between main reservoirs and non-main reservoirs on the profile

Special attention shall be paid to the location of ultra-high permeability layers—"thief layers" on the profile and their geological origin.

6) Interlayer barriers

Interlayer barrier condition is the other side of interlayer reservoir heterogeneity, and can block fluid movement. The barriers in clastic reservoirs are mainly argillaceous rocks, and also include a small amount of evaporites and other rocks. The main description content includes: (1) The rock type (lithology) of barriers; (2) The distribution (position) of barriers on the profile; (3) The thickness of barriers and its planar changes.

(II) **Microscopic heterogeneity**

The basic storage space of reservoir rocks can be divided into pores and throats. Generally speaking, a larger space surrounded by rock particles is called a pore, and the narrow part connecting two large pore spaces is called a throat. Pores are the basic storage spaces for occurrence of fluids in rocks, and throats are channels for controlling fluid seepage in rocks. Obviously, when fluids flow in a complex pore system, they need to go through a series of alternating pores and throats. When oil and gas are displaced out of porous media in the process of oilfield development, oil and gas are controlled by the smallest section (throat diameter) in a flow channel.

Microscopic heterogeneity mainly describes the size, distribution and geometry of pores and throats, clay matrix, sand grain structure, etc. These factors directly affect the oil displacement efficiency of injected fluids.

1. Shape of pore throats

The size and shape of pore throats are important factors that control reservoir properties. The size and shape of pore throats mainly depend on the size, shape, contact relationship, and cementation type of particles. The difference in the size and shape of throats leads to different capillary pressures and affects the storage properties of rocks. There are five typical throat types in sandstones: pore-shrunk throats; the variable section shrinkage part is a throat; flaky throats; bended flaky throats; tubular throats (Figure 9-5).

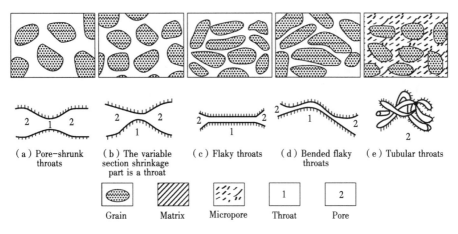

Figure 9-5　Types of pore throats

2. Distribution of pore throats

The distribution of pore throats refers to the frequency distribution and degree of homogeneity of various sizes of pore throats, pore-throat ratio, the coordination number of pore throats, etc.

1) Parameters reflecting throat size

The distribution characteristics of throats are characterized using relevant parameters based on capillary pressure data in general. Methods for measuring capillary pressure include semi-permeable diaphragm method, centrifuge method, mercury injection method, dynamic capillary pressure method, vapor pressure method, etc. Different measurement methods will have different measurement results. For each oilfield, the measurement data obtained using the best method shall prevail, and the data measured using other methods shall be adopted after unified conversion.

The semi-permeable diaphragm method is a classic method, and can simulate the actual reservoir conditions relatively approximately. However, the pressure measured using this method is relatively low, the required balance time is long, and it cannot be used widely. Therefore, oilfields select representative rock samples for the semi-permeable diaphragm method to create a set of representative curves for comparison and correction with other methods.

At present, the mercury injection method (mercury intrusion method) is often used to determine the distribution of pore throats, and the methods of calculating pore throat characteristic values using mercury injection data include the method of normal probability curve diagram and the moment method. The pore throats of various reservoirs are not completely distributed in a normal manner; moreover, the saturation value of mercury pressed into the rock pore throat volume under the highest pressure of the mercury injection instrument usually fails to reach 95% or even 84%, so the calculation parameter error is large. The data required by the moment method to calculate the characteristic parameters of pore throats can be directly obtained from analysis data, and all pore ranges are involved in the calculation; in addition, the original data can be directly input into the computer for calculation. Therefore, this method is mainly used in oilfields at present.

The following is an introduction to the detailed method of using the moment method to characterize the pore throat distribution characteristics and parameter calculation based on the capillary pressure data measured by the mercury injection method.

There are three typical methods, any of which can be selected according to different needs.

(1) Columnar frequency histogram of throat pores.

One is the uneven distribution form shown in Figure 9-6. Horizontal parallel lines are plotted along the capillary pressure curve at the same interval. The saturation at the intersection of one horizontal line with the capillary pressure curve minus the saturation at the intersection of the previous horizontal line with the capillary pressure curve is the percentage of the throat pore volume of the corresponding interval of the two horizontal lines to the total pore volume.

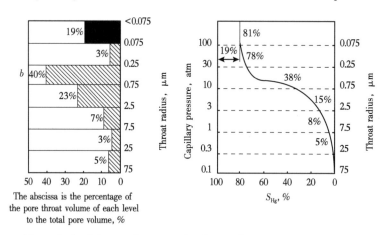

Figure 9-6 Columnar frequency distribution diagram of throat pore size

The other is a columnar frequency distribution diagram of the throat pore size divided by equal values (Figure 9-7). The throat size interval is divided into 13 intervals such as 10μm, 6.3μm, 4.0μm, 2.5μm, 1.6μm, 1.0μm, 0.63μm, 0.4μm, 0.25μm, 0.16μm, 0.1μm, 0.063μm, <0.04μm, etc. Take the throat pore radius as the abscissa, and find out the mercury saturation from the pressure values corresponding to these 13 intervals on the capillary pressure curve. The saturation difference of each interval is the percentage of the throat pore volume of the interval to the total pore volume.

This diagram has the advantages of being intuitive and convenient for comparison. By plotting the permeability contribution of each interval in the diagram, which level of throat pores is the most important in seepage flow can be judged.

(2) Frequency distribution curve and cumulative frequency distribution curve of throat pores.

This method is basically the same as the above-mentioned histogram, except that the

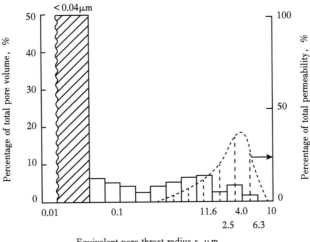

(a) The top shows Dorji-Sita sandstone: ϕ=24.0%, K=2.4mD

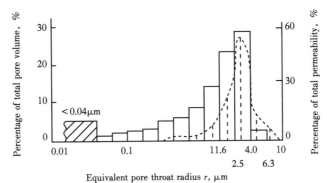

(b) The bottom shows the Permian Hoboth dolomite: ϕ=14.1%, K=5.3mD, (Rieckmann, 1963)

Figure 9-7　Columnar frequency distribution of throat pores divided by equivalent values

columnar expression is replaced by a smooth curve formed by the columnar center points. The cumulative frequency curve just superimposes the throat pore volumes of the previous intervals, as shown in Figure 9-8.

(3) Volume distribution curve and distribution function curve of throat pores.

Figure 9-9 shows the volume distribution frequency curve and cumulative volume distribution frequency curve of throat pores, where the ordinates are the volume of mercury injected and the percentage of mercury saturation. The abscissa corresponding to a point on the cumulative frequency curve is the throat radius, and the ordinate corresponding to this point is the sum of all throat pore volumes above the throat radius. A point on the frequency curve refers to the percentage of the pore volume controlled by the throat radius to the total pore volume.

2) Basic parameters for quantitative characterization of pore throat distribution characteristics

(1) Maximum connected throat radius (r_d): it refers to the largest connected throat radius in the pore system, that is, the throat radius measured when the non-wetting phase begins to enter the rock sample at the time of displacement pressure, in μm.

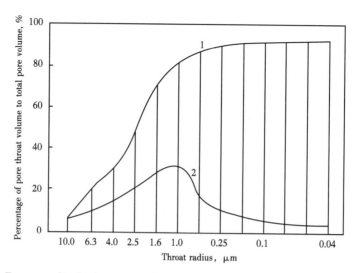

Figure 9-8　Frequency distribution curve and cumulative frequency distribution curve of throat pores.
1-Cumulative frequency distribution curve; 2-Interval frequency distribution curve

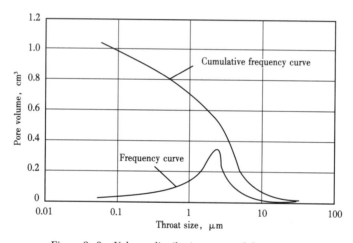

Figure 9-9　Volume distribution curve of throat pores

As shown in Figure 9-10, a tangent line BA is plotted on the capillary pressure curve along the flat part of the curve. The pressure value at which the tangent line intersects with the vertical axis is the displacement pressure p_d (or threshold pressure). The pore throat radius r_d corresponding to this pressure value is the maximum connected pore throat radius.

(2) Median pore throat radius (radius r_{50} corresponding to the median saturation): the throat radius corresponding to 50% saturation of the non-wetting phase (mercury). As shown in Figure 9-10, curve I is the mercury injection curve. Point a is the point where the mercury saturation is 50%, and the corresponding throat radius is the median pore throat radius r_{50} of the rock sample.

(3) Throat mean ($\bar{r}$): throat size mean.

$$\bar{r} = \sum_{i=1}^{n} \Delta S_i r_i / 100$$

Where: r_i—The throat radius of a certain interval in the throat radius distribution function, μm (the same below);

ΔS_i—The non-wetting phase saturation of a certain throat interval corresponding to r_i, expressed in percentage (the same below);

(4) Peak throat radius: the throat radius of the largest percentage value on the throat pore frequency distribution diagram. The throat radius corresponding to b on the pore throat frequency distribution histogram in Figure 9-10 is the peak throat radius of the rock sample.

(5) Maximum non-flowing pore throat radius: the corresponding throat radius when the permeability contribution value approaches zero (>99% is often used in actual work).

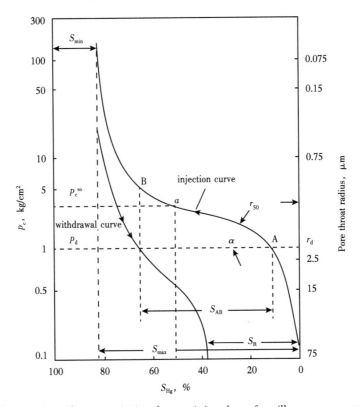

Figure 9-10 Three quantitative characteristic values of capillary pressure curve

3) Parameters reflecting throat sorting degree

(1) Standard deviation (sorting coefficient) δ: reflecting the sorting degree of throat size; the better the sorting, the smaller the value.

$$\delta = \left[\sum_{i=1}^{n} (\gamma_i - \bar{r})^2 \times \Delta S_i \right]^{1/2}$$

(2) Variation coefficient (relative sorting coefficient) C_s: reflecting the relative homogeneity of throat size distribution. The smaller the value, the more even the throat distribution.

$$C_s = \delta / \bar{r}$$

(3) Homogeneity coefficient a: it characterizes the sum of the deviation of each throat radius

(r_i) from the maximum connected throat radius (r_d) in the reservoir pore system. The value range is 0–1; the closer this value is to 1, the more uniform the throat distribution is.

$$a = \frac{\sum_{i=1}^{n} \dfrac{r_i \Delta S_i}{r_d}}{\sum_{i=1}^{n} \Delta S_i}$$

(4) Throat distribution skewness S_{KP}: it means that the throat distribution is biased to large throats or small throats relative to the mean, and is generally +2– –2. The pore skewness of good reservoirs is positive, and is mostly 0.25–1.0, while the pore skewness of all poor reservoirs is negative.

$$S_{KP} = \frac{1}{100} \delta^{-3} \sum_{i=1}^{n} \Delta S_i (\gamma_i - \bar{r})^3$$

(5) Throat distribution kurtosis K_P: it is a parameter indicating the steepness of the throat frequency distribution curve, and is also used to measure the ratio of the expanding of the throat diameter of the two tails (front and rear ones) of the frequency curve distribution to the expanding of the central part.

$$K_P = \frac{1}{100} \delta^{-4} \times \sum_{i=1}^{n} (r_i - \bar{r})^4 \Delta S_i$$

For a normal curve, K_P is equal to 1; for a flat peak (bimodal) distribution curve, K_P may be lower than 0.6; for a high and narrow peak curve, K_P may be 1.5–3.0.

4) Parameters reflecting the connectivity of pore throats and controlling fluid movement characteristics

(1) Pore throat coordination number: the average number of throats connecting pores. The method is to count a certain number (50 or 100) of pores on casting slices and the number of throats connected to them, and then to calculate the average value, which is the pore throat coordination number. The larger the coordination number, the better the reservoir properties.

$$\text{Pore throat coordination number} = \frac{\text{Statistical number of throats}}{\text{Statistical number of pores}}$$

As shown in Figure 9–11, the number of pores is 6, the number of throats is 19, and the coordination number is 19/6 = 3.17.

(2) Pore-throat ratio: the ratio of the average pore diameter to the average throat diameter in the sample. Specific method: count the diameter of a certain number of pores and the diameter of a certain number of throats on casting slices or SEM photos, and then calculate their average diameters respectively. The ratio of the average pore diameter to the average throat diameter is the pore-throat ratio. The larger the pore-throat ratio, the worse the reservoir properties.

(3) Withdrawal efficiency W_e: it refers to the percentage of the volume of mercury withdrawn from the rock sample by mercury injection method to the total volume of mercury injected before pressure reduction within the limited pressure range when the injection pressure drops from the

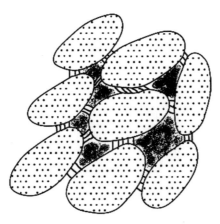

Figure 9-11　Schematic diagram of pore throat coordination number

maximum injection pressure to the minimum injection pressure. It can be calculated from the following formula:

$$W_e = \frac{S_{max} - S_R}{S_{max}} \times 100\%$$

Where: W_e—Withdrawal efficiency, %;

　　　S_{max}—Maximum saturation of injected mercury, %;

　　　S_R—Saturation of mercury remaining in the pores after withdrawal, %;

The S_{max} and S_R values can be read from the mercury injection and withdrawal curves, as shown in Figure 9-12; they can also be obtained from the corresponding experimental data of the curves.

3. Clay matrix

The clay matrix minerals filled in the pores of clastic reservoirs include detrital clay minerals and authigenic clay minerals. They have a large surface area and extremely strong activity (such as adsorption capacity, sensitivity to external fluids, etc.), so they have a very large influence on the injection capacity of various injectants, the adsorption and modification of injectants, etc. Moreover, due to their own changes, they greatly affect the displacement effects.

1) Clay content

Minerals with a particle size of less than 5μm are called clay in a particle size analysis, and their content is called the total clay content.

2) Clay mineral type

There are many types of clay minerals, typically including montmorillonite, kaolinite, chlorite, illite, and their mixed layer clay. The type and content of clay minerals occurring in different provenances and different depositional environments are different, and different types of clay minerals have different sensitivity to fluids. Therefore, different types of clay minerals and their relative content in reservoirs shall be determined separately.

Typical clay mineral analysis methods include X-ray diffraction, DTA, infrared spectroscopy, electron microscopy, etc. X-ray diffraction is often used to analyze clay minerals in domestic oilfields.

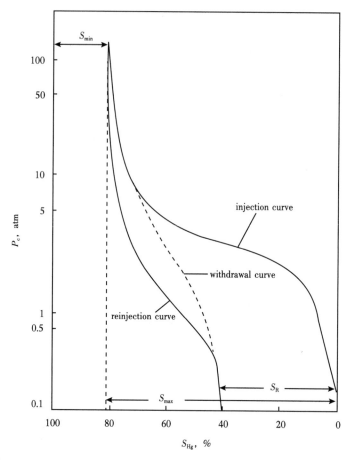

Figure 9-12 Mercury injection, withdrawal and reinjection curves (according to Wardlaw, 1976)

3) Occurrence of clay minerals

The occurrence of clay minerals has a large impact on the movement of oil and water in reservoirs. The occurrence of clay minerals is generally divided into: dispersed (filled), thin layered (lined) and bridging (Figure 9-13).

4) Research on the sensitivity of clay minerals to fluids

Clay minerals and fluids in the original reservoir are usually in equilibrium; when different fluids enter, their equilibrium will be destroyed. These external fluids do not match with reservoir fluids and reservoir minerals, resulting in a decrease in reservoir permeability. This is the sensitivity of clay minerals to fluids.

The research on the sensitivity of clay minerals to fluids includes velocity sensitivity, water sensitivity, acid sensitivity, salt sensitivity, alkali sensitivity, etc.

(1) Velocity sensitivity.

Changes in fluid flow velocity cause migration of formation fines and then blockage of throats, leading to a drop in permeability.

The intensity of velocity sensitivity is expressed by the permeability damage rate (D_K) of a rock sample:

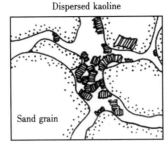

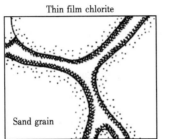 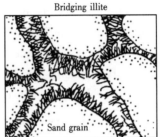

Figure 9-13 Typical chart of the occurrence of clay minerals in pores

$$D_K = \frac{K_{max} - K_{min}}{K_{max}}$$

Where: D_K—Permeability damage rate caused by velocity sensitivity;

K_{max}—The maximum value of the permeability of the rock sample before reaching the critical flow velocity, mD;

K_{min}—The minimum value of the permeability of the rock sample, mD.

Velocity sensitivity evaluation indexes: strong velocity sensitivity $D_K > 0.7$; medium to relatively strong velocity sensitivity $0.5 < D_K < 0.7$; medium to relatively weak velocity sensitivity $0.3 < D_K < 0.5$; weak velocity sensitivity $0.05 < D_K < 0.3$; no velocity sensitivity $D_K \leq 0.05$.

(2) Water sensitivity.

The clay minerals in a reservoir may be hydrated and expanded when they contact low-salinity fluids, thereby reducing the permeability of the reservoir. Water sensitivity refers to the phenomenon of permeability reduction caused by swelling, dispersion, and migration of clay after external fluids not compatible with the reservoir enter it.

Most clay minerals have different degrees of expansibility. Among the common clay minerals, montmorillonite has the strongest expansibility, followed by illite/montmorillonite and chlorite/montmorillonite mixed layer minerals, chlorite, illite, and kaolinite successively.

Water sensitivity is evaluated by the water sensitivity index I_w:

$$I_w = \frac{K_i - K_w}{K_i}$$

Where: I_w—Water sensitivity index;

K_w—Permeability of the rock sample measured with distilled water, mD;

K_i—Permeability of the rock sample measured with standard brine or formation water, mD;

Water sensitivity evaluation indexes: extremely strong water sensitivity $I_w \geq 0.9$; strong water sensitivity $0.7 \geq I_w < 0.9$; medium to relatively strong water sensitivity $0.3 < I_w \leq 0.5$; strong water sensitivity $0.05 < I_w \leq 0.3$; no water sensitivity $I_w \leq 0.05$.

(3) Acid sensitivity.

Acid sensitivity refers to the phenomenon of reservoir permeability reduction caused by the generated precipitates or the released fine particles during the reaction of acid fluids entering the reservoir with the acid sensitive minerals in it.

Acid fluids include two types such as hydrochloric acid (HCl) and hydrofluoric acid (HF). Hydrochloric acid (HCl) reacts with minerals with high iron content to generate $Fe(OH)_3$ precipitates or SiO_2 gel; Hydrofluoric acid reacts with high calcium minerals to generate CaF_2 precipitates or SiO_2 gel. Typical acid-sensitive minerals are shown in Table 9-1.

Table 9-1 Sensitive minerals and fluids that may damage formations

Sensitivity type		Sensitive mineral	Damage form
Water sensitivity		Montmorillonite, illite/montmorillonite mixed layer, chlorite/montmorillonite mixed layer, degraded illite, degraded chlorite, hydrated white mica	Lattice expansion Dispersion and migration
Velocity sensitivity		Kaolinite, hairy illite, microcrystalline quartz, microcrystalline feldspar, etc.	Dispersion and migration of fine particles
Acid sensitivity	HCl	Prochlorite, chamosite, chlorite/montmorillonite mixed layer, ferrocalcite, ankerite, hematite, pyrite, siderite,	Chemical precipitation $Fe(OH)_3 \downarrow$ SiO_2 gel $\downarrow$ Releasing of fine particles
	HF	Calcite, dolomite, anorthite, zeolite (laumonite, scolecite, calcium analcime, heulandite, stilbite)	Chemical precipitation $CaF_2 \downarrow$ SiO_2 gel $\downarrow$

Acid sensitivity index I_a:

$$I_a = \frac{K_i - K_{ia}}{K_i}$$

Where: K_i—Permeability of the rock sample measured with standard brine before acidification, mD;

K_{ia}—Permeability of the rock sample measured with standard brine after acidification, mD.

Acid sensitivity evaluation indexes: strong acid sensitivity $I_a > 0.7$; medium acid sensitivity $0.3 \leq I_a \leq 0.7$; weak acid sensitivity $0.05 < I_a < 0.3$; no acid sensitivity $I_a \leq 0.05$.

The prediction of acid sensitivity is complicated. Acid sensitivity can only be predicted qualitatively based on the change in the acid-sensitive ion content of the residual acid in the acid dissolution test.

(4) Salt sensitivity.

Salt sensitivity evaluation is to understand the process and degree of permeability change of reservoir rock samples under the condition of changing salinity in a series of salt solutions, and to find out the critical salinity where the permeability drops significantly, and the position of various working fluids in the salinity curve.

Salt sensitivity is a measure of the ability of formations to withstand salinity fluids. The critical salinity S_c is a parameter that characterizes the degree of salt sensitivity.

The critical salinity S_c refers to the corresponding salinity when the permeability of the rock sample begins to drop significantly as the salinity of the injected fluid decreases.

Salt sensitivity evaluation indexes are applicable to flocculation method salt sensitivity tests and lithologic displacement method salt sensitivity tests. Salt sensitivity evaluation indexes are as follows:

① Salt sensitivity evaluation using standard brine (composite salt) (unit: ppm):

No salt sensitivity $I_w \leqslant 0.05$

Weak salt sensitivity $S_c \leqslant 1000$

Medium to relatively weak salt sensitivity $1000 < S_c < 2500$

Medium salt sensitivity $2500 \leqslant S_c < 5000$

Medium to relatively strong salt sensitivity $5000 < S_c < 10000$

Strong salt sensitivity $10000 \leqslant S_c < 30000$

Extremely strong salt sensitivity $S_c \geqslant 30000$

② Salt sensitivity evaluation using NaCl brine (single salt) (unit: ppm):

No salt sensitivity $I_w \leqslant 0.05$

Weak salt sensitivity $S_c \leqslant 5000$

Medium to relatively weak salt sensitivity $5000 < S_s < 10000$

Medium salt sensitivity $10000 \leqslant S_c \leqslant 20000$

Medium to relatively strong salt sensitivity $20000 < S_c < 40000$

Strong salt sensitivity $40000 \leqslant S_c < 100000$

Extremely strong salt sensitivity $S_c \geqslant 100000$

(5) Alkali sensitivity.

Alkali sensitivity refers to the phenomenon of permeability reduction caused by the reaction of alkaline liquid with alkali-sensitive minerals in strata and formation fluids after entering strata.

The alkali sensitivity of a rock sample is evaluated using the alkali sensitivity index I_b.

$$I_b = \frac{K_s - K_{sb}(\min)}{K_s}$$

Where: I_b—Alkali sensitivity index;

K_s—Permeability of the rock sample measured with KCl brine, mD;

$K_{sb}(\min)$—Minimum permeability of the rock sample measured with alkaline solutions of different pH values, mD.

Alkali sensitivity evaluation indexes: no alkali sensitivity $I_b < 0.05$; weak alkali sensitivity $0.05 < I_b < 0.3$; medium alkali sensitivity $0.3 < I_b < 0.7$; Strong alkali sensitivity $I_b > 0.7$.

Section 2   Description of Remaining Oil Distribution

The final result of fine reservoir description requires the deepening of the understanding of remaining oil distribution and potential of various reservoirs, and the provision of visual and digital fine prediction geological models for quantitative remaining oil distribution.

Remaining oil refers to the part of crude oil that is still preserved in the pore space of an oil layer (or reservoir) in a certain stage of its development process, and also the crude oil that can be produced by deepening the understanding of underground geological bodies and improving oil production technology etc. Residual oil refers to crude oil that cannot be produced under certain production conditions. Therefore, residual oil is part of the remaining oil.

The study of the distribution of remaining oil with the goal of EOR after waterflooding development of oilfields has attracted widespread attention from domestic and foreign oil companies. According to the analysis by Texas Bureau of Economic Geology, the recovery ratio can reach 34% under the existing technical conditions; residual oil (unmovable) accounts for 47% (the strong water washing part accounts for 70%-80%); the movable oil of non-swept volume accounts for 16%; the recovery ratio can be enhanced by 30% by improving the oil production technology. There are still 19% reserves available for tapping potential relying on the existing secondary oil recovery technology (water injection), deeply understanding the heterogeneity of reservoirs, drilling infill adjustment wells, and improving various oil production technology measures. According to the 1997 statistical data from Yu Qitai *et al.*, in case of developing 39 waterflooding sandstone oilfields with geological reserves more than 5000×10^4t (their geological reserves account for 75.8% of the total geological reserves of the sandstone oilfields that have been put into waterflooding development), the average maximum ripple factor (corresponding to $f_w = 1.0$) of 67.6% can be achieved, so the corresponding maximum non-spread factor is 32.3%. Under the existing development method, the average maximum sweep efficiency (corresponding to $f_w = 1.0$) of 67.6% can be achieved using the existing development mode; therefore, the corresponding maximum non-sweep coefficient is 32.3%. It is not difficult to see that the non-swept remaining oil accounts for a large proportion. How to improve the recovery ratio of different types of reservoirs and carry out research on remaining oil is a worldwide research topic to which the oil companies of various countries attach great importance, and is also one of the problems that the domestic and foreign petroleum industries are eager to tackle.

Most of the oilfields in Eastern China have now entered the stage of high and ultra-high water cut development. At present, their average comprehensive water cut has exceeded 80%, and even 90%, their rude oil production has become a clear decreasing trend, and their recovery ratio is only about 30%. However, there are still 50% recoverable reserves in the reservoirs, and this part of remaining oil will be the focus of oilfield development and the main direction of tapping potential.

The distribution of remaining oil is obviously affected by geological factors and development factors in the middle and late development stages of waterflooding oilfields. Geological factors mainly include sedimentary microfacies, microstructure flow unit division, reservoir heterogeneity, etc. Development factors mainly include the perfection degree of the injection-production well pattern, injection-production ratio, production performance, etc. Under the combined influence of these static and dynamic factors, the spatial distribution of remaining oil becomes very complicated, and remaining oil is scattered and relatively enriched locally. Fine geological models can be used to quantitatively predict the spatial distribution of remaining oil in the research on fine reservoir description; moreover, remaining oil enrichment areas are determined based on the quantitative prediction results of remaining oil through numerical reservoir simulation, and targeted measures can be proposed to guide the remaining oil production measures in the middle and late stages of oilfield development and enhance the ultimate recovery ratio of oilfields.

I. Formation mechanisms of remaining oil and its controlling factors

Macroscopic remaining oil refers to the remaining oil that is studied and characterized by means of well logging, conventional core analysis, etc. and can be identified by naked eyes. The formation and distribution of macroscopic remaining oil in reservoirs is controlled by multiple factors such as macroscopic factors, microscopic factors, etc. in the process of oilfield development. There are two mechanisms for the formation and distribution of remaining oil. (1) The vertical heterogeneity of reservoirs causes the non-uniformity of vertical water-displacing-oil in reservoirs and the non-piston nature of the displacement process, thereby resulting in the alternating distribution of oil and water in reservoirs; with the continuous deepening of the development process, crude oil gradually does not dominate in the space (large and medium pores) of reservoirs. (2) Due to multiple factors, the injected water front fails to reach or displace crude oil, and crude oil occupies an advantage in the large and medium pore spaces of reservoirs, thus forming remaining oil enrichment areas.

According to the above two formation mechanisms, the macroscopic remaining oil controlling factors come down to two aspects such as reservoir heterogeneity and production non-uniformity. Reservoir heterogeneity includes structure heterogeneity, reservoir heterogeneity and fluid heterogeneity. Reservoir heterogeneity is the most important internal geological factor that controls the distribution of remaining oil, including the planar heterogeneity caused by the distribution of parameters such as reservoir scale, geometry and continuity, as well as porosity, permeability, etc. in sand bodies; interlayer heterogeneity caused by differences in the thickness, porosity, permeability, etc. of single sand layers; and intralayer heterogeneity caused by the vertical change of reservoir properties within a single sand layer, non-permeable interlayers, etc. Production non-uniformity is an external controlling factor for the distribution of remaining oil, mainly including the non-uniformity of reservoir production caused by the combination of series of strata, well pattern deployment, perforation position, injection-production correspondence, injection-production intensity, etc.

(I) **Control action of sedimentary conditions**

Clastic deposition conditions determine the sedimentary rhythm and the type of beddings, and also control the spatial distribution of sandstones, the distribution of sedimentary facies, and the heterogeneity of reservoirs. Differences in sedimentary rhythm, bedding type, sedimentary microfacies, etc. affect the distribution of remaining oil in the middle and late stages of development.

1. Sedimentary rhythm controls the oil displacement effect

The remaining oil distribution and development effects of reservoirs with different rhythmicity are quite different. The permeability of the reservoir with reverse rhythm is high in the upper part and low in the lower part. The injected water first advances along the top high permeability layers. Due to the gravity differentiation of oil and water and the capillary action, the injected water sinks along the middle and low permeability layers in the lower part under the condition of the rock's partial hydrophilicity, and the swept volume of water flooding gradually increases, thus leading to uniform longitudinal water line advancement, large water washing thickness, good oil displacement effect, and low remaining oil distribution. The permeability of the reservoir with positive rhythm is high in the bottom and decreases upwardly, the reservoir heterogeneity is strong, and the vertical distribution of permeability is quite different. Under the action of gravity, the injected water basically flows along the high-permeability layer in the bottom of the reservoir, resulting in fast water inflow at the bottom, sufficient water washing, and high water-displacing-oil efficiency. However, the swept volume of water flooding increases slowly, and generally speaking, the water displacement effect is poor. Moreover, there is no injected water in the middle and upper parts of the reservoir, and the remaining oil is concentrated in the middle and upper parts. The water displacement characteristics in the reservoir with composite rhythm are relatively complicated, and its oil displacement effect and remaining oil distribution are between those of the reservoir with positive rhythm and those of the reservoir with reverse rhythm.

2. The type of sedimentary beddings affects the oil displacement effect

Experimental results show that in the process of water-displacing-oil, the injected water quickly advances along the high-permeability laminas of sedimentary beddings so as to achieve clean water washing, and the effect of oil displacement along the low-permeability laminas is poor. There are obvious differences in the impact of different bedding types on oil displacement. Horizontal (parallel) beddings and micro-wavy beddings are stably distributed and have a large extension distance, forming relatively high permeability zones, so that oil displacement is not thorough, the oil displacement effect is poor, and the remaining oil is relatively enriched.

3. Sedimentary microfacies control the movement of oil and water

The planar distribution of sedimentary microfacies has obvious control effect on the movement of oil and water. The water absorption capacity of the reservoirs in channel edge microfacies belts is lower than that in central microfacies belts such as river channels, diaras, point bars, etc. The degree of water flooding is high, and the oil displacement efficiency is high in central microfacies belts, while both the degree of water flooding and the oil displacement efficiency are low in marginal microfacies belts, resulting in relative enrichment of remaining oil in marginal microfacies belts.

(II) **Reservoir heterogeneity controls the spatial distribution of remaining oil**

Reservoir heterogeneity is the main controlling factor for remaining oil distribution. It is generally deemed that the relative enrichment degree of remaining oil is higher in the area with a higher degree of reservoir heterogeneity; on the contrary, the relative enrichment degree of remaining oil is lower.

The existence of barriers and interlayers aggravates the heterogeneity of reservoirs, and different occurrences of barriers and interlayers have different influences on the formation and distribution of remaining oil.

1. The influence of barriers and interlayers with parallel bedding surfaces on the distribution of remaining oil

A large number of barriers and interlayers with parallel bedding surfaces are developed in the reservoirs deposited in braided rivers. Numerical simulation results show that the distribution of remaining oil enrichment areas is controlled by the planar position of interlayers in reservoirs and the perforation position of oil and water wells. When interlayers are located between water injection wells and oil production wells, they have the smallest control effect on remaining oil. If an interlayer is encountered during drilling of an oil production, the remaining oil is concentrated in the lower part of the interlayer under the influence of the barrier effect of the interlayer. If an interlayer is only encountered in a water injection well and water is injected above the interlayer, the interlayer has the greatest impact on the distribution of remaining oil.

When an interlayer is located in the mid-upper part of a reservoir with positive rhythm, it has the largest impact on the distribution of remaining oil. The larger the number of interlayers, the more obvious the impact; the larger the interlayer area, the more enriched the remaining oil.

2. The influence of barriers and interlayers with diagonal bedding surfaces on the distribution of remaining oil

Barriers and interlayers with diagonal bedding surfaces are generally developed in the reservoirs deposited in meandering river point bars and delta fronts. Numerical simulation results show that when water is injected against the dip direction of an interlayer, the sweep coefficient and recovery ratio are slightly large; when water is injected along the dip direction of an interlayer, the sweep coefficient and recovery ratio are small; especially when an interlayer is encountered in an oil production well, the sweep coefficient and recovery ratio are smaller, and the remaining oil is relatively enriched. In short, water injection along the dip direction of an interlayer tends to form a remaining enrichment area.

(III) Structures control remaining oil distribution

The degree of influence and control of structures on the formation and distribution of remaining oil is different in different development stages. The distribution of remaining oil is mainly controlled by fault block structures in the early stage of development; for example, remaining oil is enriched in the high part of a fault block structure. Anticline structures play a certain control role in the middle and late development stages of oil and gas fields, but microstructures play a major role in controlling remaining oil distribution.

Faults also control the distribution of remaining oil. Closed faults cause imperfect injection and production systems. The oil production wells near faults are effected in one direction, the water-displacing-oil effect is poor, and remaining oil enrichment areas are easily formed. During channeling of injected water along the faulting plane of an open fault, the oil in the nearby oil layer cannot be displaced so as to forms a "stagnant zone", which becomes a remaining oil enrichment area.

(Ⅳ) Injection-production well patterns control the planar distribution of remaining oil

The lithology and physical properties become worse in the edge and corner areas and pinch-out areas of reservoirs, so the heterogeneity of reservoirs is enhanced, and the injected water advances unevenly, thus easily forming remaining oil enrichment areas and affecting oilfield development effects. When oil and water wells are all in high-permeability areas and especially when oil production wells are in a transitional area from a high-permeability area to a low-permeability area, the production of each oil well is high. If water injection wells are located in a high-permeability area and oil production wells are located in a low-permeability area, the water injection effect is poor and the production of oil production wells decreases.

Remaining oil is relatively enriched near the dividual flow line of inter-well injected water and in the part with poor well pattern control.

Ⅱ. Remaining oil distribution pattern

Remaining oil distribution patterns in continental clastic reservoirs are affected and controlled by factors such as sedimentary facies belts, microstructures, fault sealing, macroscopic reservoir heterogeneity, microscopic characteristics of reservoirs, development non-uniformity, etc.; there are complex and variable types of remaining oil distribution patterns. Different oil and gas fields and different types of reservoirs have different remaining oil distribution patterns. However, a lot of research data show that there are some common remaining oil distribution laws to follow, and corresponding remaining oil distribution patterns of guiding significance can be established.

(Ⅰ) Remaining oil planar distribution patterns

The planar distribution patterns of remaining oil are comprehensively affected and controlled by the planar heterogeneity of reservoirs and the heterogeneity of injection and production, and are summarized as follows.

1. Planar changes of sedimentary microfacies and remaining oil distribution patterns

Planar changes of sedimentary facies include the transformation of sedimentary microfacies and the changes of reservoir physical properties in different parts of sedimentary microfacies. The differences in physical properties of different microfacies and the differences in physical properties of different parts of the same microfacies lead to the non-uniformity of fluid flow in underground reservoirs. The injected water always enters the high-permeability reservoir first, and then rushes along the direction of the high pressure gradient. Until the pressure gradient in this direction becomes small, the injected water expands to the edge, leading to poor water displacement effect in low-permeability reservoirs and high remaining oil saturation.

(1) The remaining oil distribution law in the reservoirs of different sedimentary microfacies in fluvial facies is different. It is generally deemed that permeability has obvious directionality. Injected water rapidly rushes along the direction of the thick sand body in the main channel, and the degree of water flooding is high, while the permeability of the thin sand body in the margin of

the channel is relatively low and the degree of water washing is poor. Therefore, the remaining oil is relatively enriched in channel edge sand bodies areally.

(2) Remaining oil distribution pattern in delta sedimentary sand bodies. Areally, the injected water first rushes along the sand body axis, and then expands to both sides in a delta front bar sand body; the sweep degree of the injected water is high and the water-flooded thickness is large in the layer. The lithology of the sand body on both sides of the sand bar and the inter-channel shoal becomes worse, and argillaceous strips increase, so the remaining oil is relatively enriched.

In spite of small sand body thickness in front sheet sands and distal bar sand bodies, the planar connectivity degree of sand bodies is high, the in–layer permeability heterogeneity is relatively weak, and the injection-production well pattern is easily controlled, so the degree of in-layer water flooding is high, and areally, the injected water advances slowly and relatively evenly. The injected water rushes along the mid-upper high permeability section of the sand body with reverse rhythm or composite rhythm longitudinally in the water washing process. Affected by the difference between oil density and water density, the injected water slowly advances towards the lower part of the sand body under the action of gravity. Therefore, the water flooding degree and water washing degree of the sand body are high vertically. However, there are still some low-permeability zones at the bottom that have not been washed with water or the sweep degree of the injected water is low, forming relatively remaining oil enrichment areas.

2. Remaining oil distribution pattern in microstructure areas

There is a close relationship between microstructures and remaining oil distribution. Generally, there is much remaining oil in the high part of a microstructure, where a remaining oil enrichment area is easily formed. When the pressure gradient and physical conditions around a water injection well in a reservoir are basically the same, under the action of gravity, the injected water firstly rushes into an oil production well in the lower part of the structure (negative-direction microstructure), a water-flooded area is firstly formed in the lower part of the structure, and the injected water reaches the area with a high degree of water flooding. In this case, the remaining oil is mainly distributed in the high part of the structure (positive–direction microstructure). Similarly, oil is displaced upwards and remaining oil is relatively enriched in all oil wells located on micro-fault noses and micro-anticline structures.

3. Fault combination and remaining oil distribution pattern

The nature and combination of faults in fault blocks have a significant impact on the distribution of remaining oil. Open faults often make it easy for oil and water to flow along faults into shallow reservoirs, while closed faults directly block the upward flow of oil and water and thus cause oil and water to stay in relatively high parts locally so as to form remaining oil enrichment areas.

Sun Mengru *et al.* (2004) established 4 remaining oil distribution patterns caused by blockage by closed faults in Tuo-30 fault block in the study of the remaining oil distribution in Shengtuo Oilfield (Figure 9-14). Pattern A shows that the remaining oil in the area sandwiched by faults is relatively enriched under the single-direction effect. Patterns B and D indicate that the area sandwiched by two faults within the fault block is actually under single-direction effect due to the impact of closed faults, and the remaining oil is relatively enriched in the high part of the faults. Pattern C indicates that in the low part of the structure controlled by a fault, the remaining oil near the fault is relatively enriched due to factors such as injection-production well patterns etc.

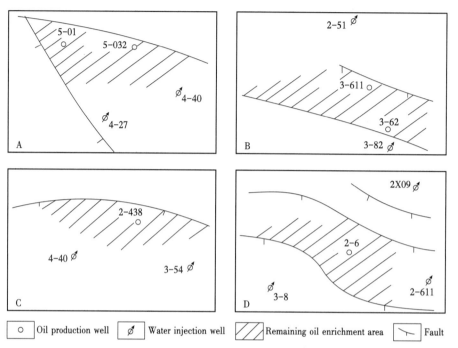

Figure 9-14 Remaining oil distribution pattern blocked by faults (according to Sun Mengru et al., 2004)

4. Remaining oil distribution pattern in imperfect injection-production well pattern areas

Areally, remaining oil is distributed near the inter-well dividual flow line and in the parts with poor well pattern control, the injection-production relationship is imperfect, and the remaining oil saturation near both sides of a production well row is generally high.

Sun Mengru et al. (2004) established 4 remaining oil distribution patterns caused by imperfect injection-production relationship in Tuo-30 fault block (Figure 9-15).

Pattern A: there is no water injection well and oil production well in the channel sand reservoir of Well 4-728 and Well 4-61 in Layer 2^1 of Sha-2 Member, the original condition of the non-produced reservoir is maintained, the reservoir is made to be in high pressure building state, the average pressure is about 30% higher than the original pressure, and the remaining oil saturation is greater than 60%. Pattern B: there is only an oil production Well 4-75 and no water injection well in Layer 2^3 of Sha-2 Member, the pressure near the oil production well is low, and the reservoirs of Well 4-728 and Well 3-65 have not been produced basically, making the remaining oil saturation greater than 66%. Pattern C: the reservoir thickness (12-20m) is large in Layer 8^3 of Sha-2 Member, and the degree of reservoir continuity is high; affected by closed faults and imperfect injection-production well patterns, some reservoirs have been fully produced, while some reservoirs have not been well produced or have not been produced basically, forming remaining oil enrichment areas. Pattern D: water injection Well 2-3 and Well 3-51 in Layer 2^1 of

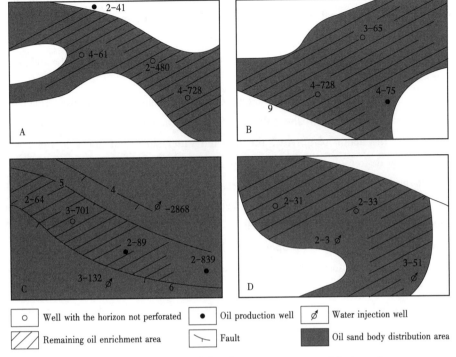

Figure 9-15 Remaining oil distribution patterns caused by imperfect injection-production well patterns (schematic)

Sha-2 Member are located in the channel margin where sand bodies are not developed, and their water injection effect is poor, while Well 2-33 is located in the main channel part; although most of the reservoirs in these well areas have not been well produced, they are remaining oil enrichment areas (where the remaining oil saturation is greater than 60%).

(Ⅱ) **Vertical remaining oil distribution patterns**

The vertical distribution of remaining oil is comprehensively controlled by the sedimentary rhythm of reservoirs and the distribution of barriers and interlayers.

1. Intralayer remaining oil distribution pattern

Intralayer remaining oil distribution is controlled by intralayer heterogeneity. According to the study of the relationship between heterogeneity and remaining oil distribution in continental clastic reservoirs in Eastern China, remaining oil is relatively enriched in the mid-upper rhythm of the reservoir with positive rhythm, and remaining oil can still be found in the mid-lower part of the reservoir with reverse rhythm.

The reservoirs deposited in fluvial facies are dominated by positive rhythms, and the water flooding degree in the lower part of the heterogeneous pattern in the simple positive rhythm layer is high. The mid-upper part is the relatively enriched part of remaining oil, and the thickness of the enriched interval is about 1/4-1/2 of the total thickness. The remaining oil in the heterogeneous pattern in the composite positive rhythm layer is enriched in multiple sections. In the relatively homogeneous rhythm pattern, the injected water advances evenly and the degree of water flooding is high. There is still a certain amount of remaining oil in the mid-upper part of the main channel sand layer, but the bottom has been severely water-flooded, and subdivision within the layer is not easily achieved. The movable remaining oil can only be produced using the technology of improving oil displacement efficiency. Relatively enriched sections of remaining oil may be found in a reservoir of over 10m thick with some interlayers by subdividing flow units and ascertaining the structure of sand bodies.

Delta front bar sand bodies mostly show reverse rhythms and composite reverse rhythms vertically, where injected water rushes along the high permeability sections in the middle and upper parts. Under the action of gravity, the injected water slowly advances to the lower part of sand bodies, so the degree of water flooding and water washing thickness of sand bodies are large vertically, but the low-permeability area of the low part fails to be washed with water or the swept degree of the injected water is low, which can become a relative remaining oil enrichment area.

2. Interlayer remaining oil distribution pattern

The differences in the physical properties of different reservoirs in development series of strata lead to differences in the water-displacing-oil process during the injection-production process in the case of commingled injection and production. The starting pressure of water injection in high permeability layers is low, the injected water is easy to advance along high permeability layers, and their degree of production is high. The starting pressure of water injection in low permeability layers is low, and their degree of production is low. Under the same or similar injection and production conditions, interlayer longitudinal sedimentary facies changes control the remaining oil distribution between reservoirs.

There is obvious interlayer heterogeneity between main layers and non-main layers vertically, which determines interlayer remaining oil distribution patterns. For example, the main Layers 3^3, 3^5, 4^2, and 4^4 of Guantao Formation in Zhongyi area of Gudao Oilfield belong to meandering river point bar microfacies deposits, which are characterized by large effective reservoir thickness, high original oil saturation, large water absorption, large liquid production, high recovery percent, and low remaining oil saturation. The non-main layers such as 3^1, 3^2, 3^4, 4^1, and 4^3, etc. have poor reservoir physical properties and relatively low producing degree, resulting in relatively high remaining oil saturation in the non-main layers. However, the thickness of the non-main oil layers is small, and their recoverable remaining oil reserves are small, so that recoverable remaining reserves are still mainly distributed in the main layers.

III. Remaining oil research methods

At present, a series of mature remaining oil research and prediction methods and technologies have been formed, but all of them have their respective limitations. Remaining oil distribution shall be qualitatively and quantitatively determined comprehensively using various methods according to the specific conditions of reservoirs.

(I) **Prediction of remaining oil distribution by comprehensive geological analysis**

Comprehensive geological analysis is one of the effective methods to study and predict remaining oil. This method is used to comprehensively study and analyze remaining oil and predict remaining oil distribution using production performance data on the basis of comprehensive analysis of geological factors such as microstructure, sedimentary microfacies, reservoir heterogeneity, etc.

1. Microstructure analysis

Due to the existence of microstructures, a reservoir is divided into multiple micro-traps, thereby affecting the movement direction and velocity of fluids in the reservoir, and controlling the formation and distribution of remaining oil in the reservoir.

According to the research results of Li Xingguo (2000), oil wells located on positive microstructures are subjected to upward oil displacement in all directions or multiple directions, the remaining oil flows to the area from all directions, and the geological conditions are favorable. In negative microstructures, oil is displaced downwards in all directions, injected water flows to the area, and the geological conditions are unfavorable.

According to the research results of Sun Mengru (2004) et al., for the wells in the same development period, the water cut of the well located in the high part of a micro-fault nose structure in the same layer is higher than that of the well located in the limb of the structure, while the oil saturation value is relatively low (Table 9-2).

Table 9-2 Statistics of microstructures and reservoir parameters in 2-442 well area of the third block in Shengtuo Oilfield

Well No.	Horizon	Elevation m	Sand layer thickness m	Porosity %	Permeability mD	Shale content %	Water cut %	Oil saturation %
2-442	1^1	-1808.6	2.7	25.4	301	20.6	91	37.4
	2^2	-1833.4	3.5	26.1	1032	14.1	87	47.1
2-412	1^1	-1813.5	4.0	24.4	370	13.7	94	23.7
	2^2	-1839.7	6.0	24.7	415	10.6	88	41.9

2. Prediction of remaining oil distribution by comprehensive heterogeneity analysis

Heterogeneity analysis includes two aspects such as reservoir heterogeneity and injection-production heterogeneity. Reservoir heterogeneity means that due to the influence of factors such as the distribution and connectivity of reservoirs, etc., reservoir properties change unevenly, resulting in low water displacement efficiency and relative enrichment of remaining oil in some areas of reservoirs. Injection-production heterogeneity means that due to imperfect injection-production well patterns or unreasonable injection-production systems, some areas in reservoirs cannot be effectively oil-displaced, forming relatively enriched areas of remaining oil.

The existence of barriers and interlayers has a very large influence on the distribution of remaining oil. The larger the area of the barriers and interlayers and the more remarkable the included angle between the occurrence of the barriers and interlayers and that of the reservoirs, the more the formed remaining oil.

3. Planar changes of sedimentary microfacies affect remaining oil distribution

The change of sedimentary facies belts is determined by sedimentary conditions. Different sedimentary conditions and hydrodynamic energy will form different sedimentary rock combinations or lithofacies, and the rock combinations in different sedimentary facies belts are very different.

The planar difference of sedimentary microfacies in the same reservoir has a large control effect on the water drive efficiency and the formation and distribution of remaining oil. Therefore, in-depth research on the change law of sedimentary microfacies of reservoirs can provide guidance to the study of remaining oil and the prediction of remaining oil distribution.

4. Influence of fault sealing property on remaining oil distribution

The research on rifted basins in Eastern China shows that remaining oil is relatively enriched near a closed fault, while remaining oil is relatively poor near an open fault. Good fault sealing property (or near the pinch-out line of sandstones) leads to poor water injection effect of oil production wells (they tend to be under single-direction effect), thus making for enrichment of remaining oil.

(II) Production logging analysis method

The production situation of reservoir profiles and the distribution of remaining oil are judged mainly using the water injection profile test data of water injection wells and the fluid production profile test data of oil production wells. Among the intervals where the reservoir thickness is opened through perforation: the main water intake interval and oil production interval shall be the intervals with high producing degree of reserves and the least remaining oil distribution; after multiple tests, the interval with no water absorption and no fluid production is the interval with the worst producing degree of reserves and enriched remaining oil; the other intervals are between the above two.

1. Injection profile logging of water injection wells

The injection profile refers to the distribution of water absorbing capacity in each perforated interval on the profile at certain injection pressure and water injection quantity of an injection well. The injection profile reflects the change in water absorbing capacity of the reservoir profile and the distribution of water absorption thickness.

Tracer logging is performed using radioisotopes in an injection profile test. Before adding tracers, an isotope baseline is measured, and then isotope tracers are added to the injected water. The two isotope curves before and after are compared, and the well interval with isotope anomaly values is added on the measured curve, which can reflect the water absorbing capacity of the corresponding interval (Figure 9-16).

The injection profiles of water injection wells determine the fluid production of oil production well in oilfields subjected to waterflooding development. Therefore, the injection profile data of water injection wells can be used to understand the water absorption situation of reservoir profiles, monitor the water drive state of reservoirs, and analyze the production situation of reservoir profiles and the distribution of remaining oil. Generally, the injection profile data of water injection wells and the fluid production profile data of oil production wells that can be measured in the same period are comparatively analyzed, which can better judge the water washing and production situation of reservoir profiles and the distribution of remaining oil on profiles.

2. Fluid production profile logging of oil wells

Fluid production profile logging refers to the logging that measures the fluid production, water cut, fluid density and other parameters longitudinally distributed along the well depth in each

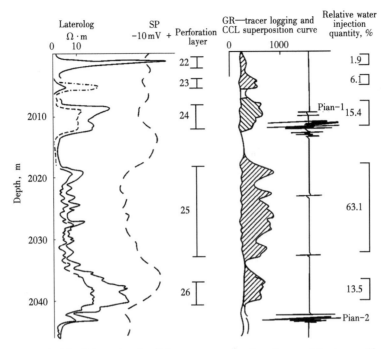

Figure 9-16 Schematic of radioisotope tracer logging of water injection profile

production interval of an oil production well under normal production conditions so as to determine the property and quantity of the produced fluids from the reservoir profile (Figure 9-17).

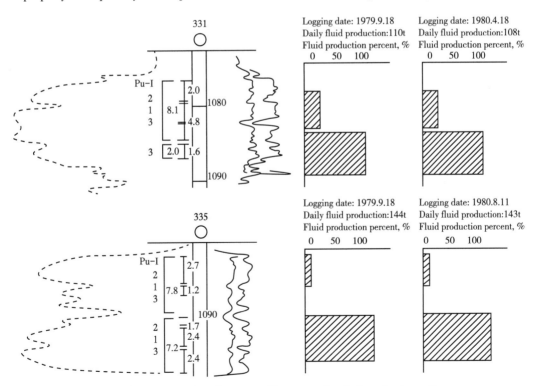

Figure 9-17 Fluid production profile test results of oil production well

The produced fluid may be a single-phase fluid of oil, gas or water, or a two-phase flow of oil-water, oil-gas, or gas-water, or a three-phase flow of oil-gas-water. While measuring the separate-layer fluid production, it is also necessary to consider parameters such as water cut or gas content, temperature, pressure, fluid density, etc. For a production well with oil-water two-phase fluid, the oil production and water production of the fluid production profile can be determined after measuring the two parameters such as liquid volume flow rate and water cut. For the three-phase flow of oil, gas and water, a density curve can be used to roughly determine the nature of the produced fluid and whether it is gas, oil or a mixture of oil and water. The producing situation of reserves and water washing degree in each interval of the reservoir profile as well as the distribution of remaining oil on the profile can be understood by regularly monitoring a single well or a single well group and comparatively analyzing the measured data.

(Ⅲ) **Water-flooded layer logging**

The water-flooded layer logging technology is an important method for understanding the distribution of remaining oil during waterflooding development or natural water flooding of reservoirs, and is also an important content of oilfield performance monitoring. China's onshore oilfields basically use the waterflooding development mode, and a large number of well pattern adjustments are carried out, thereby forming a set of water-flooded layer logging methods and interpretation technologies applicable to the development of onshore reservoirs.

1. Basic experimental research on petrophysical properties during water flooding

In order to ensure the accuracy of water-flooded layer logging interpretation, it is necessary to study the changes of petrophysical properties during water flooding of reservoirs.

1) Changes in electrical properties

During the water flooding process, the resistivity of formation mixtures changes with the resistivity of the injected water and the degree of water washing. The SP amplitude is of baseline shift in the mudstone section, the part and direction of the shift depend on the water-flooded part and the salinity of the injected water, and the shift amplitude depends on the difference between the salinity of the original formation water and that of the formation mixtures. When water of different salinity is injected, the formation resistivity changes accordingly. The dielectric constant of rocks increases with the increasing water saturation.

2) Changes in acoustic properties

Affected by long-term water injection, the radius of pores and throats in reservoirs will be increased to a certain extent, and sometimes rocks will be broken so as to generate fractures, thereby increasing the interval transit time (about $30-150\mu s/m$). When the water saturation increases, the coefficient of volume compressibility of rocks will decrease.

2. Water-flooded layer logging methods

1) Open hole logging methods

(1) SP baseline shift method The baseline of the SP curve of the water-flooded part of the sandstone reservoir developed by fresh water injection is shifted, as shown in Figure 9-18.

(2) IPP logging. When an external electric field is applied to rocks (excitation process) and then the external electric field is cancelled, a polarization potential (also known as artificial

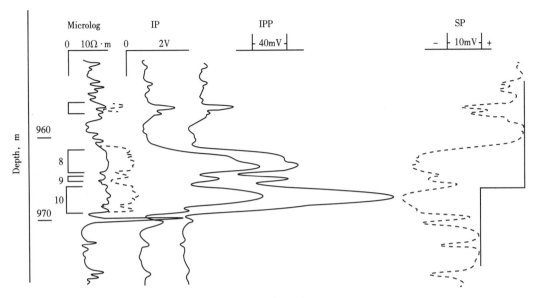

Figure 9-18 SP and IPP logging curves

potential) will be generated. In some cases, the polarization potential can also be used to determine water-flooded layers.

(3) Resistivity logging, induction logging, lateral logging, long electrode distance gradient logging and microlog are all basic methods for judging water-flooded layers.

(4) Dielectric logging. It is used to measure the conduction of electromagnetic waves in strata, determine the dielectric constant and water-flooded layers, and calculate the remaining oil saturation.

In addition, logging methods such as acoustic travel time logging, neutron gamma logging, GR logging, etc. can also be used to determine water-flooded layers in some specific cases.

2) Cased hole logging methods

(1) C/O spectral logging. The ratio of the secondary gamma-ray energy spectra generated by the indicator elements C and O of oil (hydrocarbon) and water is used to determine water-flooded layers and calculate the remaining oil saturation. C/O spectral logging is applicable to the strata with a porosity of >20%.

(2) Neutron lifetime logging. It measures the time from generation to capture of thermal neutrons in strata, and is called thermal neutron lifetime logging. It is suitable for gypsum-salt strata, and pay formations where high salinity water is injected or "wastewater" is reinjected.

3. Water-flooded layer logging series

Establishing water-flooded layer logging series is the prerequisite for logging interpretation and analysis in the development of waterflooding oilfields. Principles of establishing water flooded layer logging series: (1) Accurately measure the radial formation resistivity in wellbore; (2) Adapt to the change of formation water salinity, and accurately obtain the resistivity of mixed liquid in oil layers and water flooded layers; (3) Accurately calculate physical property parameters; (4) Adapt to changes in formation lithology; (5) Able to effectively subdivide intervals in thin layers and thick layers; (f) Adapt to changes in hole diameter and mud properties.

According to different drilling and completion processes and logging tasks, the water-flooded layer logging series are divided into two major types such as open hole water-flooded layer logging series and cased hole water-flooded layer logging series.

(1) Open hole water-flooded layer logging series include: SP baseline shift logging (applicable to reservoirs developed by fresh water injection), IPP logging, resistivity logging (induction logging, laterolog, long electrode distance gradient logging, and microlog), dielectric logging, etc. In addition, logging methods such as acoustic logging, neutron gamma logging, GR logging, etc. can also be used as water-flooded layer logging series in specific cases. For example, water-flooded layer logging examples in the main formations of Daqing Oilfield: 0.25m gradient, 0.45m gradient, 2.5m gradient electrode array, SP, high-resolution laterolog, high-resolution acoustic logging, micro-spherically focused logging, compensated density logging, inclination logging, caliper logging, etc.

(2) Cased hole water-flooded layer logging series include: C/O spectral logging, neutron lifetime logging, etc. In recent years, the C/O spectral logging technology has developed rapidly, and is applied in logging of 400-500 wells per year. Oilfields such as Daqing, Shengli and Jilin have continuously improved the accuracy of remaining oil calculation. The through-casing resistivity logging technology is currently being tested and studied, and a breakthrough will be made in its key part. Because of its simple logging process and low cost, it can be used for time-lapse logging to monitor the change process of remaining oil saturation of reservoirs. It is a new water-flooded layer logging method with broad promotion prospects.

4. Water-flooded layer logging interpretation

1) Qualitative interpretation

Methods for determining water-flooded layers in sandstone reservoirs include: SP baseline shift method, C/O ~ Si/Ca curve difference method from C/O spectral logging, IP method, flushed zone resistivity method, radial resistivity method, flowable fluid measurement method, etc. All of them can be used to judge water-flooded layers accurately to a certain extent. It is more accurate to carry out comprehensive interpretation using the above methods.

2) Quantitative interpretation

Quantitative interpretation is to calculate water saturation through and determine the flooding grade of target layers according to logging data. It is divided into: strong flooding (water cut more than 80%), medium flooding (water cut 40%-80%), weak flooding (water cut 10%-40%), and reservoir (water cut less than 10%). The methods include reservoir water-flooding model method and mathematical statistics method.

(1) Reservoir water-flooding model method. Archie's formula applicable to static conditions is extended. Reservoir water-flooding model method is summarized as the following three methods.

① Standard model method. it is used to correct the changes in the salinity, shale content, calcareous content, particle size (median), etc. of formation mixtures, and the influence caused by the inconsistency of logging itself. In case of calibration to uniform conditions, this method highlights the effect of oil-bearing property on resistivity and improves the accuracy of calculating water saturation using Archie's formula.

② Desalination coefficient method. The method uses the desalination coefficient to correct the influence of the change in formation resistivity caused by the decrease in the salinity of formation mixtures due to water injection, so as to improve the accuracy of calculating water saturation using Archie's formula.

③ Mathematical model method. The relation equations between formation resistivity and parameters such as water saturation etc. are established taking Archie's equation as a special example in the case of unchanged salinity of injected water based on rock tests and mathematical physics concepts. This model is suitable for solving water saturation at any time after water injection.

(2) Mathematical statistical methods for reservoir flooding: The methods such as discriminant analysis method, fuzzy mathematics, artificial intelligence, grey theory, artificial neural network, etc. are mainly used to interpret the water cut of water-flooded layers and divide flooding grades.

(Ⅳ) Inspection well sealed coring method

After waterflooding development of old oilfields, representative parts are selected to conduct sealed coring drilling in target intervals. This is the most direct method to obtain the remaining oil saturation of reservoirs. The oil saturation data obtained from sealed coring analysis can truly reflect the remaining oil saturation data of reservoirs, and can be used to judge the distribution of the remaining oil in the reservoir profile. In addition, the data can be combined to the location of sealed coring wells to judge the planar distribution of remaining oil, which is verified by means of staged formation testing.

(Ⅴ) Numerical reservoir simulation method

Numerical reservoir simulation is an important method for quantitative study of remaining oil distribution. The principle of this method: by solving difference equations, the values of the parameters such as pressure, remaining oil saturation, etc. of grid nodes in a reservoir are obtained and then the spatial distribution of remaining oil in each development stage is studied and predicted using the static and dynamic data of the reservoir based on geological models and the theory of fluid seepage.

This method is to study fluid evolution laws through history fitting, further simulate reservoir development indexes, and obtain parameters such as remaining oil saturation, remaining reserves, remaining movable oil saturation, etc. on the premise of accurately establishing a reservoir model (Figure 9-19).

(Ⅵ) Comprehensive analysis method of reservoir engineering

The use of reservoir engineering analysis method is to study the distribution characteristics of remaining oil in terms of statistical laws or engineering tests. Underground oil and water movement conditions and change laws are comprehensively analyzed and judged, the producing situation of reserves is understood, and the distribution of remaining oil is studied by analyzing water breakthrough and waterflooding effect of oil wells as well as planar and interlayer changes in production, pressure, water cut, and OGR based on the production performance data of oilfields combined with static geological characteristics and production logging data of reservoirs. This

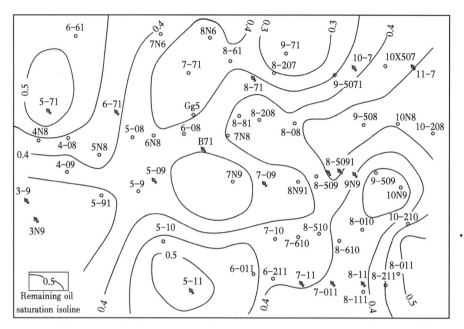

Figure 9-19 Remaining oil saturation distribution obtained from numerical reservoir simulation in the south block of Gudao Oilfield

comprehensive analysis method of reservoir engineering can use abundant production data, has the advantages such as long-term continuous tracking and analysis, low cost, etc., and is an important method widely applied in the field.

The above various remaining oil research methods have their own characteristics and limitations. The remaining oil value and distribution knowledge obtained using any method will have reliability deviations, and need to be verified and corrected using other methods. The comprehensive application of the above-mentioned methods helps to improve the accuracy of predicting remaining oil distribution.

Section 3 Natural Gas Distribution Description

The concept of oil reservoir description was proposed first. Gas reservoir description has been gradually formed under the continuous development and expansion of natural gas development business in recent years, and is closely related to oil reservoir description. There is little difference in the description of geological bodies between oil and gas reservoirs, and their research content and technical means are common. Therefore, oil reservoir description and gas reservoir description are collectively referred to as reservoir description in the previous chapters. Crude oil and natural gas differ in geophysical responses, flow mechanisms and production methods, so the essential difference between oil reservoir description and gas reservoir description is reflected in the description of pore fluids.

I. Characteristics of Natural Gas Distribution Description

(I) Comparison with Oil Reservoir Description

For a long time, oil reservoir description actually includes oil reservoirs and gas reservoirs, and is developing as a unified system, and gas reservoir description is rarely treated independently. As the development of gas reservoirs continues to deepen, a considerable number of gas fields have gone through a longer development stage. Even after entering the late stage of development, some description methods and technical means for gas reservoirs have been gradually accumulated, the description content and description goals required in gas reservoir description in different stages of gas reservoir development have been ascertained, and the method system for gas reservoir description has been gradually separated from oil reservoir description. In recent years, natural gas development targets have become increasingly complex, and especially unconventional gas fields dominated by tight gas have been developed in a large scale, so the oil reservoir description technology has shown a certain degree of inadaptability in the process of describing gas reservoirs. For example, tight gas reservoirs have low porosity and permeability and relatively low gas saturation, and most of them have high bound water saturation. Therefore, conventional oil reservoir description technologies cannot effectively solve the problems involving the identification of gas and water layers in tight gas reservoirs, the description of gas drainage scope, etc., and there is a need to develop targeted gas reservoir description technologies. In this development situation, gas field development workers continue to face new development problems, seek solutions to problems, and promote the development of gas reservoir description technologies in practice.

According to the comparison of the similarities and differences between oil reservoir description and gas reservoir description, most of the technologies for oil reservoir description and gas reservoir description are the same and have no essential differences, but there are differences in the main content of technology applications, the problems to be solved, and application goals. As far as oil reservoir description is concerned, since Daqing Oilfield was put into development in the 1960s, China has formed oil reservoir description technologies characterized by "fine correlation of layers and characterization of three major reservoir heterogeneities" for waterflooding oilfields. In addition, with the development of development technologies involving well pattern adjustment, fine water injection, deep profile control, chemical flooding, CO_2 flooding, etc., the fine oil reservoir description technology aiming at fine characterization of remaining oil has been vigorously developed. As far as gas reservoir description is concerned, the geological description parameters and methods are basically the same as those for oil reservoir description, and gas reservoir description focuses more on the combination of dynamic description with static description and the development of a gas reservoir description technology system centering on "description of reservoir-permeator unit scale and drainage boundaries".

The main difference between gas reservoir description and oil reservoir description is not the difference in geological bodies, but the difference in geological body description goals and fluid development modes resulting from different formation fluids. For oil fields, both oil and water are liquids, and the formation energy can be supplemented by water injection etc. during the development process, while gas-liquid two phases such as natural gas and formation water are developed in gas fields, gas is easy to flow, and the development process mainly relies on formation energy. At present, there is no development mode of gas fields by supplementing energy. Therefore, the main differences between gas reservoir description and oil reservoir description are reflected in the following two aspects.

1. Differences between oil reservoir description and gas reservoir description due to different fluid properties

Unlike oil reservoirs, the gas in gas reservoirs is easy to flow and has high compressibility, and gas reservoirs are more affected by formation water than oil reservoirs. Especially in low-permeability tight gas reservoirs, the increase in water production will cause gas wells to fail to produce. In view of different fluid properties, the differences between oil reservoir description and gas reservoir description are mainly as follows. (1) Gas has good fluidity and large pressure transmission range, and the well spacing is usually large during the production process. Especially for conventional gas reservoirs, the well spacing can be measured in kilometers. Even if the well pattern of tight gas reservoirs is relatively dense, the well spacing is more than 100m in most cases. The well spacing for oil reservoir development can be less than 50m. A large well spacing makes it more difficult to predict inter-well reservoirs. (2) The gas layer identification method for gas reservoirs is roughly the same as the oil layer interpretation method for oil reservoirs, but they differ to some extent in geophysical response characteristics. For conventional logging data, gas layers have some unique identification characteristics in addition to common characteristic such as high resistivity etc. For example, gas layers have an excavation effect, but oil layers do not. In terms of seismic data response characteristics, gas layer characteristics are more prominent, and technologies including gas layer identification by seismic spectrum attenuation, gas layer identification by AVO, pre-stack seismic inversion, etc. have been formed. The seismic response characteristics of oil layers are not obvious enough, and their prediction is more difficult. (3) The compressibility of gas layers is very strong, the physical properties of gas change with pressure more widely, and gas reservoir engineering parameters are more complex. (4) There is a large difference in flow mechanism between natural gas and crude oil. Gas flows by pressure difference, and has the characteristics of diffusion.

2. Differences between oil reservoir description and gas reservoir description due to different development modes

The development modes of oil reservoirs are completely different from those of gas reservoirs. Most oil fields produce little oil relying on natural energy, and are developed mainly relying on supplementary energy. Oil reservoirs can be produced using the production modes such as pumping units, supplementing energy by water injection, polymer injection, etc. Therefore, injection-production systems are the core of oil reservoir description, which determines that the heterogeneity of reservoirs at different scales and the resulting three major injection-production contradictions are the focus of description. Gas reservoirs use the depletion development mode relying on natural energy, the pressure drop sweep scope of gas reservoirs is the core of description. The pressure drop sweep scope of gas reservoirs mainly depends on the size, distribution characteristics and gas-water flow characteristics of reservoir-permeator units. For unconventional gas reservoirs, their description also has a certain relationship with the corresponding drilling and reservoir stimulation and development technologies. Therefore, gas reservoir description needs to describe the scale of gas drainage units by combining development technologies, which is different from oil reservoir description.

(Ⅱ) Main Parameters of Gas Reservoir Description

In view of some characteristics of gas reservoir development and using the key content of domestic oil reservoir description for reference, the content of gas reservoir description is summarized into 2 major parts such as static description and dynamic description, 8 characteristic elements, and 35 main parameters (Table 9-3) in response to the main problems to be solved by gas reservoir description.

Table 9-3 Parameters of gas reservoir description

Characteristic elements of gas reservoir	Static description parameters	Dynamic description parameters
Strata	Strata boundaries at different levels, thickness, lithologic composition	
Structure	Structural features of key bedding surfaces, faults	Fault sealing property
Reservoir	Lithology, reservoir space, fracture parameters, physical properties, the geometry and connectivity of reservoirs, NTG	Stress sensitivity, sand production, and multi-medium seepage characteristics
Fluid	Fluid composition, formation water occurrence	Relative permeability, phase state, physical properties of gas, water invasion mode and energy
Boundary conditions	Trap boundary, GWC, geological boundary of reservoir-permeator unit	Pressure drop boundary/flow boundary
Formation energy	Formation pressure, formation temperature, energy of edge water and bottom water	Pressure field distribution
In-situ stress field	Modulus of elasticity, principal stress orientation	
Reserves	Energy storage coefficient/abundance, undeveloped proven reserves	Dynamic reserves/EUR, producing degree of reserves and remaining reserves

The characteristic elements of a gas reservoir constitute all of the gas reservoir, including strata, structure, reservoir, fluid, boundary conditions, formation energy, in-situ stress field, and reserves. The description of these eight gas reservoir characteristic elements differs in different development stages, but basically covers the entire process of gas reservoir development.

1. Strata

The understanding of strata is the foundation of geological research, macroscopically including stratigraphic age, stratigraphic structure, and stratigraphic distribution. The key points include stratigraphic boundaries, formation thickness and lithologic composition of strata at different levels of gas layer development. The description of these three elements can lay the foundation for the study of gas reservoir distribution law. The description results are mainly reflected in the establishment of stratigraphic framework and lithologic combination.

2. Structure

The core parameters of structural description are the structural features of bedding surfaces, the distribution of faults, and their sealing property. Most gas reservoirs are affected by the developmental morphology of structures. Even for tight gas reservoirs, the gas-water distribution will be affected by local micro-scale structural changes, or by the distribution of structural fractures. Therefore, the results of structural description shall not only solve the problem of structural form and amplitude changes in a region, but also reveal the distribution of faults and their control over the distribution of gas layers. Especially for complex structural gas reservoirs, improving the accuracy of recognition of structures is an inevitable requirement for gradual deep development of gas reservoirs.

3. Reservoir

Reservoir description is mainly based on static parameters, and also involves several key dynamic parameters. Static parameters include lithology, reservoir space, fracture parameters, distribution of physical properties, the geometry and connectivity of reservoirs, and NTG. Dynamic parameters include stress sensitivity, sand production, and multi-medium seepage characteristics. Reservoir description is the foundation of gas reservoir development. In different gas reservoirs, there are various reservoirs, their distribution law is quite different, and the changes in their physical properties are large. Therefore, reservoir description is the core of gas reservoir description, and is also a difficult and strongly comprehensive gas reservoir description task with multiple methods. The data used for reservoir description include static data such as cores, logging data, seismic data, etc. as well as production performance data such as well testing data, gas testing data, etc., and involve a wide range of disciplines. The main result of reservoir description shall give gas layer enrichment areas, the continuity of gas layer distribution, and the connectivity of gas layers, and provide the basis for the determination of well pattern and well spacing.

4. Fluid

Gas reservoir fluids are mainly two phases such as gas and water; for condensate gas reservoirs, there is condensate oil in them. Fluid description mainly involves dynamic parameters, including relative permeability characteristics, phase state characteristics, physical properties of gas, and water invasion mode and energy. There are few static parameters, mainly including the properties of fluid components and formation water occurrence. For gas reservoirs, in addition to describing the distribution of gas layers, the description of gas reservoir water bodies is very important. For edge (bottom) water gas reservoirs, the coning of water bodies will cause flooding in gas reservoirs and then failure to produce them. For gas reservoirs with developed interlayer retained water, the distribution of water bodies directly affects the development effect of gas wells. At present, there is no effective method to predict the distribution of water bodies in gas reservoirs with developed interlayer retained water. The result of fluid description is mainly to solve the problem of gas-water distribution law.

5. Boundary conditions

The description of boundary conditions is a characteristic of gas reservoir description. Gas reservoirs are developed relying on their pressure. The boundary conditions determine the pressure relief range of gas and have a direct impact on the productivity of gas wells and the available reserves of the entire gas reservoir. The static parameters of boundary condition description are mainly trap boundaries, GWC and the geological boundaries of reservoir-permeator units. The key dynamic parameters are pressure drop boundaries/flow boundaries.

6. Formation energy

The core of formation energy description is the change of formation pressure. During the development of a gas reservoir, changes in pressure directly reflect the degree of gas recovery. Therefore, it can be said that for gas reservoirs, pressure description is an indispensable research content in the whole process of gas reservoir development. The static parameters of formation energy description are formation pressure, formation temperature and edge (bottom) water energy, which are especially important in the early stage of gas reservoir development. The dynamic parameter of formation energy description is pressure field distribution, which reflects the pressure change during the development of a gas reservoir, and can provide guidance to the prediction of the distribution of the undeveloped reserves of the gas reservoir.

7. In-situ stress field

In-situ stress field description is a supplement to the understanding of gas reservoirs, which is more meaningful for unconventional gas reservoirs and is closely related to the implementation of reservoir stimulation technologies. The described parameters include elastic modulus and principal stress orientation.

8. Reserves

The description of reserves is phased. The key parameters of reserves description in the early stage of development are energy storage coefficient/abundance and undeveloped proved reserves. The parameters of reserves description in the middle and late stages of development are mainly dynamic reserves/EUR, producing degree of the reserves and remaining reserves distribution. Reserves description is also a comprehensive description content, which needs to be comprehensively demonstrated using multiple parameters involving static data and dynamic data.

(Ⅲ) **Division of Gas Reservoir Description Stages**

The division of gas reservoir description stages is similar to that of oil reservoir description stages, and gas reservoirs have fewer development stages than reservoirs.

The stages of oilfield development have long been recognized by people, and some common stages and basic practices have been formed, which have similarities and minor differences. Generally speaking, after an oilfield is discovered, it can be roughly divided into evaluation stage, scheme design stage, implementation stage, monitoring stage, adjustment stage (high water cut stage), tertiary oil recovery stage, and finally oilfield abandonment stage. Each stage reflects the deepening of people's understanding of oil reservoirs. In general, these stages can be classified into three major development stages such as early, middle, and late ones, or they can be called respectively the three major development stages such as oilfield development preparation stage, main development stage and enhanced oil recovery stage. From the perspective of oil reservoir description, the oil reservoir description corresponding to the three major stages is very different, which is reflected in the fact that the degree of data possessed, the development problems to be solved, and the focus and accuracy of oil reservoir description are extremely different and the oil reservoir description technologies and methods used are also very different. Therefore, it has been currently agreed that oil reservoir description is divided into three stages such as early, middle and late ones.

The gas field production stage can be divided into evaluation stage, scheme design stage, scheme implementation stage, monitoring stage, adjustment stage, and gas field abandonment stage. The stages of a gas field are fewer than those of an oilfield. It can also be divided into productivity construction stage, stable production stage and decline stage according to production deployment. Compared with oilfields, the development of gas fields is relatively single, the stages of gas field development are not very obvious, and the boundaries between the stages are sometimes intersecting. Using the division of oil reservoir description stages for reference, gas reservoir description can also be divided into three stages, namely early, middle and late gas reservoir descriptions.

1. Early gas reservoir description

The stage from the discovery of a gas field to the completion of its development plan (before the gas field is put into development) can be called the early stage of development. The gas reservoir description performed in this stage can be called the early-stage gas reservoir description. The main task of this stage is to evaluate the feasibility of the gas reservoir development and then formulate the overall development plan. In this stage, there is few drilling data and a lack of dynamic data, the seismic data is mainly 2D data, and some gas fields may have local dense well pattern pilot areas. According to development evaluation and design requirements, determine the proven geological reserves and predicted recoverable reserves of the evaluation area, propose planned development deployment, determine the development mode and well pattern deployment, make suggestions on gas production engineering facilities, estimate the possible production scale, and evaluate the economic benefits to ensure there is no error of principle in the study of development feasibility and plan. The task of the gas reservoir description is to determine the basic framework of the gas reservoir (including structure, stratigraphy, deposition, etc.), ascertain the storage characteristics and 3D spatial distribution characteristics of the main reservoirs, and determine the type of the gas reservoir and the distribution of gas-water systems. Therefore, the gas reservoir description in this stage focuses on the establishment of a geological conceptual model based on the large framework and principle without pursuing too many details.

2. Middle gas reservoir description

After the preparation of the gas field development plan is completed and the gas field is fully put into development, the gas field is stably produced based on the production capacity target designed in the plan, that is, before the gas field enters the decline period. This stage can be called the main development stage of the gas field. The gas reservoir description performed in this stage can be called middle gas reservoir description. Once the gas field is fully put into development, drilling data and dynamic data will increase rapidly, and a variety of testing and monitoring data will gradually become available. The task of development and research in this stage is to implement the development plan, prepare well completion and perforation schemes, determine the well pattern and well spacing, carry out initial production and injection allocation, predict development performance, optimize the well pattern and well spacing, increase the producing degree of reserves, and make up for the decline in production. For this reason, the task of gas reservoir description in this stage is to carry out division and correlation of layers, further ascertain various structure and reservoir characteristics not determined in the early gas reservoir description, depict the enrichment law of gas layers, establish a static geological model, perform numerical simulation, and predict the production law of gas wells. The focus of gas reservoir description in this stage is to further ascertain the distribution of gas layers, optimize well positions, and provide support for gas field productivity construction.

3. Late gas reservoir description

After the gas field development plan is implemented and the gas field is stably produced for a certain period of time according to the production capacity construction target designed in the plan, the gas field enters a decline period, and begins to take account of the distribution of its remaining reserves so as to further improve its recovery ratio; that is, it enters the late stage of development. In this stage, a deep and objective understanding of the gas field has been achieved, and the main targets of production and potential tapping have turned to be the relatively dispersed and locally relative rich gas areas not swept by pressure drop. On the basis of early gas reservoir description and middle gas reservoir description, carry out further refinement, predict the changes of various sand bodies between wells more finely, accurately and quantitatively, and reveal the distribution of micro-faults and microstructures. The focus of gas reservoir description is to establish a fine 3D prediction model, reveal the spatial distribution of remaining reserves and enhance the gas field's recovery ratio.

II. Main Methods of Natural Gas Distribution Description

(I) Logging Identification of Gas Layers

For conventional gas reservoirs, gas layers usually have a typical "excavation effect", and their interpretation difficulty is not large. However, with the discovery and large-scale development of unconventional gas reservoirs such as tight gas reservoirs etc. in recent years, new difficulties in the identification of gas layers have been brought. Tight reservoirs have low porosity and permeability, and the fluids contained in them have a relatively small impact on their typical characteristics. Therefore, the logging response characteristics of gas layers are not obvious, and some characteristic parameters need to be built to identify gas and water layers. When strata contain gas, acoustic logging, density logging, and neutron logging all have a good indication of gas layers, which is manifested in increased interval transit time values and decreased neutron logging and density logging values. Therefore, acoustic, acoustic logging, density logging, and neutron logging can be used to build gas layer sensitivity parameters and then identify gas layers, gas-water layers, and gas-bearing water layers. Here the P-wave slowness ratio method and the

AK crossplot combination method are introduced.

1. P-wave slowness ratio method

The "excavation effect" of natural gas is quantified as a parameter DT, and layers are identified according to the magnitude of DT. Synthesize the neutron logging value into the interval transit time Δt_1, $\Delta t_1 = (1-\phi_n)\Delta t_{ma}+\phi_n\Delta t_f$; Calculate DT, $DT = (\Delta t-\Delta t_1)/\Delta t$, where, Δt, Δt_{ma}, and Δt_f are strata interval transit time, skeleton interval transit time and fluid interval transit time respectively, in μs/m, and ϕ_n is the neutron porosity of strata.

When strata contain gas, Δt increases, ϕ_n decreases, and Δt_1 decreases, so $DT>0$. When strata are not gas layers, $DT\leq 0$.

2. AK crossplot combination method

The slope of the line connecting the skeleton point with the fluid point on the neutron-density crossplot is defined as:

$$A = \frac{\rho_{ma} - \rho_f}{\phi_f - \phi_{ma}} = \frac{\rho_b - \rho_f}{\phi_f - \phi_n}$$

The slope of the line connecting the skeleton point with the fluid point on the neutron-acoustic crossplot is defined as:

$$K = \frac{\Delta t_f - \Delta t_{ma}}{\phi_f - \phi_{ma}} \times 0.01 = \frac{\Delta t_f - \Delta t}{\phi_f - \phi_n} \times 0.01$$

Where: ρ_b, ρ_{ma}, ρ_f—rock bulk density, skeleton density and fluid density respectively, g/cm³;

Δt, Δt_{ma}, and Δt_f—strata interval transit time, skeleton interval transit time and fluid interval transit time respectively, μs/m;

ϕ_n, ϕ_{ma}, and ϕ_f—neutron porosity of strata, hydrogen index of skeleton, and hydrogen index of fluid, respectively.

A and K are parameters that reflect the characteristics of rock skeleton and pore fluids without being affected by porosity. When strata contain gas, both ρ_b and ϕ_n decrease, and Δt increases, so both A and K decrease. The ρ_b and ϕ_n of oil layers and water layers are relatively large, and their Δt is small, so both A and K are relatively large. Magnify the difference between gas layers and water layers, select $\sqrt{A^2+K^2}$ as the parameter for gas layer identification, and define the identification parameter as AK, which is expressed by the following formula:

$$AK = \sqrt{\left(\frac{\rho_b - \rho_f}{\phi_f - \phi_n}\right)^2 + \left(\frac{\Delta t_f - \Delta t}{\phi_f - \phi_n} \times 0.001\right)^2}$$
$$= \frac{1}{\phi_f - \phi_n}\sqrt{(\rho_b - \rho_f)^2 + [(\Delta t_f - \Delta t) \times 0.01]^2}$$

Taking the western area of Sulige Gasfield as an example, the analysis of the single-layer gas test intervals shows that the P-wave slowness ratio method and the AK crossplot combination method have good effects in identifying gas and water layers. Therefore, by working out the $AK-DT$ crossplot (Figure 9-20) as per the data of the single-layer gas test points, the ranges of the AK

and DT values of gas layers, gas-water layers, and gas-bearing water layers have been determined. Gas layers: $AK<4.55$, $DT>0.065$; gas-water layers: $4.55<AK<4.64$, $0.02<DT<0.065$; Gas-bearing water layers: $AK>4.64$, $DT<0.02$. Neutron logging, density logging and acoustic logging require relatively high hole conditions. Therefore, when the hole diameter is enlarged, neutron logging, density logging and acoustic logging cannot reflect the properties of the strata and fluids around the hole very well. Therefore, 92 single-layer gas test points have been analyzed for hole diameter, and 21 hole diameter enlargement sections have been excluded. The 71 single-layer gas test points with good hole conditions have been interpreted using the $AK-DT$ method. The interpretation result shows that 7 points are in line with the gas test conclusion. Compared with the interpretation accuracy of the original interpretation conclusion, the interpretation accuracy of this interpretation conclusion has increased by 17% and is up to 90%.

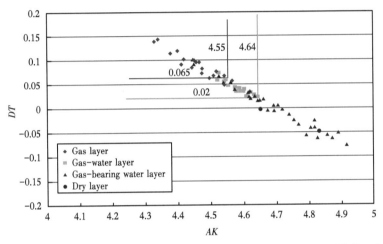

Figure 9-20　$AK-DT$ crossplot of single-layer gas test in the western area of Sulige Gasfield

(Ⅱ) Seismic Gas Bearing Detection

1. Wavelet absorption coefficient method

The main principle is that when a seismic wave propagates in strata, its energy will continue to attenuate as the propagation distance increases. When there is gas in strata, energy attenuation will be accelerated at the high-frequency components of the wavelet. A seismic trace is the convolution of the seismic wavelet and the reflection coefficient of strata, the reflection coefficient does not absorb energy, and the information of energy absorption is contained in the wavelet. Conventional absorption analysis is direct calculation using seismic traces, so that the strong amplitudes in non-absorbent anomalous seismic traces are "detected". Moreover, convolution suppresses the effective information of the wavelet spectrum and increases the unreliability of reservoir prediction. The advantage of this method is that the influence of the reflection coefficient of strata and the amplitude anomaly caused by the difference between upper and lower strata are eliminated using this method, so it is more reliable and has very high sensitivity.

2. AVO analysis technology

Based on the gas-bearing sandstones from logging interpretation, the AVO gas-bearing sandstone combination is determined through the determination of geophysical parameters of rocks and logging AVO forward modeling; that is, the amplitude is enhanced as the angle of incidence increases. Thus the variation law of the amplitude with offset is identified for the sandstone section on the pre-stack fidelity-preserved dynamic correction gathers. In addition, the gas-bearing properties of sandstones are qualitatively predicted using the technologies such as separate-offset stacking, AVO attribute stacking and extraction, etc.

3. Multi-wave seismic technology

The pre-stack AVO analysis of multi-wave Z-component with large angles is different from conventional seismic analysis. The difference is that the offset of about 7000m maintains the AVO information of nearly 35° incidence angle. In addition, very good gather data are obtained through the flexible combination of small trace intervals, and the original fidelity of single-trace detection makes AVO phenomena be more clearly reflected. The high-precision analysis of Z-component instantaneous wavelet energy attenuation by absorption is based on the relationship between the high-frequency component energy of seismic waves and gas bearing properties. According to the analysis of near-well seismic traces, the signal energy of 10-15Hz at high production well points is relatively strengthened, while the energy of 30-40Hz (main frequency) is relatively weakened. This indicates that the absorption coefficient of the instantaneous wavelet obtained by this change has a good corresponding relationship with the gas-bearing properties of reservoirs. However, the prerequisite for the application of the technology must be that a sufficiently wide frequency band is reserved in the complete fidelity and processing of field signals. The medium-offset stack profile with the best quality of the multi-wave Z-component ensures the effective application of this technology. The joint interpretation of P-wave and S-wave is the calibration of P-wave and PSV in their respective time domains for known wells with P-wave and S-wave logging data; then the PSV-wave profile in the PSV-wave time domain is compressed to the P-wave time domain to obtain the Poisson's ratio and calculate the amplitudes of the target horizons of P-wave and PSV-wave (attributes such as maximum amplitude, root mean square amplitude, average amplitude, etc.); next, the effectiveness of reservoirs can be evaluated relatively accurately through the quantitative calculation of the amplitude ratio.

(Ⅲ) Formation Pressure Evaluation

Formation pressure is particularly important for gas field development. It can be said that the oil production is the production of oil saturation, and the natural gas production directly produces formation pressure. The formation pressure is obtained through formation pressure evaluation of gas wells. The main methods include the following.

1. Dynamic monitoring method

The static pressure test before a new well is put into production, the shut-in test of a production well and the pressure monitoring of an observation well are the direct methods to determine the formation pressure of a gas well. Restricted by production management, technical conditions and economic factors, the shut-in time of a production well is often limited. When it is difficult to completely shut in the well till reaching stable conditions, the well test analysis method can be used to estimate the formation pressure.

2. Pressure drop calculation method

The pressure drop method is another name of the material balance analysis method for constant-volume closed elastic gas drive gas reservoirs. The pressure drop method is often used to calculate the dynamic reserves of gas reservoirs, and also to evaluate the formation pressure of gas wells. For constant-volume closed elastic gas drive gas reservoirs, the ratio of formation pressure to gas deviation coefficient has a linear relationship with cumulative production. The pressure drop of gas wells also has this relationship when the inter-well interference effect is not significant or the supply scope of a single well continues to be in dynamic equilibrium without the influence of water invasion. The ratio of formation pressure to gas deviation coefficient is calculated by using the existing data, and the linear relationship between the ratio and cumulative production is established. Then the ratio of unknown wells can be calculated according to the established linear relationship and the current cumulative production. Thus, the formation pressure of unknown wells can be determined by trial and error iterative calculation according to the relationship between pressure and deviation coefficient. The application condition of this method is that gas reservoirs have no abnormal effect that changes the linear relationship on the pressure drop diagram, e.g. water invasion, late low-permeability recharge, special laws of pressure attenuation in abnormal high-pressure gas reservoirs, etc. In addition, there is no change in the supply scope of a single well, and some accurate formation pressure data corresponding to different cumulative gas production times have been obtained through testing. When the above condition is met, the reliability of the method can be guaranteed.

3. Wellhead stable static pressure conversion method

There are very few methods of running a pressure gauge to the bottom hole to measure the formation pressure in the actual production process of a gas well. Under certain conditions, the formation pressure can be quickly and approximately calculated using the wellhead pressure data during the shut-in period under certain conditions. In the case of simplified consideration of ideal gas, the hydrostatic column pressure calculation formula is $p_{ws}=p_{wh}+DH$, where p_{ws} is the bottom hole static pressure, p_{wh} is the wellhead static pressure, D is the average pressure gradient in the wellbore, and H is the vertical depth in the middle part of the gas well's pay formation. In the case of a single static gas column, the average pressure gradient in the wellbore is directly proportional to the average density of the gas in the wellbore, and also to the average pressure in the wellbore. According to the arithmetic average method, the average pressure in the wellbore is calculated, and it can be known through deduction that the average pressure gradient in the wellbore is also directly proportional to the wellhead static pressure, and the proportional coefficient is related to the well depth. For gas fields with little difference in gas composition and gas well depth, the empirical relationship between the static pressure gradient and the wellhead pressure can be established using the wellbore static pressure gradient test data of gas wells. As long as the relationship between pressure gradient and wellhead pressure is grasped, it is very easy to determine the bottom hole pressure of a gas well according to the formula. The applicable condition of this method is that the shut-in pressure and wellhead pressure are stable, the fluid in the wellbore is single-phase gas, and the relative density and depth of the gas well-application object are approximately the same as the corresponding parameters of the sample gas well for establishing the empirical relation.

4. Binomial productivity equation estimation method

The formation pressure can be inversely calculated according to the binomial productivity equation for steady seepage of gas wells. The production data of gas wells are easy to obtain. If the productivity equation coefficient of a gas well has been obtained through test analysis, the formation pressure at the corresponding time can be calculated as long as the bottom hole flowing pressure is measured. The applicable condition of this method is that there is no change in the productivity equation coefficient of a gas well. Bottom hole purification improvement or blockage pollution during the production process, water production of a gas well, and large difficulties in reaching the stable state of seepage flow in the supply area of a low-permeability formation gas well all make this method in applicable.

(Ⅳ) **Evaluation of Gas Drainage Scope**

Gas drainage scope is an important part of gas reservoir description and can also be referred to as gas drainage radius, drainage radius, pressure relief sweep scope, etc. The core of gas drainage scope refers to the distribution area of gas around the wellbore that can participate in flowing during the entire life cycle of a gas well. In terms of geological evaluation, the scope of effective gas communication is predicted by analyzing the morphology, scale and connectivity of effective reservoirs. Dynamic evaluation is the main method for the study of gas drainage scope of a gas well, including well test interpretation, production instability method, curve integration method, pressure drop method, elastic two-phase method, pressure buildup method, decline analysis method, etc. Different development stages have different dynamic data and information, and the applicable dynamic evaluation methods are different. Here the first three typical methods are introduced.

1. Well test interpretation method

The well test interpretation method is the main test means available in the early stage, which can be used to preliminarily determine the production capacity of a gas well, and obtain the seepage capacity and pressure relief sweep scope of underground reservoirs. Transient well testing is based on strict seepage theory and is a scientific and effective method for detecting the pressure relief boundary of a gas well and judging its geometry. According to the transient well testing analysis of the 12 gas wells put into production in the early stage of Sulige Gasfield, the geometry of the effective sand body controlled by a single well mainly has three forms such as two-zone combination, parallel boundary and rectangle, the gas supply scope is small, and the single-well controlled reserves are low. The inner area scope of the two-zone composite model is 50-70m; the river channel width predicted by the parallel boundary model is 60-110m; the river channel width and length are 60-180m and <1000m respectively as predicted by the rectangle model. Because tight gas reservoirs are tight and have weak seepage capacity, pressure transmission is slow, and short-term gas supply is mainly in near-well zones. The pressure drop is large, away from the location of the well point, so the pressure drop temporarily cannot sweep the boundary, forming a pressure drop funnel. As the production time increases, the pressure drop scope will gradually expand. According to the physical characteristics of tight gas reservoirs, the pressure relief scope obtained from early well testing is relatively small, but it also indirectly reflects the development laws of Sulige Gasfield, such as small scale and small connectivity scope of effective sand bodies, etc.

2. Production instability analysis method

The production instability analysis method is suitable for gas wells that have been produced for a long period of time and have at least 2–3 years of production history. The prediction result from this method has high reliability. This method is subject to few restrictions. This method mainly requires the actual production data of a gas well, including production and pressure data, and the physical property data of the reservoirs encountered in the gas well. This method has undergone rigorous mathematical derivation, and can be used to effectively evaluate the gas well's drainage scope, dynamic reserves, etc. Its principle basis: fluids flow from the reservoir to the wellbore in two stages including the unsteady flow stage at the beginning of well opening and the boundary flow stage in the late period. In the unsteady flow stage, the pressure drop does not sweep the boundary, and the boundary does not affect flowing, which is the flow stage usually described by transient well testing. When the pressure drop spreads to the boundary and affects the flow, the flow in the reservoir enters the boundary flow stage (including the pseudo-steady flow stage in the case of constant production and the boundary flow stage in the case of variable production). The traditional Arps production decline curve describes the production decline trend under boundary flow conditions. The "modern production performance analysis method" includes the entire flow process from the unsteady flow stage to the boundary flow stage. In the unsteady flow stage, the functional relationships are established between the introduced parameters such as new production (pseudo-pressure normalized production), pressure (production-normalized pseudo-pressure) and time (material balance pseudo-time) functions and the dimensionless parameters in the interpretation of transient well testing, so that the characteristic chart of the unsteady flow stage of the gas well is established. In the boundary flow stage, the traditional Arps production decline curve is non-dimensionalized, and the introduction of new production, pressure and time functions makes the later Arps decline curve converge into an exponential decline curve or a harmonic decline curve. Thus, the characteristic chart of gas well production curve is established. The first half of the chart is a set of characteristic curves of the unsteady flow stage that represent different dimensionless well control radii (r_e/r_{wa}). This set of curves converges into an exponential decline curve (or a harmonic decline curve) in the boundary flow stage. In order to improve the accuracy of curve analysis, in addition to the pressure-normalized production, the integral form and the derivative form of the production-normalized pressure are also used to assist in the analysis. The skin factor, reservoir permeability, fracture length, etc. of the gas well can be calculated by fitting the unsteady flow stage of the chart. The controlled reserves (dynamic reserves) of the gas well can be calculated by fitting the boundary flow stage of the chart.

The traditional Arps decline curve method is an empirical method. Its advantage is that it does not require reservoir parameters and can be used to predict production and calculate recoverable reserves only by using the production variation trend. The applicable conditions of the method are as follows. (1) The gas well is produced at a constant bottom hole flowing pressure. (2) From the perspective of strict flow stage, the decline curve represents the boundary flow stage and cannot be used to analyze the unsteady flow stage during early production. (3) In the analysis, the production time of the gas well (field) is required to be long enough to find the trend of production decline. (4) Reservoir parameters and gas well production measures will not change.

The production curve characteristic chart fitting method established by Fetkovich, Blasingame, Agarwal-Gardner, *et al.* on the basis of the Arps decline curve takes account of the impact of flow pressure changes on production by utilizing the pressure-normalized production, so that the curve can better reflect the flow characteristics of the reservoir itself, and make the analysis method suitable for both constant production and variable production. In addition, by introducing the pseudo-time function and material balance pseudo-time function, the PVT properties of the gas that change with the formation pressure and the reservoir stress sensitivity are taken into account. In terms of the flow stage, the production curve characteristic chart includes both the early unsteady flow stage and the late boundary flow stage.

The production curve characteristic chart fitting method needs only production data and flowing pressure data, and does not require shut-in pressure measurement data except for the original formation pressure. As long as the flow in the reservoir achieves pseudo-steady flow (under constant production conditions) or boundary flow (under variable production conditions), the analysis can be performed. Reservoir permeability, skin factor, and well-controlled reserves are calculated and the relationship between wells and water drive characteristics, etc. can be qualitatively analyzed by fitting the typical chart of gas well production curves.

The production curve characteristic chart includes traditional Arps and Fetkovich methods, modern Blasingame, Agarwal-Gardner (AG), Normalized Pressure Integral (NPI), Transient, Flowing-Material-Balance (flowing material balance) methods, etc. In addition, the analytical method combined with gas well production history fitting based on the chart can be used to comprehensively obtain the dynamic control reserves and drainage scope of the gas well.

3. Curve integral method

The curve integral method can be used to evaluate the dynamic reserves of a gas well. The principle of the curve integral method: mainly using the production performance curve of the gas well and through function fitting of the curve, the integral of the production time is calculated by the function; moreover, according to the abandoned production conditions of the gas well, the ultimate cumulative gas production of the gas well is obtained (Figure 9-21). This method makes full use of the production decline law of actual production wells. Especially for gas wells and blocks with a long production time, the method is more accurate. Block Su-36-11 was put into production in 2006, where most of the gas wells have been produced for more than 5 years, and the decline in the production of the gas wells has stabilized. The curve integral method can reflect the decline in the production of gas wells in this block very well and evaluate the accurate ultimate cumulative gas production. Curve fitting and function integration have been performed this time mainly in the way of combined analysis of the wells put into production in a separate year, i.e. the combination of the gas wells put into production in the same year, so as to finally obtain their average ultimate cumulative gas production.

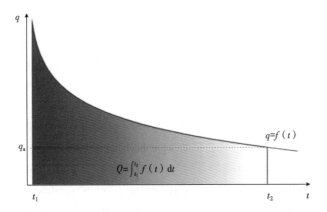

Figure 9-21 Schematic diagram of calculating the gas well's dynamic reserves using the curve fitting integration method

(V) Interference Testing Evaluation

Whether there is interference between gas wells and the degree of interference in the process of gas field development are key factors for reasonably designing development well patterns, enhancing the gas field's recovery ratio, and improving gas field development benefits. For conventional massive integral structural gas reservoirs, it is appropriate to use the development mode of sparse wells with high production, maximize the well spacing, reduce the degree of interference between wells, and ensure the high and stable production of gas wells. For lithologic gas reservoirs with poor connectivity and especially for lenticular sandstone gas reservoirs represented by Sulige Gasfield, the complete avoidance of inter-well interference can ensure a high cumulative production of a single well, but the gas field's recovery ratio is low, thus making against full utilization of resources. Therefore, it is necessary to scientifically optimize the well pattern and reasonably increase its density by optimizing the well pattern and finding the juncture of the degree of interference between wells with the gas field development benefits, so as to ensure that the gas field's recovery ratio is maximized on the premise of ensuring beneficial development. The inter-well interference evaluation study in Sulige Gasfield shows that inter-well interference experiments in 42 well groups have covered different well spacings and row spacings. With the increase in the experimental well spacing and row spacing, the proportion of well groups with inter-well interference gradually decreases. When the well spacing is 500-600m and the row spacing is 600-700m, the proportion of well groups with inter-well interference is less than 25%. When the well spacing is greater than 600m and the row spacing is greater than 800m, there is basically no well group with inter-well interference. According to the development characteristics such as multi-layer continuous superimposition of effective sand bodies, the main effective sand body in Sulige is smaller than 600m×800m in scale. Correspondingly, when the well pattern is reduced to 600m×800m, inter-well interference hardly occurs, so it is feasible to increase the well pattern density.

Chapter 10　Geological Models of Reservoirs

The ultimate goal of fine reservoir description is to establish a geological model of reservoirs, including the skeleton model of reservoirs, e.g. sand body model, sedimentary facies model or flow unit model, etc., as well as various physical parameter models and fluid distribution models of reservoirs, e. g. porosity model, permeability model, oil saturation model, remaining oil distribution model, etc. The key to establishing a geological model of reservoirs depends on two aspects: (1) qualitative and quantitative description of the internal structure of reservoirs, such as the shape, occurrence, scale, and combination law of effective sand bodies, barriers and interlayers; (2) selection of modeling methods and modeling parameters. There are a variety of modeling methods, and commercial software has been formed at present. For general reservoirs, description workers only need to master the operation of software to achieve geological modeling, and the selection of modeling parameters is mainly based on the understanding of reservoir structure. The reservoir description work carried out in the previous chapters is mainly to deeply understand the characteristics of reservoirs and fluids, and to provide important parameter basis for the establishment of the final geological model of reservoirs.

Section 1　Classification of Geological Models of Reservoirs

The geological model of reservoirs shall be able to meet the needs of different development stages of oilfields, reflect the porosity, permeability, fluid characteristics and dynamic characteristics of reservoirs, and also meet the needs of prediction of different levels and scales of geological bodies. It is essentially a comprehensive reflection of the static and dynamic characteristics of reservoirs in 3D space. Different scholars have different research levels of geological bodies, as well as different focus on the study of reservoir target parameters, so there are different classification methods for geological models of reservoirs. Among various classification methods, it is appropriate to classify geological models of reservoirs according to the tasks in the development stage and the accuracy of model establishment. Different oilfield development stages have different workloads, there are differences in the data and understanding degree of reservoirs, and the development tasks to be solved are also different. As the production degree of reservoirs increases, the modeling work advances gradually from shallow to deep. The geological models of reservoirs to be established in different development stages are correspondingly different. On the whole, with the progress of oilfield development stages and the increasing of reservoir production degree, the requirements for geological models of reservoirs have also changed from simple to fine and from coarse to precise. Qiu Yinan (1991) divided a geological model of reservoirs into three types: conceptual model, static model and prediction model. This classification method is a typical representative of geological model classification methods.

I. Conceptual model

For a certain sedimentary type or genetic type of reservoir, its representative characteristics (heterogeneity, continuity, etc.) are abstracted, typified and conceptualized so as to establish a reservoir geological model that is universally representative of such reservoir in a research area (oilfield), which is called a conceptual model. The conceptual model is not a geological model of a reservoir or a set of specific reservoirs, but it represents the basic appearance of a certain type of reservoirs in a certain area (oilfield).

The conceptual model is widely used in the early development of an oilfield. The conceptual model of reservoirs is mainly used to study various development strategy issues from the discovery of the oilfield to its evaluation stage and development design stage. In this stage, the oilfield has only a few exploration wells and appraisal wells with large well spacing. Limited by data conditions, it is impossible to make a detailed description of the entire reservoirs, and only a conceptual model of the reservoirs in the research area can be established referring to theoretical sedimentary models, diagenetic models and prototype models of similar sedimentary reservoirs in adjacent areas according to a small amount of information. This conceptual model is crucial to the determination of development strategy and can avoid strategic mistakes. For example, in the deployment of well patterns, large well spacing can be used for sheet sand bodies, small well spacing is required for channel sand bodies, and horizontal wells are better for massive bottom water reservoirs.

The conceptual model shall generally be established using core data as directly as possible based on reservoir sedimentology so as to avoid relying on indirect data such as logging interpretation data etc. This is because the accuracy of quantitative logging interpretation is not high enough in the early evaluation stage of reservoirs.

The conceptual model is very important in the development feasibility and development design research stage. Instructive decision-making research on various development strategies can be carried out through the numerical simulation of reservoirs. For example, the conceptual model can be used to study all the following issues: the technical and economic feasibility of the investment in development, the optimization of development modes, series of strata and well patterns, the estimation of the recovery ratio in each stage, the prediction of the main problems that may occur in the production process, the strategic issues that must be correctly decided before development, etc.

II. Static model

For one reservoir or a set of reservoirs in a specific oilfield (or development area), the geological model established by truthfully describing the changes and distribution of the reservoir characteristics in 3D space is called the static model of the reservoir.

The truthful description of the entire reservoirs generally requires a dense well pattern, that is, it can only be carried out after the development well pattern is established. The static model is mainly intended for the implementation of the oilfield development plan (that is, the determination of injection and production well types, the implementation of the perforation scheme, etc.), daily oilfield development performance analysis, operation construction, proration production and injection schemes, and local adjustment services.

Since the 1960s, China's oilfields have established such static models after put into development, but most of them were compiled manually, such as various layer plan views, reservoir profiles and grid diagrams. Individual oilfields have also worked out solid models to show reservoirs more intuitively. These static models of reservoirs have played an essential role in the development of water flood fields in China.

Since the 1980s, using the computer technology, foreign countries have gradually developed a 3D static model that relies on computer storage and display, that is, after reservoirs are gridded, 3D data volumes are established using the parameters of each grid block according to the 3D spatial distribution position. Thus, 3D display of reservoirs can be achieved, reservoir models of different horizons and different sections can be displayed by means of any slices and cross sections, and various other calculations and analyses can be performed, and more importantly, the models can be directly connected with numerical simulation.

The static model only describes the appearance of reservoirs revealed by multi-well patterns, and does not pursue the interpolation accuracy and extrapolation prediction of inter-well parameters.

Static models are widely used in the practice of waterflood development in China. From the daily management of production wells to various adjustment measures of oilfields, static models are all essential geological foundations.

III. Prediction model

The prediction model is the result of the in-depth development of oilfields. Compared with the static model, the prediction model not only emphasizes the description of multiple single wells, but also pays more attention to the prediction of inter-well reservoirs; and the established reservoir model is more accurate than the static model. The prediction model is the predictive interpolation or extrapolation of parameters of reservoirs between control points and in regions beyond control points. The prediction model requires richer well pattern information so as to obtain more reliable reservoir distribution rules and provide more reliable prediction parameters.

The prediction model is of great significance to tapping the potential of remaining oil. After waterflood development of reservoirs, there is still a large amount of remaining oil underground, and there is a need for development adjustments, well pattern density increasing, or tertiary oil recovery. Therefore, a highly accurate reservoir model and remaining oil distribution model need to be established. Tertiary oil recovery technologies have achieved rapid development in the past 20 years, but with the exception of thermal recovery of heavy oil, other technologies have not reached the level of universal industrial application. One of the important reasons is that the accuracy of reservoir models cannot meet the needs of establishing a high-precision remaining oil distribution model. The distribution of reservoir parameters is extremely sensitive to the distribution of remaining oil, so reservoir characteristics and their subtle changes are much more sensitive to tertiary oil recovery injection agents and oil displacement efficiency than to water injection efficiency, and the reservoir model is required to have higher accuracy. In order to meet the needs of the middle and late stages of waterflood development and tertiary oil recovery for remaining oil production, it is necessary to predict the changes and their absolute values of parameters of reservoirs of tens of meters or even several meters between wells under the conditions of development well patterns (generally under the condition of 100 meters); that is, fine reservoir prediction models or fine reservoir geological models are established.

Section 2 Geological Modeling Methods

The purpose of reservoir description is to establish a quantitative reservoir geological model. This model is the basis of research work involving reservoir simulation, reservoir engineering, oil production technology, etc. With the rapid development of computer technology, the traditional geological work methods (e.g. compiling various 2D maps) have been gradually replaced by 3D quantitative geological models established and displayed by computer technology.

Reservoir modeling is actually to characterize the spatial distribution and change characteristics of reservoir structure and reservoir parameters. How to interpolate and extrapolate the characteristics of reservoirs between and beyond data points based on the data of known control points is the key point in the technology of establishing a geological model of reservoirs. According to this feature, the methods for establishing a geological model of reservoirs can be divided into two types: deterministic modeling and stochastic modeling.

I. Deterministic modeling method

Using the deterministic modeling method, it is deemed that the interpolation result between data control points is unique and deterministic, that is, the definite, unique and true reservoir parameter between well points is inferred by trying to start from deterministic control points (such as well points). The core issue of modeling is inter-well reservoir prediction. Given the data, the main method to improve the precision of the reservoir model is to improve the accuracy of inter-well prediction.

There are mainly several deterministic geological modeling methods below.

(I) **Traditional geological mapping methods**

Linear interpolation according to geological trend, including simple linear interpolation, trend surface drawing method, linear interpolation under facies belt control, etc.

These methods are mature for structural phenomena and parameters with weak heterogeneity, such as formation pressure, temperature field, fluid saturation, porosity, etc. Sometimes, they are even useful for the permeability distribution of stable sedimentary bodies, such as delta front mouth bars and sheet sands.

(Ⅱ) **Development seismic inversion methods**

1. 3D seismic

With the development of 3D seismic technology, seismic technology has evolved from being applied only to structural interpretation to reservoir description, it is possible to use 3D seismic data for high-resolution reservoir parameter inversion, and this new development seismic technology has been gradually formed.

The advantages of 3D seismic such as high planar coverage rate and high lateral acquisition density just make up for the deficiency of a too sparse well pattern such as insufficient control points. Therefore, the development seismic technology has become an indispensable technology in the description of reservoirs. In recent years, new acquisition, processing, interpretation and inversion technologies have emerged continuously, e.g. thousand-component to multi-component seismic, four-dimensional seismic, cross-well seismic, etc. At present, the relationship of seismic attributes (such as amplitude, etc.) and inverted formation attributes (such as interval transit time, acoustic impedance, etc.) with core (or logging) porosity is established, and the porosity inversion is carried out and then the permeability is calculated with the porosity. This method has been widely used. 3D seismic data volumes are converted into 3D reservoir attribute data volumes, thus directly realizing 3D modeling.

The biggest shortcoming of 3D seismic data is the low vertical resolution, generally 10-20m, and the vertical resolution of high-precision 3D seismic data of shallow layers can reach about 5m. Therefore, using the conventional 3D seismic technology, it is very difficult to distinguish the scale of a single sand body, and only the scale of a sand group can be identified; in addition, the accuracy of the predicted reservoir parameters (such as porosity, fluid saturation) is low, which is only the average value of large intervals.

3D seismic methods are mainly used in reservoir modeling in the exploration stage and early evaluation stage, and are used to determine the sequence stratigraphy framework, structural traps, fault characteristics, the macroscopic framework of sand bodies and the macroscopic distribution of reservoir parameters.

2. Cross-well seismic

Because of using downhole seismic sources and multi-trace receiver spreads, cross-well seismic has more advantages than surface seismic.

(1) Both seismic sources and geophones are in a well, so that the attenuation of seismic wave energy by near-surface weathering layers is avoided and the SNR can be improved.

(2) High-frequency seismic sources are used and inter-well sensors are very close to targets, so this is favorable for improving the resolution of seismic data.

(3) Using the first arrival of seismic waves, cross-well seismic tomography of P-waves and S-waves is achieved, velocity fields can be accurately established, and the interpretation accuracy of inter-well reservoir parameters can be greatly improved.

Admittedly, seismic attributes are not only controlled by physical properties of rocks, but also affected by other factors; in addition, limited by seismic resolution, the development seismic inversion result still has high uncertainties, and its effective application must be closely combined with geology. The geological laws and patterns of the region shall be fully utilized in the process of processing, interpretation, and inversion, and the application of the development seismic inversion result shall also be closely combined with geological understanding.

Since the development seismic inversion result is unique, it also belongs to the category of deterministic modeling.

(Ⅲ) Kriging method

The current popular computer geological mapping software still belongs to the category of deterministic modeling as long as it is based on interpolation technology. There are many typical interpolation methods, which can be roughly divided into traditional statistical valuation methods and geostatistical valuation methods (mainly including Kriging method). The traditional mathematical statistics interpolation method (such as the inverse distance square weighting method) only takes account of the distance between the observation point and the point to be estimated, without regard to the interrelation between the known point positions, i.e. the spatial correlation of reservoir parameters caused by geological laws, so the interpolation accuracy is relatively low. In order to improve the estimation accuracy of reservoir parameters, the Kriging method is used for inter-well interpolation.

The Kriging method is a smooth interpolation method and actually a special weighted average method. It is difficult to characterize the subtle changes and discreteness of inter-well parameters (such as complex changes of inter-well permeability). Moreover, the Kriging method is a partial valuation method, which does not take into account the overall structure of parameter distribution to a sufficient degree. When the reservoir continuity is poor, the well spacing is large, and the well spacing distribution is uneven, the estimation error is large. Therefore, the inter-well interpolation point given by the Kriging method is a certain value, but it is not the true value, but close to the true value. The magnitude of the error depends on the applicability of the Kriging method and objective geological conditions. As far as inter-well interpolation is concerned, the Kriging method can reflect objective geological laws better than the traditional mathematical statistics method, and has relatively high estimation accuracy, and is a powerful tool for quantitative reservoir description.

(Ⅳ) Other methods

The above three methods are deterministic modeling methods with universal significance and can be applied in many cases. In addition, there are some modeling methods for special cases. For example, with the widespread application of horizontal wells in China, a horizontal well modeling method has been gradually formed. A horizontal well is drilled along the strike direction or dip direction of reservoirs, and the lateral change parameter data of reservoirs or logging interpretation results are directly obtained, thereby establishing a deterministic reservoir model. Especially in the modeling of inclined barriers and interlayers, horizontal wells have played an important role (Figure 10-1).

The modeling of field outcrops and modern sedimentary bodies is also a deterministic modeling method. For field outcrops, carry out intensive section drilling and sampling, obtain the heterogeneity of outcrop section reservoirs, and establish reliable geological models. This is an important method for studying reservoir prototype models.

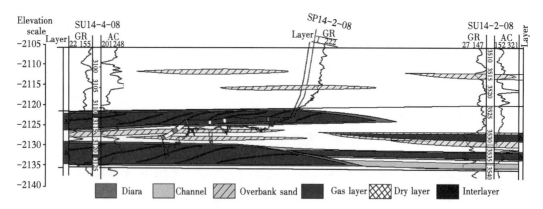

Figure 10-1　Identification of interlayers in horizontal wells in Sulige Gasfield

II. Stochastic modeling methods

With the rapid development of computer technology, geostatistics has developed into a relatively complete theoretical system. Because of its strong practical advantages, it has been widely regarded and applied as an important tool to solve problems. Especially the stochastic simulation technology formed on this basis has received great attention because of its great advantages in analyzing the heterogeneity and spatial uncertainty of oil and gas reservoirs. At present, almost all modeling commercial software is mainly based on stochastic modeling methods.

(I) **Overview of stochastic simulation**

An underground reservoir itself is definite, and has definite properties and characteristics at every location point. However, the underground reservoir is also very complex. It is the result of the combined effects of many complex geological processes (sedimentation, diagenesis, and tectonism) and has a complex structure (reservoir facies), and the reservoir parameters vary greatly in space. The data used to describe reservoirs are always incomplete in the process of reservoir description, so it is difficult for people to grasp the definite and true characteristics or properties of reservoirs at any scale. Especially for continental reservoirs with poor continuity and strong heterogeneity, it is more difficult to accurately characterize their characteristics. In this case, due to the lack of knowledge, the reservoir description is uncertain. These reservoir properties that need to be determined by "guessing" are the stochastic nature of reservoirs.

The stochastic modeling technology of reservoirs is a high and new technology for reservoir description that emerged in the late 1980s. It came into being to meet the needs of deepening of oil and gas field development and further enhance the oil and gas recovery ratio by using advanced secondary and tertiary oil recovery technologies. This provides reservoir geologists with a reservoir geological model that is quantitative and refined and can realistically reflect "uncertainty". The stochastic modeling technology not only respects the inherent geological law of reservoirs and reflects some objective randomness, but also quantitatively describes the uncertainty of the reservoir geological model caused by the lack of data and the limitations of people's understanding. It provides an important basis for risk analysis in strategic decision-making for oil and gas field development. This is the reason why this technology was highly valued by the petroleum industry as soon as it appeared.

Haldorson (1990) proposed six reasons for applying the stochastic simulation technology to describe deterministic reservoirs:

(1) The information on the spatial distribution and internal (geometric) structure of reservoirs and rock properties is incomplete at various scales;

(2) Complex spatial arrangement of reservoirs and facies;

(3) It is difficult to grasp the changes and change forms of rock properties relative to the spatial position and direction;

(4) No understanding of the relationship between rock physical properties and the rock volume used to obtain the average value (proportion problem);

(5) There are more static data than dynamic data;

(6) Convenient and fast.

Due to the randomness of reservoirs, their prediction result is of multiplicity. The only prediction result obtained using the deterministic modeling method has a certain degree of uncertainty, and using this as the basis for decision-making is risky. For this reason, stochastic simulation methods are widely used to model and predict reservoirs. Therefore, there are multiple reservoir models instead of one reservoir model to be built using stochastic simulation methods; i.e., several possibilities are realized within a certain range, thus meeting the requirements of correctness of oilfield development decisions within a certain risk range. This is an important difference from deterministic modeling methods. For each implementation (i.e. model), the statistical theoretical distribution features of the simulated parameters are consistent with the statistical distribution features of control point parameter values, i.e., so-called equal probability. The difference between different implementations is direct reflection of reservoir uncertainty. If all implementations are identical or of very small difference, this indicates that there are few uncertainty factors in the model. If there is relatively large difference between implementations, the uncertainty is large. It can be seen that one of the important purposes of stochastic modeling is to evaluate the uncertainty of reservoirs.

In addition, with the development of reservoir geology, the improvement of seismic reservoir prediction accuracy and the increasing of well pattern density, stochastic simulation methods are widely used in fine reservoir modeling in the late development stage. By increasing constraint conditions, the accuracy of reservoir modeling is improved, and a single model with high reliability can be obtained as the geological model of the research area. Under this kind of thinking, only a single model is generated using a stochastic simulation method.

The stochastic simulation method is quite different from the interpolation method such as Kriging method, mainly involving the following three aspects.

(1) The interpolation method only considers the accuracy of the local estimated value, and strives to provide the optimal (minimum estimated variance) and unbiased (the mean of the estimated value is the same as that of the observed point value) estimation of the unknown value of the estimated point, without regard to the spatial correlation (discreteness) of the estimated value. The stochastic simulation method first considers the overall nature of the result and the statistical

spatial correlation of the simulated value, and secondly the accuracy of the local estimated value.

(2) The interpolation method gives a smooth estimate between the observations (for example, a smooth curve chart of the research object), which weakens the discreteness of the observation data and ignores the subtle change between wells. Conditional stochastic simulation systematically adds "stochastic noise" to the interpolation model, so that the generated result is much more realistic than that using the interpolation model. "Stochastic noise" is just the subtle change between wells. For each local point, the simulated value is not completely true, and the estimated variance is even greater than that using the interpolation method, but the simulated curve can better represent the fluctuation of the real curve (Figure 10-2).

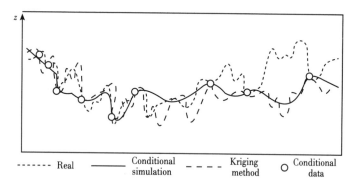

Figure 10-2 Comparison of stochastic simulation method with Kriging method

(3) The interpolation method (including the Kriging method) only generates one reservoir model. In stochastic modeling, many optional models are generated, and the difference between various models is just the reflection of spatial uncertainty.

Stochastic modeling has greater advantages for the study of reservoir heterogeneity, because a stochastic model can better reflect the discreteness of reservoirs, which is particularly important for oilfield development and production. The Kriging method is actually a linear smoothing low-pass filter and an estimation of conditional mathematical expectation, and thus has a smoothing effect. The interpolation method conceals the degree of heterogeneity (i.e., discreteness), and especially the degree of heterogeneity of reservoir parameters (such as permeability) with obvious discreteness, so it is not suitable for the characterization of permeability heterogeneity.

(Ⅱ) **Stochastic simulation algorithm**

Stochastic simulation is based on random function theory. A random function is characterized by the distribution function and covariance function (or variogram) of a regionalized variable. A random function $Z(X)$ has countless possible implementations $(A_s(X), s=1,2\cdots,\infty)$. The basic idea of simulation is to extract multiple possible implementations from a random function. Conditional restriction can be carried out in the simulation, that is, the experimental data of the observation point are used to restrict the simulation process so that the simulated value of the sampling point is the same as the measured value, which is conditional simulation.

At present, a variety of random simulation algorithms have been formed. The following main methods are highlighted here.

1. Marked point process simulation

The basic idea of marked point process simulation: based on the law of probability of point process and the distribution of geometric objects in space, generate space distribution of the central points of these objects, and then mark the object properties (such as the geometric form, size and direction etc. of the objects) on each point; i.e., generate the attribute information of these space points through stochastic modeling and match it with the existing condition information. From the perspective of geostatistics, marked point process simulation is to simulate the joint distribution of object points and their properties in 3D space.

1) Basic principle

Assume that U is a vector of spatial coordinates, X_K is a random variable describing the geometric size (shape, size, direction) of type K, and the geometric size can be defined by a parameterized analytical expression. The geometric shape, size and other attributes of the center point here are determined through the joint distribution of $X_K(u), I_K(u,k)$ ($k=1,2\cdots,K$). Where, $I_K(u,k)$ is a random function indicating whether the Kth type of geometric attribute appears at the position U. When U belongs to X_K, it is 1; otherwise it is 0. Thus by simulating the position of the target in space, the relevant attributes of the target are simulated. When the known condition information is met, a simulation implementation can be achieved.

According to different point process theories, the distribution of object center points in space can be independent, interrelated or exclusive. In practical applications, the location of the target points is determined by the following rules.

(1) Density function (the proportion and distribution trend of each facies). The spatial distribution of the target point density can be uniform, or a certain distribution trend can be given according to geological laws;

(2) Correlation function (whether the wells are connected);

(3) Repulsion principle (the minimum distance that can not be contacted between objects in the same facies or objects in different facies);

(4) The principle of gradual change of facies (gradual change relationship between different facies). The determination of the marked point process is a "step-by-step approximation process". Carry out iteration with multiple combinations of various parameter distributions and interactions until a satisfactory image is finally obtained.

2) Simulation steps

Take the simulation of distributary channel sand bodies as an example (Figure 10-3).

(1) Determine a lithofacies as the background facies. For example, interdistributary mudstones can be selected as the background facies, and distributary channel sand bodies are used as the simulation target bodies in the simulation of the lithofacies of delta plains;

(2) For the target body to be simulated, randomly select some position points, and give its shape to meet the appropriate size, anisotropy and direction;

(3) Check the position points, and simulate the form to match the previous conditional information (such as well data or seismic data) through the process of adding, canceling or replacing multiple times;

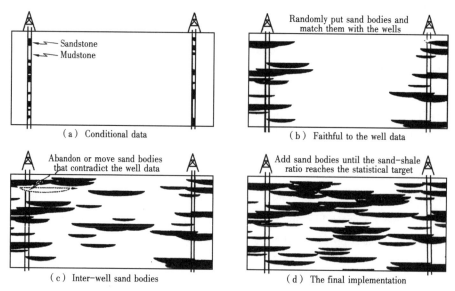

Figure 10-3 Schematic diagram of marked point process simulation (according to Srivastava, 1994)

(4) Check whether the distribution of various facies reaches the known proportion (or objective function). If it reaches the known proportion, the simulation process is approved; otherwise, go back to the previous step to continue.

The marked point process simulation method has its unique advantage: flexible application, i.e., some geologic data (such as facies percentage, width-to-thickness ratio of sand bodies, spatial distribution rules of various facies, etc.) can be easily added in the model as the conditional information, so that the recognition from geologists can be generalized as much as possible. This is equivalent to a human-computer interactive modeling process. In addition, mathematically speaking, spatial data do not need to obey a certain distribution.

3) Scope of application

The marked point process simulation method is suitable for the simulation of the target with background facies, e.g. channels and crevasse splays in alluvial systems (the background facies: flood plains), delta distributary channels and mouth bars (the background facies: interchannel and lacustrine mudstones). In addition, non-permeable argillaceous interlayers, calcareous cementation zones, faults, and fractures in sand bodies can all be simulated by this method.

In the mid-1980s, the marked point process simulation method was mainly used in few oilfields in the early development stage, to describe the continuity and connectivity of reservoirs, and to provide a basis for the design of oilfield development plans. In the 1990s, in-depth research on this method was carried out by Norwegian Computing Center as a representative, and the shortcomings of this method were overcome, e.g. being difficult to faithful to well data, simplification of the beginning of target objects, being only suitable for low density well patterns, etc. After continuous improvement of the method, the channel and general lithofacies module of the stochastic simulation software "STORM" was finally developed and launched. It is applicable to the establishment of a fine geological model under multi-well conditions to meet the needs of water injection in the middle and late stages of oilfield development. By the mid‑1990s, the research team headed by Professor Wang Jiahua of Xi'an Shiyou University succeeded in determining the remaining oil distribution in the late stage of waterflood development of the GD oilfield in Eastern China using the marked point process simulation method.

2. Simulated annealing

Simulated annealing is different from other stochastic modeling methods. The main feature of simulated annealing: the units of the original data points that the model needs to meet, the multivariate statistical relationship, the variogram relationship, the geological knowledge, etc. are combined into a component optimization problem, and the modeling result is obtained by solving this nonlinear optimization problem.

Simulated annealing was originally used for composition optimization problems. It is necessary to find the optimal ranking in a system of many components to minimize the overall energy or objective function of the system. Simulated annealing is similar to thermodynamic balance.

1) Basic principle

Under high temperature conditions, molecules can move freely, and their distribution is disordered. As the temperature drops slowly, molecules are ordered to form crystals (representing the lowest energy state of the system). Boltzmann's probability distribution $P\{E\} \sim e^{-E/(K_b T)}$. It expresses the probability distribution of energy in the thermal equilibrium state system at temperature T. K_b is Boltzmann's constant, which is a constant that relates temperature with energy.

Metropolis *et al.* used this principle to simulate the motion of molecules. The probability that a system changes from energy state E_1 to energy state E_2 is $p = e^{-(E_2-E_1)/K_b T}$. If $E_2 < E_1$, the system is always changing; a favorable direction is taken in general and sometimes a disadvantageous direction is also taken. This principle is called Metropolis principle. Any optimization method similar to the annealing thermodynamic process can be called a simulated annealing method.

The basic idea of simulated annealing: an initial image is continuously disturbed until it matches some predefined features included in the objective function. There are two key issues in simulated annealing: (1) objective function; (2) how to decide whether to accept or reject a disturbance.

The objective function is similar to the Gibbs free energy in the real annealing process and is called an energy function, which expresses the difference between the spatial characteristics achieved in each simulation and the desired those. The spatial characteristics can be as follows: (1) univariate distribution map (such as histogram); (2) variogram or indicator variogram; (3) the correlation between primary variables and secondary variables (such as seismic data) or the condition distribution between them; (4) the geometry, volume content, vertical sequence, cross bedding, etc. of lithofacies (or other discrete variables), and any combination of the above. The objective function is the difference between the simulated variogram and the model variogram, and has the expression below:

$$O = \sum_h \frac{[r^*(h) - r(h)]^2}{r(h)^2}$$

Where: $r^*(h)$—Simulated variogram;

$r(h)$—Predefined variogram;

O—Expressing their difference, that is, energy. When the energy is 0, it means that the simulated variogram is faithful to the predefined variogram.

The second key issue of simulated annealing is how to decide whether to accept or reject a certain disturbance. The probability distribution of accepting the disturbance is given by the Boltzmann probability distribution:

$$P\{accpt\} = \begin{cases} 1 & ,O_{new} \leq O_{old} \\ e^{-(O_{new}-O_{old})/t} & ,O_{new} > O_{old} \end{cases}$$

In this distribution, all ideal disturbances ($O_{new} \leq O_{old}$) are accepted; non-ideal disturbances ($O_{new} > O_{old}$) are accepted with an exponentially distributed probability. The parameter t in the exponential distribution is similar to the temperature in simulated annealing. The higher the t, the greater the probability of accepting an undesirable disturbance. t cannot drop too fast; otherwise the simulation implementation will fall into local optimization and no longer converge; But t cannot drop too slowly; otherwise the convergence speed will be too slow. The process of determining how to control t is called making an annealing plan. An empirical method of selecting an appropriate t is proposed by Di Longshi et al.

2) Simulation steps

(1) Generate an initial parameter field. It can be generated by other simulation methods, or is formed by randomly taking values from the univariate distribution function and then placing them on network nodes. If there are secondary variables, the initial value can also be extracted from the conditional distribution in the scatter diagram.

(2) Establish the objective function, and set the initial temperature and annealing plan.

(3) Disturb the initial parameter field, such as exchanging parameter values on two different network nodes.

(4) If the objective function decreases, accept the disturbance; If the objective function increases, the disturbance is accepted with a certain probability (the Boltzmann probability distribution of the real annealing process).

(5) Continue the disturbance process and reduce the probability of accepting undesirable disturbances (reducing the temperature parameter of the Boltzmann probability distribution) until the objective function is sufficiently low and there is no improvement in subsequent iterations.

The advantage of simulated annealing is that any desired statistics combination can be entered into the energy function. The disadvantage is that it has just developed and is not yet mature. When the energy function is complicated, the algorithm converges very slowly and theoretically there is a lack of a unified mathematical tool.

3) Scope of application

In reservoir description and modeling, the simulated annealing method can be used directly in stochastic simulation, and can also be used in post-processing of simulation implementation.

3. Sequential simulation

The general idea of sequential simulation: sequentially obtain the cumulative conditional distribution function (CCDF) of each node along a random path, and extract the simulated value from CCDF. The conditional data used to extract CCDF include both the original sample points and the points that have been simulated. The purpose of this simulation algorithm is to make full use of more conditional data to restore the spatial correlation of variables. Because this method can estimate local conditional probability distribution, it is a very flexible method and has a wide range of applications.

1) Sequential principle

A simulation implementation $Z^{(1)}(u)$ is obtained from the simulation steps. At $u=1$, the CCDF of the variable is obtained from n original sample data. Then randomly extract a quantile from CCDF as the value $Z_1^{(1)}$ of the node. At the next node ($u=2$), the simulated value of the previous node is added to the original conditional data, so that the conditional data for obtaining CCDF are increased from the original n to $n+1$. Take $Z_2^{(1)}$ from CCDF and add this value to the simulation of the next node. Then the capacity of the conditional information is increased by 1 again so as to become ($n+2$).

Thus all N nodes are randomly simulated in sequence, and a simulation implementation $Z^{(1)}(u)$ can be obtained. In this sequential simulation process, N cumulative conditional distribution functions need to be determined:

$$P\{Z_1 \leq z_1 \mid (n)\}$$
$$P\{Z_2 \leq z_2 \mid (n+1)\}$$
$$P\{Z_3 \leq z_3 \mid (n+2)\}$$
$$\ldots$$
$$P\{Z_N \leq z_N \mid (n+N-1)\}$$

The sequential simulation method can be used in Gaussian stochastic simulation and indicator stochastic simulation, and their main difference is that the method of obtaining CCDF is different. In the sequential Gaussian simulation method, each CCDF is assumed to be Gaussian distribution, and its mean and variance are given by simple Kriging equations. The sequential Gaussian simulation method is mainly used in stochastic simulation of continuous variables (such as porosity). In sequential indicator simulation, CCDF is directly given by indicator Kriging equations. The sequential indicator simulation method is mainly used in stochastic simulation of permeability. In addition, Markov-Bayesian simulation method and indicator component simulation method also use the sequential simulation idea.

2) Simulation steps

All "sequential" simulation methods use the following steps (Figure 10-4):
(1) Randomly select a node to be simulated;
(2) Estimate the cumulative conditional distribution function (CCDF) of the node;
(3) Randomly extract a quantile from CCDF as the simulation value of the node;
(4) Add the new simulation value to the conditional data group;

(5) Repeat steps (1)-(4) until all grid nodes have been simulated.

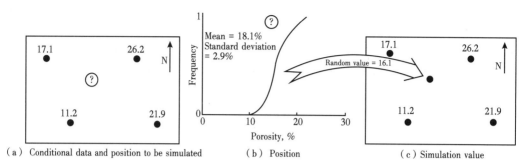

Figure 10-4 Simple diagram of sequential simulation (according to Srivastava, 1994, modified)

3) Scope of application

If the principle of sequential simulation is strictly followed in computer implementation, there will be more and more data for determining CCDF. This is because the capacity of conditional information increases from n to $(n+N-1)$, and the calculation process will become more and more complicated. In practical applications, closer data often shield the influence of farther data, so only the closer data are retained as the conditional information for obtaining CCDF. But the search radius cannot be too small, and the range of conditional data must be large enough to reflect the variogram. One solution is to use the concept of multi-level grid, that is, to simulate N nodes with two or more steps. The first step is to use a coarse grid to reflect the variogram of a large range; The second step is to simulate the remaining grid with a small range of conditional data. In sequential simulation, the sequence of simulating N nodes is preferably random. When N nodes are visited by rows, there will be artificial effects along the rows.

4. Truncated Gaussian simulation

The truncated Gaussian domain is a discrete random model, which is used to study discrete variables or type variables. In the simulation process, 3D continuous variables are truncated through a series of threshold values and truncation rules to establish the 3D distribution of type variables.

A continuous 3D variable distribution is built through a Gaussian domain model, in which the continuous variables (e.g. grain size median) are converted into Gaussian distribution (normal distribution), and then the distribution of 3D continuous variables is built by using a continuous Gaussian simulation method (e.g. sequential Gaussian simulation). In the process of establishing the Gaussian domain distribution, geological trends can be applied to make the distribution of 3D continuous variables reflect geological laws. The threshold value can be obtained through statistics of actual data. The threshold value can be a constant, or a threshold trend given by geological laws, such as a function of threshold value and depth, or a function of threshold value and plane.

Truncation rule: suppose that n lithofacies can be described by an indicator function of each lithofacies. For the ith lithofacies, its indicator value can be defined by Gaussian random function $Y(x)$:

$$I(a_{i-1}<y(x)\leqslant a_i)=\begin{cases}1,\text{ if }y(x)\in(a_{i-1},a_i)\\0,\text{ other}\end{cases}$$

Therefore, point X belongs to the ith lithofacies and $y(x)\in(a_{i-1},a_i)$. Where, a_i is the truncated value, and then the function can be defined:

$$F(x)=\sum_{i=1}^{n}\text{cod}(i)I(a_{i-1}<y(x)\leqslant a_i)$$

Therefore $y(x)$ is taken as i at position x, if and only if the position belongs to facies i, that is, $I(a_{i-1}<y(x)\leqslant a_i)=1$

$$F(u)=i,\text{ if }t_{i-1}(u)<z(u)<t_i(u)$$

Where: $F(u)$—Type variable (or facies);

$Z(u)$—Gaussian domain;

$t_i(u)$—Threshold value at position u.

The truncated Gaussian domain can be extended to the multivariate truncated Gaussian domain, and its discrete nature is defined by the truncated linear combination of N Gaussian domains, so it can simulate the distribution of geometrically complex type variables.

In truncated Gaussian simulation, the distribution of the target objects depends on the truncation of continuous variables by a series of threshold values; therefore, the Facies distribution in the simulation implementation will be ordered, that is, the order of the simulated type variables is fixed. As shown in Figure 10-5, Facies 1, Facies 2, and Facies 3 are distributed in sequence. Facies 1 is in contact with Facies 2, and Facies 2 is in contact with Facies 3, while Facies 1 cannot be in direct contact with Facies 3. It can be seen that this method is suitable for the simulation of sedimentary facies with ordered distribution of facies belts, such as the stochastic simulation of deltas (including delta plains, delta fronts and prodeltas).

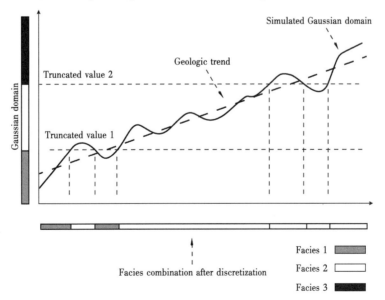

Figure 10-5 The truncation of the continuous Gaussian domain in the truncated Gaussian simulation

5. Fractal simulation

It is a mathematical method proposed by Mendelbert (1983) to describe many complex and irregular shapes in nature. Any infinitely complex and non-differentiable form or structure has a certain self-similarity within it, that is, the part is similar to the whole. The main method of using fractal geometry method to determine the distribution of inter-well reservoir parameters is the fractal simulation method. Generally, the simulated value shall be consistent with the true value where there is a measured value at the well point; the main trend of the inter-well parameter changes shall be consistent with the trend of smooth interpolation such as Kriging interpolation etc.; the heterogeneous characteristics of inter-well parameters require that the predicted value should be consistent with the true value.

1) Simulation theory

The generated fractal geometric parameter field must be equal to the given true value at the measurement point, and shall meet the requirement of the given fractal geometric characteristics in the statistical sense. In order to achieve the above goal, the following method of superposition of random field with Kriging field is proposed.

(1) First, generate a fractal random field $r(x,y)$, which satisfies the fractal characteristics specified by Hurst index H and noise variance $O(L)$.

(2) Use the Kriging method to generate a modified field $u(x,y)$. The final fractal Kriging field is defined as:

$$u(x,y) = r(x,y) + v(x,y)$$

At the well point $u(x,y)$, the fractal Kriging field is required to be equal to the given well point value $u_j(j=1,2,\cdots,n)$, that is, $u(x_j,y_j) = u_j(j=1,2,\cdots,n)$.

According to the above definition, the value of the $v(x,y)$ field at the well point can be first obtained as $v_j = v(x_j, y_j) = u_j - r(x_j, y_j)$ $(j=1,2,\cdots,n)$.

Therefore, the following Kriging method can be used to obtain the value at any point $v(x_o,y_o)$:

$$V(x_o,y_o) = \sum_{j+1}^{n} \lambda_j y_j$$

Where, v_j is the known value of n adjacent wells. The weight coefficient λ_j satisfies the following equation:

$$\sum_{j+1}^{n} r(L_{ij})\lambda_j + a = r(L_{io}) \quad i=1,2\cdots,n \qquad \sum_{j+1}^{n} \lambda_j = 1$$

Where: $r(L_{ij})$—Variogram between two points i and j.

The above equation can also be extended to the addition of more data sources and more known conditions.

(3) The generation method of a fractal random field. The Fourier filtering method is mainly introduced here, which has the best effect. The implementation steps are as follows:

① Generate Gaussian noise series (i.e. Gaussian distribution series);

② Perform Fourier transformation;

③ Multiply the coefficient of each frequency wavelet by $\dfrac{c}{f^{\beta/2}}$, $\beta = \begin{Bmatrix} 2H+1 \text{ vs. } f\beta m \\ 2H-1 \text{ vs. } fGn \end{Bmatrix}$

Where: $f\beta m$—Fractional Brownian motion;

fGn—Fractional Gaussian noise.

④ Take the inverse Fourier transform.

In the application, the boundary part can be cut appropriately to reduce the boundary effect. The above method can be easily extended to 2D or 3D.

2) Scope of application

Fractal simulation has the characteristics such as flexibility etc. and a wide range of application.

(1) The pore structure of sandstone reservoirs has strong self-similarity and is a fractal structure. The MIFA method of calculating fractal dimension using mercury injection capillary curve proposed by Shen Pingping *et al.* is simple and accurate, and can quantitatively describe the characteristics and heterogeneity of sandstone pore structure.

(2) The pores of most oilfield sandstone samples can be divided into three parts: pores without fractal characteristics; large pores with fractal characteristics, whose fractal dimension has a good correlation with the recovery ratio in the water-free production period; small pores with fractal characteristics, whose fractal dimension has a good relation with bound water saturation.

(3) First determine the type of fractal curve using the spectrum analysis method; then determine the Hurst index according to R/S analysis or variogram analysis method. It is an effective method to determine the heterogeneity of reservoir parameters.

(4) The porosity and permeability of most oilfields have fractal characteristics, and their Hurst indexes are 0.9 and 0.8, respectively. In comparison with the traditional Kriging method, the fractal simulation method of reservoir parameter modeling is more advantageous in reflecting the heterogeneity distribution and statistical characteristics of reservoir parameters. The result of simulation with the fractal simulation method is closer to reality than that with the traditional Kriging method, and history fitting is more convenient.

(5) Research on the distribution of fracture networks with the fractal simulation method: some scholars believe that the fractal dimension of 3D fractures is around 2.5 (Hew-ett, 1994), and the "box counting method" is generally used for the distribution dimension of fractures.

Section 3 Fine Geological Modeling Steps

The purpose of reservoir description (characterization) is to establish a reservoir geological model. The final result of modern reservoir description technology is the establishment of a 3D geological model. The general trend is to develop from qualitative to quantitative, from macroscopic to microscopic, and from a single geological discipline to a multi-disciplinary and multi-specialized integrated direction.

Ⅰ. Modeling strategy

Reservoir characterization is to maximize the integration of multiple data and minimize the uncertainty of reservoir prediction. In the 1990s, three-level and two-step modeling procedures and steps were established, namely the three levels such as single-well geological model, 2D geological model, and 3D overall model, as well as the two steps such as establishment of reservoir architecture and the filling of physical parameters. This achievement has not only promoted reservoir geological models to be standardized and programmed quickly in China, but also determined the basic content, parameters and research methods of geological models in each step, and has been recognized as a standard modeling procedure in China and adopted (Figure 10-6). For a long time, single-well geological model and 2D geological model have been widely applied, which are equivalent to single-well interpretation, profile correlation and planar distribution laws, and will be used in the stratigraphic division, sedimentary facies analysis, single sand body description and reservoir parameter description described in the previous chapters. Therefore, the establishment of the two models is very common, and they are the key to the understanding of geological bodies and an important basis for the establishment of 3D models.

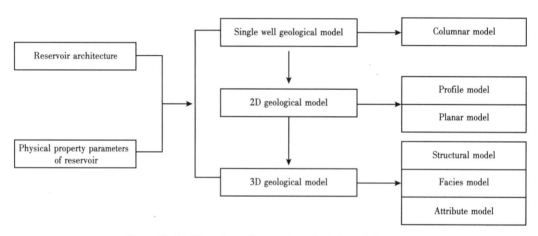

Figure 10-6　Flow chart of reservoir geological modeling steps

With the development of 3D geological modeling technology, the emphasis on the concept of 1D and 2D geological models is becoming less and less. They are carried out as conventional geological work of single wells, profiles and planes, and the geological models of single wells used for numerical simulation are often derived from the shearing of a 3D overall model, so 3D modeling has gradually become a normal work.

Ⅱ. 3D overall model establishment

Describing the distribution of reservoir geological bodies and reservoir parameters in 3D space is to quantitatively give the spatial distribution of various attribute parameters in the reservoir skeleton model. The core issue of establishing a geological model is to predict inter-well parameters and how to reasonably interpolate and extrapolate the same parameter value of the non-drilled area (prediction point) between wells based on the parameter values of existing well points (control points, original sample points).

A reservoir parameter model is characterized by various contour lines in traditional geological methods. The smaller the contour line spacing, the finer the model. Fence diagrams are generally used in 3D characterization; of course, solid models can also be used. With the development of modern computer technology, geological models of reservoirs can be displayed by 3D data volumes. A reservoir is gridded, and each grid is assigned a parameter value to characterize its spatial attribute changes. The smaller the grid size, the finer the geological model.

The accuracy of the geological model is determined by the error of the parameter interpolation value. The smaller the error, the higher the accuracy. Factors affecting the accuracy of a parameter model:

(1) The finer the division of reservoir units, the more difficult it is to improve accuracy;

(2) The stronger the heterogeneity of the attribute itself, the more difficult it is to improve the accuracy. Permeability is the parameter with the relatively largest degree of heterogeneity, so permeability modeling is also the most difficult;

(3) The accuracy is directly proportional to the degree of understanding of geological laws; that is, whether there is a rich geological knowledge database is an important factor in improving accuracy.

(I) **Establishment of reservoir skeleton model**

The reservoir skeleton model is established on the basis of describing the spatial distribution of reservoir structure, faults, strata and lithofacies. It mainly characterizes the 3D spatial distribution of discrete variables of reservoirs and belongs to the category of discrete models.

The facies model is the 3D spatial distribution of different facies types inside reservoirs. The reservoir facies model can quantitatively characterize the size, geometry and 3D spatial distribution of reservoir sand bodies. Development practice shows that the distribution of facies belts strongly affects the flow of underground fluids. The changes of rock physical properties are closely related to facies types, and reasonable facies-controlled modeling is a necessary prerequisite for the accurate establishment of physical property models of rocks.

The establishment of the reservoir skeleton model is based on a database, including coordinate data, individual-layer data, fault data, and reservoir data. Reservoir data are the most important data, including core data and logging interpretation data (single well sedimentary facies analysis and research, sand bodies, barriers and interlayers, porosity, permeability, oil saturation, etc.), which are the most reliable data and are usually called hard data. Seismic data are mainly velocity data, wave impedance data, frequency data, etc. These data have relatively low reliability and are usually called soft data. Well test (including formation test) data reflect the accuracy and reliability of reservoir connectivity information, but the accuracy of the average permeability reflected by the data within a certain range around the wellbore is relatively low.

The reservoir skeleton model is composed of fault model and horizon model. The fault model actually reflects fault plane in 3D space, and fault distribution in 3D space is established mainly based on the fault documents from seismic interpretation and corrected by well data. The horizon model reflects the 3D distribution of formation boundary, and a superposed horizon model is the formation skeleton model. The basic data for modeling are mainly individual-layer data, etc. Generally, the top and bottom horizon models of each isochronous layer are generated by applying individual-layer data with the interpolation method (or stochastic simulation method). Then the horizon models are spatially superimposed to establish the reservoir skeleton model.

(II) Attribute modeling

Attribute modeling is to establish the 3D distribution of reservoir attributes on the basis of the reservoir skeleton model. Reservoir attributes include discrete reservoir properties such as sedimentary facies, reservoir structure, flow units, fractures, etc., as well as continuous reservoir parameters such as reservoir porosity, permeability, oil saturation, etc. First, perform 3D gridding of the reservoir skeleton model (structural model). Then, assign values to each 3D grid using well data and seismic data according to a certain interpolation (or simulation) method to establish 3D data volumes of reservoir attributes.

The key to modeling is the accuracy of the assignment. The factors that affect the accuracy of the model mainly include the following.

(1) Data richness and interpretation accuracy. The degree of data richness is different, and then the accuracy of the model built is also different. The richer the data available in the modeling area, the higher the accuracy of the model built. The interpretation accuracy of the existing original data also seriously affects the accuracy of the model (such as the physical parameters of reservoirs from logging interpretation).

(2) Modeling methods. There are many modeling methods. In terms of inter-well interpolation (or simulation), there are traditional interpolation methods (such as median method, distance square weighting method, etc.), as well as the currently widely applied geostatistical methods. Modeling with different modeling methods will generate reservoir models with different accuracy. The choice of modeling methods is the key to reservoir modeling.

(3) The technical level of modeling personnel, including the theoretical level of reservoir geology, the degree of understanding of geological conditions in the survey area, the computer application level, and the degree of mastery of modeling software.

The result of 3D space assignment forms a 3D data volume, which can be transformed into graphs and displayed in various forms of graphs. The modern computer technology can provide perfect 3D graphic display function. By arbitrarily rotating slices in different directions, the external form and internal characteristics of reservoirs can be displayed from different angles. These graphs can be used very conveniently for 3D reservoir heterogeneity analysis and reservoir development management.

Section 4   Deterministic 3D Reservoir Modeling

For late-stage oilfields with intensive well data, obtaining a deterministic reservoir model is the core of reservoir modeling. Regardless of whether a deterministic modeling strategy or a stochastic modeling strategy is adopted, the final result must be the most reliable geological model under current conditions. Taking Xingerzhong typical area of Daqing Oilfield as an example, this section describes the modeling example of the dense well pattern area based on the deterministic modeling idea.

I. Modeling idea

The modeling area is a late-stage oilfield with intensive well data, where the degree of understanding of the characteristics of underground reservoirs is high. Based on the characteristics of data and information of the research area, a deterministic modeling approach is adopted. First of all, carry out reservoir sedimentary microfacies modeling on the basis of the analysis of sedimentary microfacies, the description of typical single sand bodies, and the study of the reservoir geological knowledge database fully referring to the existing research results and using the human-computer interaction mode; then check and analyze the facies model by means of well pattern vacuating; finally, taking the facies model as the constraint, the distribution of physical parameters of reservoirs is simulated.

There are six geological modeling horizons including P111, P112, P12, P12-1, P13, and P13-1. The input data are mainly well data, including single well logging curves, individual-layer information, microfacies interpretation data, porosity, permeability, etc.

II. Reservoir skeleton model

(I) Structural model

There are no faults in the modeling area, the structural relief amplitude is not large, and there are dense well data. Therefore, a reliable structural model can be established through interpolation of individual-layer data of single wells. The single layer thickness in the modeling interval is not large, and is only 3-5m. Therefore, in order to ensure that no structural surface penetration phenomenon occurs, formation thickness interpolation is performed for control. Figure 10-7 is the 3D structural model showing the top and bottom surfaces. It indicates that the bedding

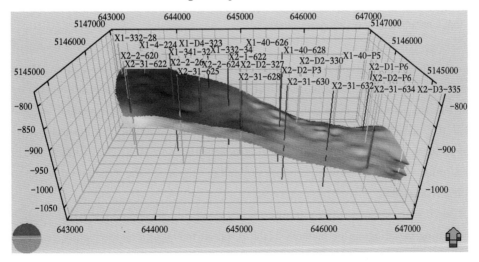

Figure 10-7 3D display of top P11 and bottom P142 (some wells are displayed; the vertical scale is enlarged by 5 times)

plane structure in the area is a monoclinic with a dip angle of about 2°, a dip direction of NE, and a maximum structural amplitude difference of about 100m.

(Ⅱ) Sedimentary microfacies model

The purpose of 3D facies modeling is to obtain the 3D distribution of different types of microfacies in reservoirs and to lay a foundation for reservoir parameter modeling. Facies modeling is to carry out inter-well 3D prediction (simulation or interpolation) using well data (single well facies profiles) and then establish the 3D distribution of facies under the guidance of the sedimentary model on the basis of the structural model. At present, the mainstream modeling software includes several mature modeling methods such as indicator simulation, marked point process, truncated Gaussian simulation and indicator Kriging. Different modeling methods are suitable for different data conditions and sedimentary facies types. The stochastic modeling method has been introduced in the preceding part of the text, and the indicator Kriging method is mainly adopted here.

The indicator Kriging method is a Kriging interpolation method, and belongs to the category of deterministic modeling methods. The indicator Kriging method is actually to perform Kriging estimation of the indicated transformation value of the original data. This method is characterized by considering the existence of outliers in the form of probability. Other Kriging methods have a smooth overall estimation result due to the weighted average, while the indicator Kriging method helps to overcome this shortcoming. On the other hand, the structural model of the indicator Kriging method is transformed from indicator transformation data instead of the original data, so this method is not affected by outliers, the result is very robust, and the method is suitable for processing the spatial estimated values of physical property parameters with large spatial changes. Moreover, the indicator Kriging method is a deterministic interpolation method and requires the variogram parameters of different facies to be given, and a facies model with good continuity and smooth boundaries can be obtained. This is very effective for deterministic reservoir modeling in dense well pattern areas.

1. Input parameters and variogram analysis

1) 3D grid setting

In order to perform 3D reservoir modeling, reservoirs must be 3D gridded on the 3D structural background, that is, an actual geological body is divided into a series of grids in X, Y, and Z directions. In the late stage of oilfield development, the thickness of the studied single layer is small and the microfacies is finely divided, so the requirements for the accuracy of reservoir modeling is high. The grid size in this modeling is set to around $10m \times 10m \times 0.5m$, and the total number of grids is up to 2474160. There are no faults in the area, the structure is gentle, and the grid setting is regular.

2) Setting of input parameters of variogram

The setting of variogram parameters in the reservoir modeling process mainly comes from two ways. (1) The variogram is fitted to the modeling data by giving an appropriate bandwidth and search radius, and the corresponding parameter settings of the variogram are obtained. This kind of fitting is a pure data analysis process, and reflects the spatial variability of data, and the obtained variogram usually does not conform to the development law of geological bodies. The secondary range of the channel sand body obtained by the variogram fitting this time is much larger than the main range, that is, the channel width is much larger than the channel length, which is not in line with geological laws. (2) The variogram parameters are directly given on the basis of geological knowledge. This parameter setting mode depends on the depth of geological understanding. The first few chapters of this book focus on in-depth anatomy and analysis of combined microfacies and typical single sand bodies in the modeling area; in addition, the corresponding reservoir geological knowledge database has been established. Therefore, the reasonable input parameters of the variogram can be given according to geological laws and adjusted through modeling experiments. The final input parameters of the variogram are shown in Table 10-1.

Table 10-1 Input parameters of the variogram for each layer

Layer	Microfacies	Main range, m	Secondary range, m	Vertical range, m
P111	Channel	600	300	3
	Branch channel	500	200	2
	Overbank	1000	800	2
	Flood plain	2000	2000	3
P112	Channel	1200	500	3
	Branch channel	600	300	2
	Overbank	1000	1000	1
	Flood plain	1000	1000	4
P12	Channel	1200	500	3
	Branch channel	600	300	2
	Overbank	1000	1000	1
	Flood plain	1000	1000	4
P12-1	Channel	2000	600	5
	Branch channel	600	300	2
	Overbank	500	500	1
	Flood plain	1000	1000	5
P13	Channel	2000	600	5
	Branch channel	600	300	2
	Overbank	500	500	1
	Flood plain	1000	1000	5
P13-1	Channel	2000	600	5
	Branch channel	600	300	2
	Overbank	500	500	1
	Flood plain	1000	1000	5

2. Modeling result analysis

A 3D sedimentary facies model of the reservoirs in the modeling area has been established using the indicator simulation method. The microfacies types include channel, branch channel, overbank and flood plain. The modeling interval is divided into 6 layers. The sand bodies in P111, P112, and P12 are less developed, the overall sandstone content is less than 5%, and overbank deposits and branch channel sand bodies predominate; the developed channel sand bodies are small in scale, most of which are scattered in potato shape; the channel sand bodies are continuous only in the P12 single layer, forming an obvious single channel form (Table 10-2, Figure 10-8). The sand bodies in P12-1, P13, and P13-1 are extremely developed, the overall sandstone content is more than 50%, channel sand body deposits predominate, and branch channel and overbank sand bodies are less developed. The channel sand bodies are superimposed and connected, and the overflow bank sedimentary sand bodies are distributed among them (Figure 10-9). As shown in the profile (Figure 10-10), the lower part of the modeling interval has obvious characteristics of continuous sand bodies, while the upper part is dominated by mudstone deposits with scattered sand bodies. This is completely consistent with the reservoir development characteristics of the modeling area.

Table 10-2 The distribution proportion of different microfacies in each single layer

Single layer / Microfacies	P111	P112	P12	P12-1	P13	P13-1
Channel	0	1.99	4.45	48.75	48.56	70.96
Branch channel	0	0.24	0.2	0.48	1.36	0.85
Overbank	0.55	2.72	0.21	0.25	0.46	0.69
Flood plain	99.44	95.05	95.14	50.52	49.43	27.5

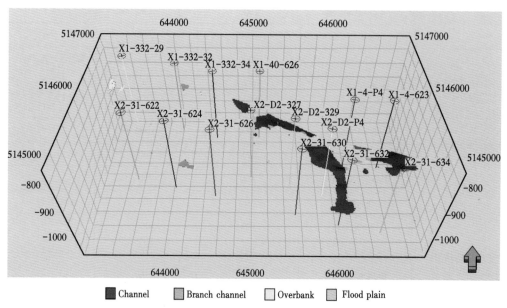

Figure 10-8 Hollowed-out display of sand bodies of P12 single layer (some wells displayed)

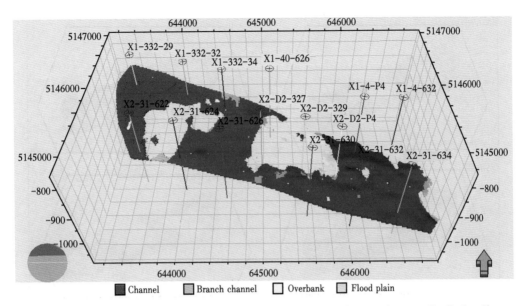

Figure 10-9 Hollowed-out display of sand bodies of P12-1 single layer (some wells displayed)

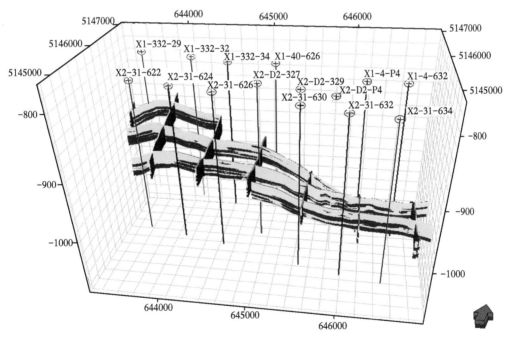

Figure 10-10 3D display of grid slices

In order to verify the reliability of the sedimentary microfacies model, the method of well pattern vacuating was adopted for verification. Well pattern vacuating was carried out three times, and the number of wells was reduced from 268 to 159, 85 and 40 successively. Keep the input parameter settings of the variogram unchanged, carry out the modeling of sedimentary microfacies at different well pattern densities, and compare the modeling results (Figure 10-11, Figure 10-12).

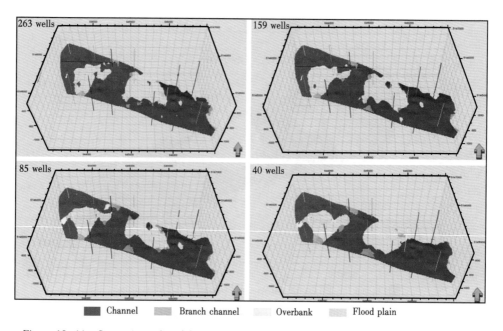

Figure 10-11 Comparison of modeling results after well pattern vacuating for P12-1 single layer
(hollowed-out display of sand bodies, display of some wells)

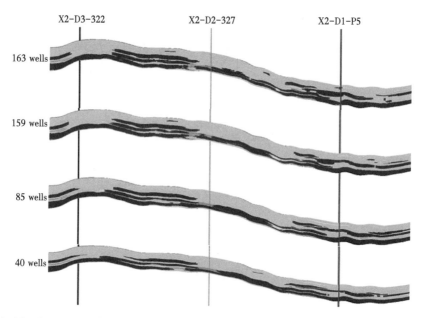

Figure 10-12 Comparison of the cross-well profiles of the modeling results through well pattern vacuating

In general, the sedimentary microfacies models established for each single layer at different well pattern densities have good consistency, and the distribution position and development scale of the sand bodies are basically the same. In particular, the model established using 159 wells has the highest similarity with the original model established using 263 wells, and the similarity of the models established using 85 wells and 40 wells to the original model becomes worse. Some microfacies sand bodies controlled by single wells no longer appear, and the continuity of some sand bodies controlled by adjacent wells in some areas becomes better. This change is the inevitable result of modeling relying solely on well data. The development of sand bodies depends on logging interpretation results. During well pattern vacuating, some wells with well developed sand bodies were removed, which directly caused absence or continuous distribution of some microfacies sand bodies. Therefore, as the well pattern density decreases, the similarity of models will gradually become worse. Nevertheless, the models established after well pattern vacuating still have a high similarity to the original model, and some major sand bodies can be well simulated in each model. The coincidence rate of the models after well pattern vacuating is above 70%, indicating that the selected method and parameter settings in this modeling are suitable for this area, and the modeling results are reliable.

III. Physical property parameter modeling of reservoirs

The 3D distribution models of reservoir porosity and permeability are mainly established here. The physical parameter modeling of reservoirs is carried out using the sequential Gaussian simulation method under the guidance of the facies-controlled modeling idea.

(I) Parameter modeling method

The traditional reservoir modeling method is mainly one-stage modeling method; that is, inter-well interpolation is performed directly according to the reservoir parameters of each well to establish a 3D distribution model of reservoir parameters. This method is simple and is mainly suitable for modeling of reservoir parameters with a single sedimentary facies distributed. In this case, the reservoir parameters of the target area have a uniform statistical distribution. For reservoirs with multi-facies distribution or complex structure, the distribution of reservoir parameters of different facies is quite different, and the application of this method will seriously affect the accuracy of the built model. For the case of multi-phase distribution, the facies-controlled modeling idea is very practical. That is, firstly build the model of sedimentary facies, reservoir structures or flow units; then according to the quantitative distribution law of reservoir parameters of different sedimentary facies (sand body types or flow units), carry out inter-well interpolation or stochastic simulation by facies (sand body or flow unit) to build a reservoir parameter distribution model. Facies (or lithology) controls the distribution of physical property parameters in 3D space; therefore, the idea of the facies-controlled parameter simulation method is more in line with geological laws and can avoid the strict requirements of most continuous variable models for stability and homogeneity.

The sequential Gaussian simulation method was used in this modeling. This method is a stochastic simulation method using Gauss probability theory and sequential simulation algorithm to generate continuous spatial variable distribution. The simulation process is carried out sequentially from one pixel to another. In addition to the original data, all the data that have been simulated are also considered in the calculation of the conditional data of a certain pixel conditional probability distribution function. This method is fast and robust, and is mainly used for stochastic simulation of continuous variables (such as porosity). It is currently the main method of physical property parameter modeling of reservoirs.

(Ⅱ) **Parameter setting and model establishment**

The input parameters of physical property parameter modeling of reservoirs mainly include variable statistical parameters (mean, standard deviation), variogram parameters (range, nugget effect, etc.), conditional data, etc. For facies-controlled modeling, a 3D facies model shall be input, and the corresponding variable statistical parameters and variogram parameters shall be input for each type of facies.

1. Distribution of physical property parameters of reservoirs

The statistics of the distribution of porosity and permeability data of the modeling area show that there is little difference in porosity and a large difference in permeability as a whole. The porosity of reservoirs in the area is mainly 20%–32%, with an average of 23.6%; the permeability varies widely, and is mainly 1–400mD, and the local permeability can be up to more than 1500mD. The distribution statistics of physical property parameters of different microfacies are shown in Figure 10-13.

2. Variogram analysis

Based on the facies-controlled modeling idea, the distribution of physical properties of reservoirs is restricted by the facies model, and the variogram parameters must also be given according to the variogram of the facies model. The channel sand bodies in the modeling area are widely distributed, the data volume of physical property parameters is large, and variogram fitting can be performed. Figure 10-14 shows the variogram fitting diagrams of porosity and permeability of channel sand bodies. The main range of porosity is 461m, the secondary range 287m, and the vertical range 3m; the main range of permeability is 300m, the secondary range 239m, and the vertical range 3.7m. This is the fitting of the variogram of all single layers. The variogram shall be set for each single layer and each microfacies type in the process of physical property parameter modeling. According to the overall variogram fitting characteristics and microfacies modeling variogram parameter settings, the variogram parameters for physical property modeling of reservoirs are given (Table 10-3).

3D models of physical property parameters of reservoirs have been established through the setting of the above parameters. Figure 10-15 shows the distribution models of porosity and permeability of a single grid layer obtained through facies-controlled interpolation. As shown in the figure, the distribution of porosity and permeability is obviously controlled by facies, which conforms to geologic laws. As shown in Figure 10-16, the lower strata have good physical properties and features such as high porosity and high permeability, and the physical properties of the upper strata are significantly worse. The models meticulously reflect the change characteristics of reservoir physical properties within a single-layer sand body, can better reveal the heterogeneity of reservoirs, and provide reliable geological models for the numerical simulation of remaining oil in the late development stage.

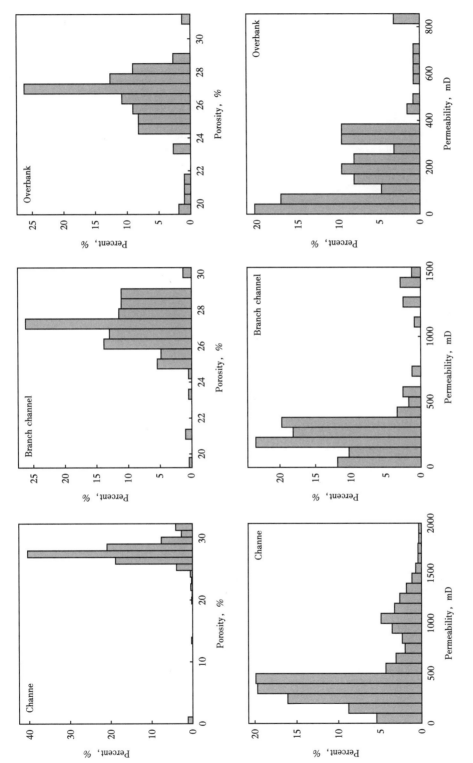

Figure 10-13　Distribution of porosity and permeability of main microfacies sand bodies

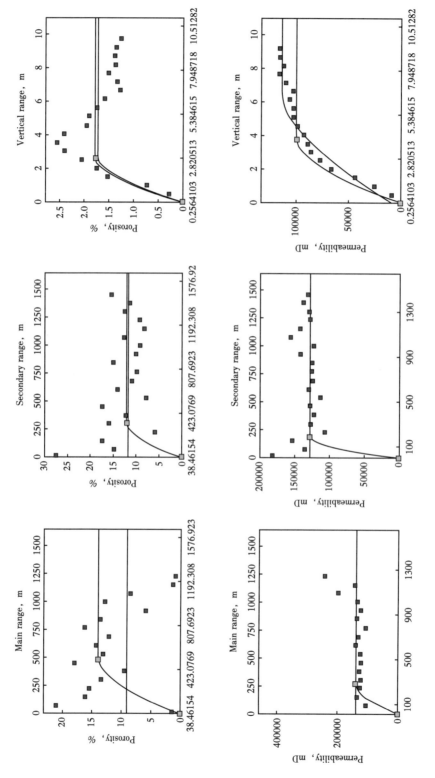

Figure 10-14　Variogram fitting diagrams of porosity and permeability of main channel microfacies sand bodies

Table 10-3 Variogram parameters of porosity and permeability of each microfacies

Layer	Microfacies	Main range, m	Secondary range, m	Vertical range, m
P111	Channel	50	50	2
	Branch channel	50	50	2
	Overbank	100	100	1
P112	Channel	200	100	2
	Branch channel	200	100	2
	Overbank	200	200	1
P12	Channel	300	100	2
	Branch channel	200	100	2
	Overbank	200	200	1
P12-1	Channel	300	200	3
	Branch channel	300	200	2
	Overbank	200	200	1
P13	Channel	400	250	3
	Branch channel	300	200	2
	Overbank	200	200	1
P13-1	Channel	400	250	3
	Branch channel	300	200	2
	Overbank	200	100	1

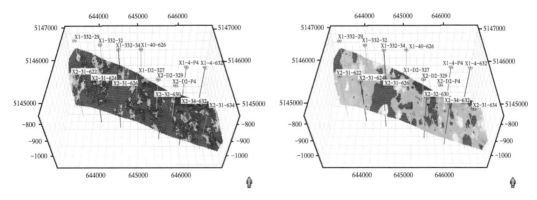

Figure 10-15 Distribution models of porosity and permeability of single grid layer

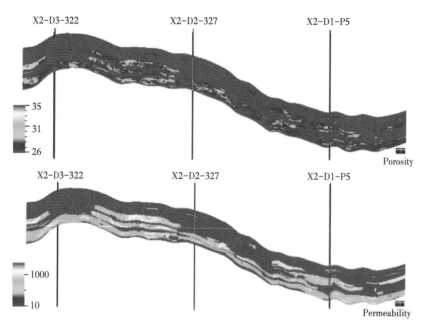

Figure 10-16 Display of the through-well profile of reservoir physical property model

Section 5 Well-Seismic Joint Stochastic Simulation and 3D Reservoir Modeling

The modeling area is Z1 well area in the northwest of Maqiao uplift of Moxizhuang structure in Junggar Basin. The current development degree of the area is low, and the well spacing is generally greater than 2km. The modeling horizon is sand group 2 ($J_1s_2^2$) of Member 2 of Jurassic Sangonghe Formation, which is mainly composed of sandstone deposits and is the main pay formation. The reservoirs have an average porosity of 9.5% and an average permeability of 16.8mD, and are characterized by superimposition of multiple sand bodies, independent accumulation of a single sand body, etc.

According to actual needs, the target layer is divided into two facies types: channel sand bodies and mudstone interlayers. In addition, through seismic profile interpretation and seismic attribute analysis, the reservoir geological model and sand body distribution parameters have been systematically analyzed and abundant modeling parameters have been obtained. The specific input information includes: single-well sedimentary microfacies interpretation data, logging constrained inversion wave impedance data volumes, SP field 3D data volumes, and geostatistical characteristic parameters.

I. Modeling idea

Due to the large well spacing in the modeling area, the idea of stochastic modeling of reservoirs by combining well data with seismic data is adopted in order to improve the credibility of the model. That is, the 3D data volumes obtained by seismic inversion are used as constraints, multiple model implementations are established, and the data volume closest to the geological knowledge is selected as the final result.

At present, the main method of seismic constrained reservoir modeling is to establish a 3D reservoir model by using the collocated co-simulation algorithm and the probability correlation between seismic inversion data and reservoir parameters. Scholars at home and abroad have carried out a lot of research work in this aspect and have obtained many successful application examples, but there are still some problems worth exploring. Due to the limited vertical resolution of seismic data and the multiplicity of logging constrained inversion itself, the correlation between wave impedance data and lithofacies probability is often poor in practical applications. As shown in Figure 10-17, with the change of the wave impedance value, there is no obvious distinction between the probability values of the two facies within the effective value distribution range, and the inversion data constrained facies modeling cannot be achieved based on this. In the case of few well data, seismic data, as the main source of cross-well information, still need to be fully utilized.

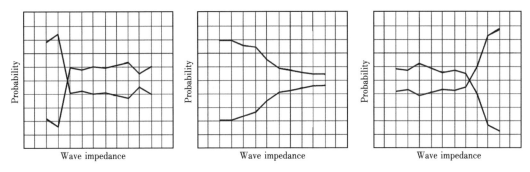

Figure 10-17 Schematic diagram of the relationship between imperfect wave impedance and lithofacies probability (two facies types)

Due to the poor probability relationship between wave impedance data and lithofacies data, facies modeling cannot be performed directly through wave impedance constraints. Consider selecting the geophysical characteristic curve that can reflect the change of lithofacies, take the inversion data volume as the constraint, and perform the stochastic simulation of the parameter field of the geophysical characteristic curve through the functional relationship between the two. Then take the constructed parameter field as the constraint, and carry out reservoir facies modeling by means of the facies probability relationship. Finally, the optimized facies model and the geophysical characteristic curve parameter field are used as constraints to establish a reservoir parameter model, thereby forming a set of comprehensive level-by-level constrained multi-information and multi-level modeling methods. Considering that the selected geophysical characteristic curve has a good corresponding relationship with lithofacies, the facies model established with this constraint is relatively definite. The functional relationship between wave impedance data volumes and geophysical characteristic curves is not strictly one-to-one, so the establishment of the geophysical characteristic curve parameter field based on the wave impedance data volumes is of uncertainty. This link is the key to stochastic simulation.

The specific steps are as follows (Figure 10-18).

(1) Obtain reliable seismic inversion data volumes through well-seismic joint inversion; carry out time-depth conversion through the establishment of the velocity field so as to correspond to the logging data in the depth domain.

(2) With seismic inversion data volume as the constraints, stochastic simulation is performed using the collocated co-Kriging method to establish multiple random parameter fields of geophysical characteristic curves.

(3) With each geophysical characteristic curve parameter field as the constraint, reservoir facies models are established by using the collocated co-Kriging method and facies modeling method. The facies models are simulated and optimized, and the facies model closest to the real situation of the underground reservoirs is obtained.

(4) Taking the optimized facies model as the control and the geophysical characteristic curve parameter field as the constraint, a reservoir parameter model is established using the collocated co-Kriging method and the sequential Gaussian simulation method.

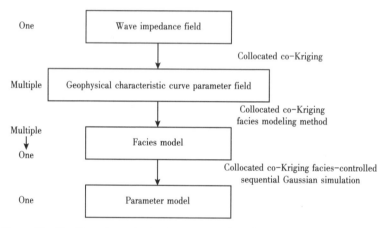

Figure 10-18 Flow chart of seismic-constrained stochastic modeling of reservoirs

II. Reservoir skeleton model

According to the exploration and development situation of the modeling area, corner point grids are set, and the grid size is about 100m×100m×0.5m. The main purpose of modeling is to reveal the distribution of sand bodies and mudstone barriers and interlayers. Therefore, the sedimentary facies are divided into two types such as sand bodies and mudstone barriers and interlayers. The modeling method uses the target-based stochastic simulation method of marked point process.

(I) **Statistics of reservoir geological parameters**

Different modeling methods require different geostatistical characteristic parameters. The modeling parameters required for reservoir facies modeling using the marked point process method mainly include the direction of the main flow line, the shape, distribution of length/width/thickness and volume content of channel sand body microfacies, etc. These parameters are mainly obtained through core data statistics, well-seismic joint profile interpretation, seismic attribute analysis, and hole data statistics. See Table 10-4 for specific parameters.

Table 10-4 Geostatistical characteristic parameters

Layer Geologic parameters	$J_1s_2^{2-1}$	$J_1s_2^{2-2}$	$J_1s_2^{2-3}$	$J_1s_2^{2-4}$
Composite channel width, m	2500	2500	2500	2500
Composite channel thickness, m	10	15	15	15
Composite channel amplitude, m	1000	1000	1000	1000
Composite channel curvature	1.02	1.02	1.02	1.02
Number of single channel(s) in the composite channel	3±2	3±2	3±2	3±2
Single channel width, m	500±100	500±100	500±100	500±100
Single channel thickness, m	10±5	10±5	10±5	10±5
Single channel amplitude, m	500±100	500±100	500±100	500±100
Single channel curvature	1.02	1.02	1.02	1.02
Channel direction, (°)	50±5	50±5	50±5	50±5
Sand content, %	17±5	82±5	65±5	77±5

(Ⅱ) **Stochastic simulation of SP field constrained by wave impedance**

Perform logging-constrained inversion of the 3D seismic data of the target interval in the modeling area to obtain high-resolution 3D wave impedance data volumes. The wave impedance data of near-well traces are extracted from the 3D wave impedance data volumes for the probability correlation analysis with the interpretation data of the sedimentary microfacies of wells. The results show that the corresponding relationship between the wave impedance inversion data volumes and the single-well sedimentary microfacies interpretation results is not ideal, and the obtained probability relationship is poor (Figure 10-19a). In addition, when the wave impedance value is greater than $1.05\times10^7 kg/m^3 \cdot m/s$, it is not easy to make a distinction between the channel sand bodies and the mudstone interlayers, while this range is the main range of the wave impedance values of the target interval. Therefore, facies modeling cannot be performed directly through wave impedance constraints. Studies have shown that SP curves can make a good distinction between the channel sand bodies and the mudstone interlayers, have a very good probabilistic relationship with sedimentary microfacies (Figure 10-19b), and can be used as the constraint conditions for sedimentary facies modeling. Therefore, a randomly simulated SP field is firstly established by means of seismic constraints in the modeling process; then a 3D facies model is established using this as the constraint.

The collocated co-simulation method is adopted to integrate well data with seismic wave impedance data to realize SP field simulation.

The basis of collocated co-simulation is collocated co-Kriging. Its realization formula is:

$$Z(u) = \sum_{i=1}^{n} \lambda_i(u) Z(u_i) + \lambda_j(u) Y(u)$$

Where, $Z(u)$ is the estimated value of the random variable; $Z(u_i)$ is i sampling points of the primary variable (hard data); $Y(u)$ is the secondary variable (seismic data); λ_i and λ_j are co-

Kriging weighting coefficients to be determined.

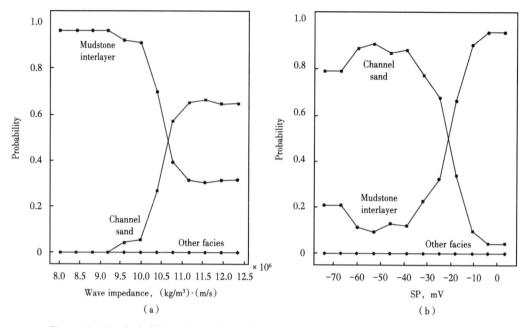

Figure 10-19　Probability relation charts of sedimentary facies vs. wave impedance and SP

The equations of the collocated co-Kriging system can be derived from the unbiased and minimum variance conditions of the Kriging method. In the simulation process, the SP logging data after baseline drift correction is the main variable, and the logging constrained inversion data is the secondary spatial constrained variable. The two have a certain correlation, and their correlation coefficient is 0.5 (Figure 10-20). The variogram parameters are determined according to seismic

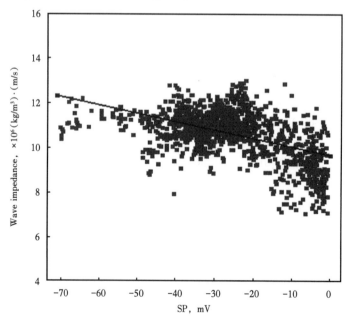

Figure 10-20　The correlation between SP and wave impedance

attribute analysis and geological model research; the azimuth angle is set to 50° according to the provenance direction. The main range of $J_1s_2^{2-1}$ layer is 3500m, and the secondary range 2500m; the main range of other layers is 5000m, and the secondary range 2500m.

Using the above simulation parameters, 10 SP field implementations have been generated through stochastic simulation (Figure 10-21). Although their overall characteristics are similar, there are differences between the implementations, representing a possible prediction result of the real situation of geological bodies.

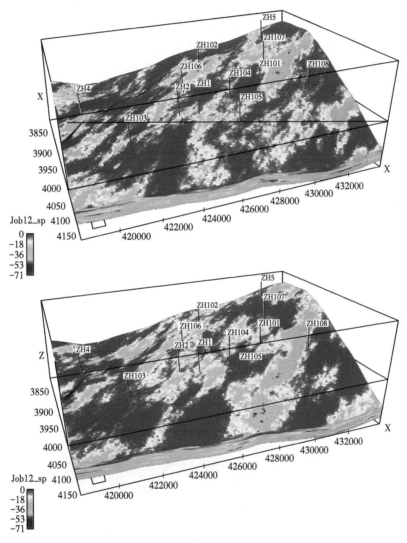

Figure 10-21 Two implementations of stochastic simulation of SP field

(Ⅲ) **Establishment of 3D facies model and parameter model**

The 3D facies model of channel sand bodies and barriers and interlayers has been established using the collocated co-simulation method and marked point process method as well as the probabilistic relationship between the SP field and sedimentary facies respectively taking each implementation of the SP field as the constraint. 10 facies implementations have been worked out in this simulation, each of which is faithful to the well point information and seismic information, and represents a possibility of the distribution of sand bodies and interlayers in the area. The comparison of the 10 sets of implementations shows that the seventh implementation can better reveal the distribution characteristics of oil and water in the survey area (Figure 10-22). Sand bodes are not developed in $J_1s_2^{2-1}$, where mudstone interlayers predominate. There are many sand bodies in the Well Z5 and Well Z107 areas close to the provenance direction, where channels are superimposed and cut mutually. All other 3 layers are characterized by the general development of sand bodies in the whole area. At first glance, they seem to be superimposed and connected, but after hollowed-out display, it is found that the barriers and interlayers are also widely distributed in them and they have a certain degree of separability, so that the generally developed sand bodies become independent sand bodies or semi-connected sand bodies. This feature accords with the geological understanding of the superimposed and continuous sand bodies in the area and the independent accumulation of single sand bodies, and also explains the fluid distribution characteristics such as multiple oil-water systems in wells and a non-uniform OWC in the same layer between wells. The cross-well section slices of the reservoir intervals from Well Z103 to Well Z101 and from Well Z102 to Well Z105 also show this feature (Figure 10-23).

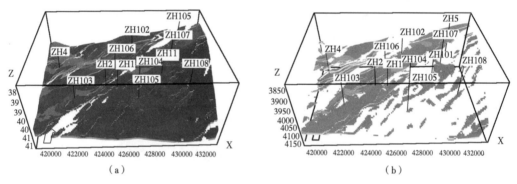

Figure 10-22 Hollowed-out display of the channel sand body (a) and mudstone interlayer (b) for the seventh implementation of the 3D facies of a layer

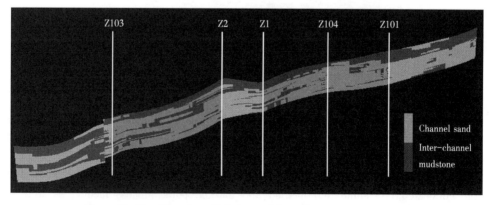

Figure 10-23 Through-well facies model profile of reservoir interval

III. 3D reservoir parameter modeling

(I) Modeling method

The sequential Gaussian simulation method is adopted, and the algorithm of this method is robust and commonly used. The Gaussian random field is the most classic random function model. The biggest feature of this model is that random variables conform to Gaussian distribution (i.e., normal distribution). This method is mainly used in stochastic simulation of continuous variables (such as porosity).

If the continuous spatial random function $\{Z(u), (u) \in A\}$ is the sum of some random functions $\{Y_k(u), (u) \in A\}$, $k = 1, 2, \cdots, K$ with similar spatial distribution, the spatial distribution of this function can be simulated by a multivariate Gaussian random function model. This principle is not that each component has the same distribution and number, but that the components are independent of each other.

The random function $\{Z(u), (u) \in A\}$ obeys the multivariate normal distribution if and only if:

(1) All subsets $\{Z(u), (u) \in B \subset A\}$ of the random function are also multivariate normal distributions;

(2) All linear combinations of the random variable components of $Z(u)$ are univariate normal distributions;

(3) The covariance function or correlation coefficient is 0, which ensures complete independence;

(4) Given any subset, the conditional distribution of any other subset of the random function $Z(u)$ is a multivariate normal distribution.

People are particularly interested in that when $k = 1$ and $u' = u_0$, the random function $Z(u)$ simulates the uncertainty of an unsampled value. Given n known data points, the conditional probability distribution function CCDF of $Z(u_0)$ is normal and has the following characteristics:

(1) The mean value or conditional expectation is consistent with the simple Kriging estimate;

(2) The variance or conditional variance is the simple Kriging variance.

Therefore, for Gaussian simulation, CCDF can be obtained by simple Kriging. Due to the normality of CCDF, the entire simulation process is greatly simplified, and sequentially determining a series of CCDF is simplified to solving a series of Kriging equations.

Of course, most geological data are not of symmetric Gaussian distribution. In practical applications, the regionalized variables (such as porosity, permeability) can be firstly subjected to normal score transformation (transformed into Gaussian distribution); after simulation, the simulation results are inversely transformed into regionalized variables.

A variety of algorithms can be used in Gaussian simulation, such as sequential simulation, error simulation, probability field simulation, etc. Sequential simulation is often used in practice, that is, sequential Gaussian simulation.

The sequential Gaussian simulation process is carried out sequentially from one pixel to another. In addition to the original data, all the data that have been simulated are also considered in the calculation of the conditional data of a certain pixel conditional probability distribution function. Simulation implementations can be obtained from random extraction of quantiles from CCDF.

The conditional simulation steps of continuous variable $Z(u)$ are as follows:

(1) Determine the univariate $F_z(z)$ that represents the full modeling area (including Z sample data). If the Z data is unevenly distributed, it shall be deserialized first, or it may also need extrapolation and smoothing;

(2) The Z data is subjected to normal score transformation using CDF $F_z(z)$ and is transformed into the Y data with standard normal distribution;

(3) Check whether the Y data after the normal score transformation meets bivariate normality (that is, the bivariate cdf of any data pair must be normal). If yes, this method can be used; otherwise, other random models shall be considered;

(4) If the multivariate Gaussian model is suitable for Y variable (sample data after normal score transformation), perform sequential simulation, namely:

① Determine a random path, and visit each grid node once every time (not necessarily regular). For each node u, a certain amount of neighborhood conditional data is retained, including the original Y data and the previously simulated grid node Y value.

② Determine the parameters (mean and variance) of the CCDF function of the random function $Y(u)$ at the node using the simple Kriging method and the variogram model of normal score, and obtain ccdf.

③ Randomly extract a quantile from ccdf, which is the simulated value $Y(l)(u)$ of the node.

④ Load the simulated value $Y^{(l)}(u)$ into the existing data group.

⑤ Simulate the next node u' along the random path until every node is reached. Once u at all positions is simulated, a random simulated image can be obtained.

(5) The entire sequential simulation process can repeat the above steps according to a new random path to obtain a new implementation.

The input parameters of sequential Gaussian simulation mainly include variable statistical parameters (mean, standard deviation), variogram parameters (range, nugget effect, etc.), conditional data, etc. For facies-controlled modeling, a 3D facies model shall be input, and the corresponding variable statistical parameters and variogram parameters shall be input for each type of facies.

(Ⅱ) **Parameter modeling result analysis**

Multiple facies implementations have been obtained through collocated co-simulation. Similarly, the porosity and permeability parameter fields can also be simulated using the idea of collocated co-simulation and the SP field as a secondary variable under the constraints of the facies model. The main parameters to be determined for simulation are as follows.

1. Facies model

The facies model is used as input to achieve the purpose of facies-controlled modeling. The petrophysical parameters will be faithful to the selected phase distribution, and the parameter assignments will vary with facies. A reservoir parameter model is established taking the selected seventh facies model as the constraint.

2. Data transformation

The following data transformation is mainly carried out.

Data truncation transformation: that is, truncate some abnormal low values and abnormal high values from logging interpretation, so that the parameters of wells and simulation implementations conform to geologic laws.

Logarithmic transformation: permeability does not show a normal distribution in general, but after logarithmic transformation on permeability, its distribution can be close to a normal distribution. Therefore, before modeling, it is generally necessary to perform logarithmic transformation on permeability; after modeling, inverse transformation is performed.

Standard normal distribution transformation: through transformation, each parameter conforms to Gaussian distribution, so that the Gaussian simulation method can be applied for modeling; after modeling, inverse transformation is performed.

3. Variogram

The variogram reflects the spatial correlation of reservoir parameters. The parameters of the variogram can be obtained by calculation. But when there are few well points for obtaining the variogram by facies, there will be a large error in obtaining the variogram. Therefore, in the actual modeling process, a geological concept model can generally be used to estimate the parameters of the variogram. The main range reflects the relative changes of parameters along the channel direction, and is equivalent to the channel length; the secondary range reflects the relative changes of parameters along the direction transecting the channel, and is roughly equivalent to the channel width; the vertical range is roughly equivalent to the channel thickness.

Referring to the previous seismic attribute information and the determined geological concept model, some parameters of Layer 1 have been determined as follows: the direction of the main range is 50° east by north, the length of the main range is 3000m, the length of the secondary range is 1500m, and the vertical range is 10m. In addition, some parameters of other layers have been determined as follows: the direction of the main range is 50° east by north, the length of the main range is 5000m, the length of the secondary range is 2500m, and the vertical range is 10m.

4. Standard deviation

It reflects the numerical variability of the petrophysical parameters of each facies. A standard normal distribution can be described by the mean and standard deviation. The standard deviation can be obtained through histogram statistics of multi-well data.

5. Correlation

It reflects the correlation degree of different types of petrophysical parameters. In order to perform collaborative modeling between parameters, the correlation between parameters can be input as constraints. The correlation can be obtained through regression analysis between parameters.

The 3D data fields of porosity and permeability parameters of each layer have been obtained through the above parameter setting and sequential Gaussian simulation (Figure 10-24 and Figure 10-25). Each parameter field obtained is strictly faithful to the well point data. Layer 1 is dominated by mudstone facies, so the porosity and permeability are both very low. In contrast, the porosity and permeability of other layers are high, and under the control of facies, are distributed in a NE-trending channel pattern. On the whole, the porosity simulation result shows mainly low porosity, the permeability simulation result shows mainly medium and low permeability, and only a very small local area has high permeability. This is in line with the geological characteristics of this area such as low porosity and medium and low permeability.

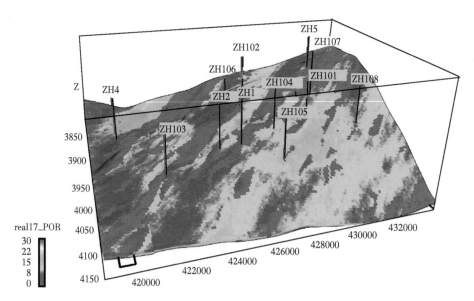

Figure 10-24　Facies-controlled 3D porosity model of $J_1 s_2^{2-4}$

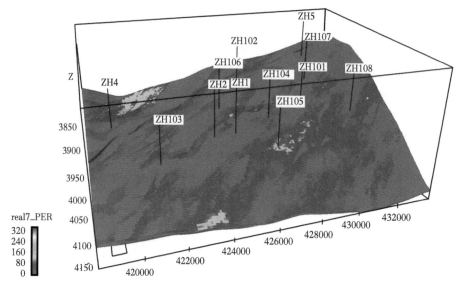

Figure 10-25　Facies-controlled 3D permeability model of $J_1 s_2^{2-4}$

After upscaling, the parameter fields can be directly input to the numerical simulation software such as Eclipse, VIP, etc.

References

Al-Henshiri M Y, Arisaka K, Al-Hassani H, et al. 2005. Integration of Dynamic & Geostatic Data Improve Sreservoir Characterization[J]. SPE 93475, 1-8.

Anderson R N, Boulanger A, He Wei, et al. 1997. 4D Seismic: The Fourth Dimension in Reservoir Management, Part 1: What Is 4D and How Does It Improve Recovery Efficiency[J]. World Oil, 218 (3): 43-49.

Andre Picarelli, Jorge Arguello, Israel Nieves, et al. 2001. High - resolution Sequence Stratigraphy and Reservoir Characterization Applied to Mature Fields: Example From Eastern Venezuela Basin[J]. SPE 69602, 1-12.

Brian Rothkopf, Steve Fredrickson, Jeff Hermann. 2003. Case Study: Merging Modern Reservoir Characterization with Traditional Reservoir Engineering[J]. SPE 84414, 1-6.

Cao Hui. 2002. Current Status of Crosswell Seismic Technology [J]. Progress in Exploration Geophysics, 25 (6): 6-10.

Chen Cheng, Jia Ailin, et al. 2000. Microfacies Architecture Patterns and the Distribution of the Remaining Oil within Thick Pays[J]. Acta Petrolei Sinica, 21 (5): 99-102.

Chen Cheng, Sun Yimei, et al. 2006. Development and Application of Geological Knowledge Database for Fan-Delta Front in the Dense Spacing Area[J]. Acta Petrolei Sinica, 27 (2): 53-57.

Chen Shijun, Liu Hong, Zhou Jianyu, et al. 2003. The Current Situation and Prospect of Crosswell Seismic Technique[J]. Progress in Geophysics, 18 (3): 524-529.

Chen Xiaohong, Mou Yongguang. 1998. 4D Seismic Reservoir Monitoring Technology and Its Application[J]. Oil Geophysical Prospecting, 33 (6): 707-715.

Chen Xiaohong, Yi Weiqi. 2003. Research on Time - lapse Seismic Reservoir Monitoring Technology[J]. Progress in Exploration Geophysics, 26 (1): 1-6.

Chen Yefei, Peng Shimi, Song Guiru. 2003. Inter-Well Prediction of Flow Units and Remaining Oil Distribution[J]. Acta Petrolei Sinica, 24 (3): 74-77.

Chitale D V, Charlotte Sullivan. 2004. Standard Workflows to Integrate Borehole Images with Other Openhole Logs for Reservoir Characterization[J]. SPE 90705, 1-9.

Corbeanu M R, Soegaard K, Szerbiak R B, et al. 2001. Detailed Internal Architecture of A Fluvial Channel Sandstone Determined from Outcrop, Cores, and 3D Ground - Penetrating Radar: Example from the Middle Cretaceous Ferron Sandstone, East-Central Utah[J]. AAPG Bulletin, 85 (9): 1565-1582.

Dai Qide, et al. 1999. Oilfield Development Geology [M]. Dongying: China University of Petroleum Press.

Deng Hongwen, et al. 2002. High-resolution Sequence Stratigraphy—Principle and Application [M]. Beijing: Geological Publishing House.

Deutsch C V, Journel A G. 1992. GSLIB: Geostatistical Software Library and User'S Manual [M]. New York: Oxford University Press.

Dou Zhilin. 2000. A Study on Flow Unit Model and Distribution of Remaining Oil in Fluvial

Sandstone Reservoirs of the Guantao Formation in Gudong Oil Field[J]. Petroleum Exploration and Development, 27 (6): 50-52.

Du Shitong. 2004. Seismic Attribute Analysis[J]. Oil And Gas Geophysics, 2 (4): 19-30.

Feng Qihong, Qi Junluo, Yin Xiaomei, et al. 2009. Simulation of Fluid-Solid Coupling During Formation and Evolution of High - Permeability Channels [J]. Petroleum Exploration and Development, 36 (4): 498-502, 512.

Flint S S, Bryant I D. 1993. The Geological Modeling of Hydrocarbon Reservoirs and Outcrop Analogs[M]. International Association of Sedimentologists Special Publication, 15.

Gao Boyu, Peng Shimi, Chen Yefei. 2005. Reservoir Dynamic Flow Unit and Remaining Oil Distribution[J]. Journal of Jilin University (Earth Science Edition), 35 (2): 182-194.

Guo Jianlin, Chen Cheng, et al. 2006. Predicting Effective Sandbody Distribution in Well Sulige 6 Area[J]. Journal of Oil and Gas Technology, 28 (5): 54-57.

Guo Jianlin, Jia Ailin, et al. 2007. Sequence Stratigraphy of Upper Jurassic-Lower Cretaceous Fan-Delta Outcrops in Luanping[J]. Geology in China, 34 (4): 628-635.

Hamman J G, Buettner R E, Caldwell D H, et al. 2003. A Case Study of A Fine Scale Integrated Geological, Geophysical, Petrophysical, and Reservoir Simulation Reservoir Characterization with Uncertainty Estimation[J]. SPE 84274, 1-7.

Han Dakuang. 1995. An Approach to Deep Development of High Water-Cut Oil Fields to Improve Oil Recovery[J]. Petroleum Exploration and Development, 22 (5): 47-55.

Hao Jianming, Wu Jian, Zhang Hongwei. 2009. Research on Fine Reservoir Modeling and Remaining Oil Distribution Using Horizontal Well Data [J]. Petroleum Exploration and Development, 36 (6): 730-736.

He Wenxiang, Wu Shenghe, Tang Yijiang, et al. 2005. Detailed Architecture Analyses of Debouch Bar in Shengtuo Oilfield, Jiyang Depression [J]. Petroleum Exploration And Development, 32 (5): 42-46.

He Wenxiang, Wu Shenghe, Tang Yijiang, et al. 2005. The Architecture Analysis of the Underground Point Bar—Taking Gudao Oilfield as An Example[J]. Journal Of Mineralogy and Petrology, 25 (2): 8186.

Huang Changwu. 2008. The Key to the Success of Digital Oilfields[J]. Petroleum Exploration and Development, 35 (4): 436.

Huang Xuri. 2003. A review of Time - Lapse Seismic Techniques Abroad [J]. Progress in Exploration Geophysics, 26 (1): 7-12.

Hu Xiangyang, Xiong Qihua, et al. 2001. Advancement of Reservoir Modeling Methods [J]. Journal of China University of Petroleum (Natural Science Edition), 25 (1): 107-112.

Hu Xuetao, Li Yun. 2000. Study of Microcosmic Distribution of Residual Oil with Stochastic Simulation in Networks[J]. Acta Petrolei Sinica, 21 (4): 46-51.

Jackson S R, et al. 1991. Application of Outcrop Data for Characterization Reservoir and Deriving Grid - Block Scale Values for Numerical Simulation [C]. Third International Reservoir Characterization Technical Conference. Tulsa, Oklahoma, U.S.A., 15.

Jia Ailin, Guo Jianlin, et al. 2007. Perspective of Development in Detailed Reservoir Description

[J]. Petroleum Exploration and Development, 34 (6): 691-695.

Jia Ailin, He Dongbo, et al. 2003. Application of Outcrop Geological Knowledge Database to Prediction of Inter-Well Reservoir in Oilfield[J]. Acta Petrolei Sinica, 24 (6): 51-53, 58.

Jia Ailin, He Dongbo, et al. 2004. Sequence Evolution of Fan-Delta Outcrops and Its Controlling on the Sandstone Reservoirs[J]. Petroleum Exploration and Development, 31 (B11): 103-105.

Jia Ailin, Huang Shiyan, et al. 2000. Approach for Detailed Study of Reservoir Outcrop[J]. Acta Petrolei Sinica, 21 (4): 105-108.

Jia Ailin, Huang Yanshi, et al. 2000. A Study on Sedimentary Simulation of Fan-Delta Outcrop [J]. Acta Petrolei Sinica, 21 (6): 107-110.

Jia Ailin. 1995. Steps to Build Reservoir Geological Models[J]. Earth Science Frontiers, 2 (4): 221-225.

Jia Ailin, et al. 2002. Techniques and Methods for Establishing Geological Models in the Evaluation Stage of Reservoirs[M]. Beijing: Petroleum Industry Press.

Jiang Hanqiao, et al. 2001. Principles and Methods of Reservoir Engineering[M]. Dongying: China University of Petroleum Press.

Jiang Xiangyun, Wu Shenghe, Yu Diyun, et al. 2007. Fluvial Reservoir Architecture Modeling and Remaining Oil Analysis[J]. SPE 109175.

Jiang Zaixing. 2003. Sedimentology[M]. Beijing: Petroleum Industry Press.

Jia Zhenyuan, Cai Zhongxian. 1992. Research Methods of Architecture[J]. Geological Science and Technology Information, 11 (4): 63-68.

Justice J H, Woerpel J C, Watts G P, et al. 2000. Geostatistical Reservoir Characterization Using Interwell Seismic Data[J]. SPE 62973, 1-8.

Justice J H, Woerpel J C, Watts G P, et al. 2000. Interwell Seismic Data for Reservoir Characterization[J]. SPE 59695.

Justice J H, Woerpel J C, Watts G P, et al. 2000. Interwell Seismic for Reservoir Characterization and Monitoring[J]. SPE 62588, 1-5.

Liao Guangming. 2006. Analysis on Sedimentary System and Reservoir Architecture of Delta-Shallow Lake Facies of Fanzhuang in Jinhu Depression[J]. Journal of Southwest Petroleum Institute, 28 (2): 4-7.

Li Bohu, et al. 2004. Fine Geology Research And Application Technology For Daqing Oilfield [M]. Beijing: Petroleum Industry Press.

Li Jianrong, Wang Lei, Wang Zhui. 2003. Some Key Problems in Production Seismology[J]. Geophysical Prospecting for Petroleum, 42 (2): 279-284.

Li Lanbin. 2000. Development Seismic Technique and Its Trend[J]. Henan Petroleum, 5: 25, 28-31.

Liu Huiqing. 2001. Special Topic on Numerical Reservoir Simulation Methods[M]. Dongying: China University of Petroleum Press.

Li Xingguo. 1994. Application of Microstructures And Reservoir Sedimentary Microfacies to Study Remaining Oil distribution in Reservoirs[J]. Petroleum Recovery Efficiency Technology, 1

(1): 68-80.

Li Xingguo. 1995. New Insights into the Ministructure[J]. Petroleum Exploration and Development, 22 (1): 64-67.

Li Xingguo. 1987. The Control Effect of Oil Layer Microstructures on Oil Well Production—Taking Shengtuo and Gudao Oilfields as Examples[J]. Petroleum Exploration and Development, 14 (2): 53-59.

Li Yang. 2003. Flow unit Mode and Remaining Oil Distribution in Reservoir[J]. Acta Petrolei Sinica, 24 (3): 52-55.

Lu Jianlin, Li Guoqiang, Fan Zhonghai. 2001. Research on Remaining Oil Distribution in Oilfields During High Water Cut Period[J]. Acta Petrolei Sinica, 22 (5): 48-52.

Ma Zaitian. 2004. Some Thoughts on Petroleum Developmect Seismology [J]. Natural Gas Industry, 24 (6): 43-46.

Mezghani M, Fomel A, Langlais V, et al. 2004. History Matching and Quantitative Use of 4D Seismic Data for An Improved Reservoir Characterization[J]. SPE 90420, 1-10.

Miall A D. 1985. Architectural Element Analysis: A New Method of Facies Analysis Applied to Fluvial Deposits[J]. Earth Science Review, 22 (2): 261-308.

Miall A D. 1983. Basin Analysis of Fluvial Sediments, In: Moden and Ancient Fluvial Systerms [M], Hoboken: Wiley-Blackwell, 279-286.

Miall A D. 1978. Lithoracies Typs and Vertical Profile Models in Braided River Deposits: A Summary in Miall, A. D., Eds. Fluvial Sedimentology [M], Canadian Society of Petroleum Geologists Memoir 5, 597-604.

Márquez L J, González M, Gamble S, et al. 2001. Improved Reservoir Characterization of A Mature Field Through An Integrated Multi-Disciplinary Approach.LL-04 Reservoir, Tia Juana Field, Venezuela[J]. SPE 71355, 1-10.

Mu Longxin, et al. 2003. Research on Fan Delta Sedimentary Reservoir Model and Prediction Method[M]. Beijing: Petroleum Industry Press.

Mu Longxin, Jia Ailin, Chen Liang, et al. 2000. Reservoir Fine Research Method—Research on Outcrop Reservoir, Modern Deposition and Fine Geological Modeling at Home and Abroad[M]. Beijing: Petroleum Industry Press.

Mu Longxin. 1999. Some Development Trends in Reservoir Description Technology[J]. Petroleum Exploration and Development, 26 (6): 42-46.

Mu Longxin. 2000. Stages and Characteristic of Reservoir Description[J]. Acta Petrolei Sinica, 21 (5): 103-108.

Pankaj P, Pathak V. 2005. Reservoir Monitoring and Characterization for Heavy-Oil Thermal Recovery[J]. SPE 97860, 1-9.

Qin Jishun, Li Aifen. 1988. Reservoir Physics [M]. Dongying: China University of Petroleum Press.

Qiu Yinan, Chen Ziqi. 1996. Reservoir Description[M]. Beijing: Petroleum Industry Press.

Qiu Yinan, Jia Ailin, et al. 2000. Development of Geological Reservoir Modeling in Past Decade [J]. Acta Petrolei Sinica, 21 (4): 101-104.

Qiu Yinan, Xue Shuhao, *et al.* 1997. Reservoir Evaluation Technology[M]. Beijing: Petroleum Industry Press.

Ran Qiyou. 2003. Current Status and Development Trend of Remaining Oil Research [J]. Petroleum Geology and Recovery Efficiency, 10 (5): 49-51.

Ran Qiyou. 2003. Research on Remaining Oil Formation Conditions[J]. Fault-Block Oil and Gas Field, 10 (6): 23-26.

Ravenne, *et al.* 1989. Heterogeneity and Geometry of Sedimentary Bodies in A Fluvial Deltaic Reservoir[J]. SPE Formantion Evaluation, 4 (2): 239-246.

Scott R Reeves, Shahab D Mohaghegh, John W Fairborn, *et al.* 2002. Feasibility Assessment of A New Approach for Integrating Multiscale Data for High-Resolution Reservoir Characterization [J]. SPE 77759, 1-12.

Shen Pingping. 2003. A New Method of Modern Reservoir Description[M]. Beijing: Petroleum Industry Press.

Soto B, Bernal M C, Silva B, *et al.* 2000. How to Improve Reservoir Characterization Using Intelligentsy Stems—A Case Study of: Toldado Field in Colombia[J]. SPE 62938, 1-14.

Sun Huanquan, Sun Guo, Cheng Huiming, *et al.* 2002. The Simulation Models of Remaining Oil Distribution at Super-High Water-Cut Stage of Shengtuo Oil Field[J]. Petroleum Exploration and Development, 29 (3): 66-67.

Sun Huanquan. 2002. Reservoir Dynamic Model and Remaining Oil Distribution Model[M]. Beijing: Petroleum Industry Press.

Sun Mengru, *et al.* 2004. Fine Geological Research on Shengtuo Oilfield[M]. Beijing: China Petrochemical Press.

Tingting Yao. 2000. Integrating Seismic Attribute Maps and Well Logs for Porosity Modeling in A West Texas Carbonate Reservoir: Addressing the Scale and Precision Problem[J]. Journal of Petroleum Science and Engineering, 65-79.

Tingting Yao. 2000. Integration of Seismic Attribute Map into 3D Facies Modeling[J]. Journal of Petroleum Science and Engineering, 69-84.

Vincent Kretz, Mickaele Le Ravalec, Frederoc Roggero, *et al.* 2002. An Integrated Reservoir Characterization Study Matching Production Data and 4D Seismic[J]. SPE 77516, 1-9.

Wang Naiju, *et al.* 1999. General Reservoir Development Models in China [M]. Beijing: Petroleum Industry Press.

Wang Shaohua, Zhang Baiqiao, Shu Zhiguo, *et al.* 2002. Using Outcrop Information of Kuqa Depression to Predict Reservoir Sand Bodies in Tuziluoke Gas Field[J]. Petroleum Exploration and Development, 29 (5): 50-52.

Wang Zhizhang, Shi Zhanzhong. 1999. Modern Reservoir Description Technology[M]. Beijing: Petroleum Industry Press.

Weber K J. 1982. Influence of Common Sedimentary Structures on Fluid Flow in Reservoir Models [J]. JPT, 34 (3).

Wei Bin, Chen Jianwen, Zheng Junmao. 2000. Utilization of Reservoir Flow Unit to Study Remaining Oil Distribution in High Water Containing Oilfield[J]. Earth Science Frontiers, 7

(4): 403-410.

Wu Jian, Li Fanhua. 2009. Prediction of Oil-Bearing single sand body by 3D Geological Modeling Combined With Seismic Inversion[J]. Petroleum Exploration and Development, 36 (5): 623-627.

Wu Shenghe, Liu Ying, Fan Zheng, et al. 2003. 3D Stochastic Modeling of Sedimentary Microfacies with Geological and Seismic Data[J]. Journal of Palaeogeography, 5 (4): 439-448.

Wu Shenghe, Xiong Qihua, et al. 1998. Hydrocarbon Reservoir Geology[M]. Beijing: Petroleum Industry Press.

Wu Shenghe, et al. 1999. Reservoir Modeling[M]. Beijing: Petroleum Industry Press.

Xiong Qihua, Wang Zhizhang, Ji Fahua. 1994. Modern Reservoir Description Technology and Its Application[J]. Acta Petrolei Sinica. 15 (Supplement): 1-9.

Xiong Qihua, Wang Zhizhang, Zhang Yiwei. 1995. New progress in the Study of Reservoir Description[J]. Journal of China University of Petroleum (Natural Science Edition), 19 (3): 96-101.

Xu Shouyu. 2005. Principles of Reservoir Description Methods[M]. Beijing: Petroleum Industry Press.

Yang Qinyong. 2003. Advances in Time-Lapse Seismic Technology[J]. Progress In Exploration Geophysics, 26 (5-6): 339-348.

Yang Zhanlong, Peng Licai, et al. 2007. Seismic Attributes Analysis and Lithological Reservoir Exploration[J]. Geophysical Prospecting for Petroleum, 46 (2): 131-136.

Yin Taiju, Zhang Changmin, Fan Zhonghai. 1997. Founding Subsurface Geological Data Bank for Shuanghe Oil field[J]. Petroleum Exploration and Development, 24 (6): 95-98.

Yin Taiju, Zhang Changmin, Fan Zhonghai, et al. 2002. Establishment of the Prediction Models of Reservoir Architectural Elements[J]. Journal of Xi'an Petroleum Institute (Natural Science Edition), 17 (3): 7-10, 14.

Yin Taiju, Zhang Changmin, Tang Jun, et al. 2000. Analysis On Reservoir Hierarchical Structure in Machang Oilfield[J]. Journal of Jianghan Petroleum Institute, 23 (4): 19-21.

Yin Taiju, Zhang Changmin, Zhao Hongjing, et al. 2001. Remaining Oil Distribution Prediction Based on High-Resolution Sequence Stratigraphy[J]. Petroleum Exploration and Development, 28 (4): 79-82.

Yin Xingyao, Liu Yongshe. 2002. Methods and Development of Integrating Seismic Data in Reservoir Model-Building[J]. Oil Geophysical Prospecting, 37 (4): 423-430.

Yin Xingyao, Zhou Jingyi. 2005. Summary of Optimum Methods of Seismic Attributes[J]. Oil Geophysical Prospecting, 40 (4): 482-489.

Yi Weiqi, Li Qingren, Zhang Guocai, et al. 2003. Study on Time-Lapse Seismic and Prediction of the Distribution of Residual oil in TN Oilfield, Daqing [J]. Progress in Exploration Geophysics, 26 (1): 61-65.

Yuan Shiyi. 2006. Progress in Reservoir Engineering Technology [M]. Beijing: Petroleum Industry Press.

Yue Dali, Wu Shenghe, Tan Heqing, *et al.* 2008. An Anatomy of Paleochannel Reservoir Architecture of Meandering River Reservoir-A Case Study of Guantao Formation, the West 7th Block of Gudong Oilfield[J]. Earth Science Frontiers, 15 (1): 101-109.

Yu Qitai. 2000. Oil Field Development Three Major Rich Areas of "Large Scale" Unswept Remaining Oil in Water Flooded Bedded Sandstone Reservoirs[J]. Acta Petrolei Sinica, 21 (2): 45-50.

Zhang Changmin. 1992. Analytic Hierarchy Process in Reservoir Research[J]. Oil and Gas Geology, 13 (3): 344-350.

Zhang Jinliang. 2008. Reservoir Sedimentary Facies[M]. Beijing: Petroleum Industry Press.

Zhang Yiwei, Xiong Qihua, *et al.* 1997. Continental Reservoir Description [M]. Beijing: Petroleum Industry Press.

Zhao Chenglin *et al.* 2001. Sedimentary Petrology (Third Edition) [M]. Beijing: Petroleum Industry Press.